AMERICAN MATHEMATICAL SOCIETY TRANSLATIONS

Series 2

Volume 113

TWELVE PAPERS IN LOGIC AND ALGEBRA

by

A. F. Lavrik
E. M. Levič
A. I. Mal′cev
A. G. Postnikov
N. P. Romanov
R. L. Šapiro
B. M. Schein [Šaĭn]
M. I. Semenenko
N. Ja. Vilenkin

ISSN 0065-9290

Published by the
AMERICAN MATHEMATICAL SOCIETY
Providence, Rhode Island
1979

AMS (MOS) subject classifications (1970). Primary 02B05, 02C15, 08A05, 10G05, 10H15, 10H20, 10J15, 16A26, 20E35, 20L05, 20M10, 20M10, 20M30, 20N10, 22E45, 33A75; Secondary 20A50, 20E25, 20F25, 20K05

Library of Congress Cataloging in Publication Data

Main entry under title:

Twelve papers in logic and algebra.

(American Mathematical Society. Translations; ser. 2, v. 113 ISSN 0065-9290)

Translated from Russian.

Bibliography: p.

1. Logic, Symbolic and mathematical—Addresses, essays, lectures. 2. Algebra—Addresses, essays, lectures. I. Lavrik, A. F. II. Series.

QA3.A572 Ser. 2, vol. 113 [QA9.2] 510'.8s [511'.3]

ISBN 0-8218-3063-5 79-9994

Printed in the United States of America

TABLE OF CONTENTS

RUSSIAN TABLE OF CONTENTS*

Семененко, М. И. *Свойства некоторых подлогик классического и интуиционистского исчислений высказываний*, Z. Math. Logik Grundlagen Math. 16 (1970), 201–238.

Лаврик, А. Ф. *Об одном предложении аддитивной теории чисел*, Усиехи Мат. Наук 14 (1959), в. 1 (85), 197–198.

Лаврик, А. Ф. *К теории распределения простых чисел на основе метода тригонометрических сумм И. М. Виноградова*, Труды Мат. Инст. Стеклов. 64 (1961), 90–125.

Постников, А. Г. и Романов, Н. П. *Упрощение элементарного доказательства А. Сельберга асимптотического закона распределения простых чисел*, Усиехи Мат. Наук 10 (1955), в. 4 (66), 75–87.

Шайн, Б. М. *К теории обобщенных групп и обобщенных груд* Теория Полугруии и её Приложения. I. Издат. Саратов. Унив., Саратов, 1965, стр. 286–324.

Шайн, Б. М. *Лекции о полугруппах преобразований*, Издат. Саратов. Унив., Саратов, 1970.

Шаииро, Р. Л. *Неприводимые представлсния группы $SU(n)$ класса* 1 *относительно $SU(n-1)$*, Успехи Мат. Наук 22 (1967), в. 1 (133), 180–182.

Виленкин, Н. Я. и Шапиро, Р. Л. *Неприводимые представления группы $SU(n)$ класса* I *относительно $SU(n-1)$*, Изв. Высш. Учебн. Завед. Математика 1967, в. 7 (62), 9–20.

Шапиро, Р. Л. *Специальные функции, связанные с представлениямн группы $SU(n)$ класса* I *относительно $SU(n-1)$ $(n \geq 3)$*, Изв. Высш. Учебн. Завед. Математика 1968, в. 4 (71), 97–107.

Левич, Е. М. *О неприводимых представлениях полициклической группы над абсолютно алгебраическим полем простой характеристики*, Латвийский Государственный Университет Ученые Записки 151 (1971), 104–124.

Левич, Е. М, *О группах без кручения, все неприводимые представления которых над некоторым полем конечномерны*, Латвийский Государственный Университет Ученые Записки 151 (1971), 125–139.

Мальцев, А. И. *Симметрические группоиды*, Мат. Сб. 31 (73) (1952), 136–151.

*The American Mathematical Society scheme for transliteration of Cyrillic may be found at the end of index issues of *Mathematical Reviews.*

Amer. Math. Soc. Transl.
(2) Vol. 113, 1979

PROPERTIES OF SOME SUBSYSTEMS OF CLASSICAL AND INTUITIONISTIC PROPOSITIONAL CALCULI*

M. I. SEMENENKO

Preface

In both classical and intuitionistic propositional calculi the following formulas are derivable:

$$A \to (B \to A), \tag{1}$$

$$A \,\&\, B \to A. \tag{2}$$

But if (2) is derivable in some calculus(1), then the derivability of (1) is generally speaking redundant, since whenever we use the derivability of (1), we can take advantage of the derivability of (2), *if* in the critical place we require that the hypotheses *are given* as connected by conjunction. In deductions it is necessary to have hypotheses connected with each other by conjunction, and this is the price we have to pay for the nonderivability of (1), if (2) is derivable. This gives a new sense to conjunction; there is some factual connection in addition to the formal connection between the conjuncts. This has consequences for the sense of implication and disjunction as well.(2)

Even these simple changes show that if from the classical (intuitionistic) propositional calculus we omit the formula (1) as derivable, we will obtain calculi that have interesting properties. In this paper we investigate the properties of such calculi.

In spite of the fact that the final version of this paper is presented purely formally, all basic results (including all theorems about relevance) were suggested by informal, epistemological considerations. In this respect the results presented in this paper are a continuation of investigations inaugurated and worked out by Ackermann [1], Anderson and Belnap [2]–[4], McCall [5] et al.

In the last 10–15 years a trend has clearly become visible in mathematical logic, characterized by the work of Ackermann [1] on an aspect of implication, the work of Anderson and Belnap [2]–[4], McCall [5], Zinov′ev, Dončenko [6] et al. All these papers in different ways and using different means try to justify the statement, as well as to solve, essentially the same problem: the formal apparatus of mathematical logic must reflect meaningful connections.

Each author in his own way tries to resolve the same problems. The striking

---

AMS (MOS) subject classifications (1970). Primary 02B05, 02C15.

*Translation of Z. Math. Logik Grundlagen Math. **16** (1970), 201–238. MR **44** #29.

(1) By "calculus", in this paper, we shall always mean "propositional calculus".

(2) Negation will be defined in terms of implication; see §1 of Chapter I.

common feature of all of these papers is the abandonment of the principle (1).

As we shall show in this paper, this agreement is not accidental, but is based on deep fundamental reasons. But this is not all. Most of the other logical principles rejected in the aforementioned papers are abandoned in consequence of eliminating (1) as a theorem. An idea is naturally suggested: to construct and investigate formalisms, from a purely formal standpoint, that satisfy the following requirements:

a) The logical principle described by the formula $A \longrightarrow (B \longrightarrow A)$ is not valid.

b) Retain as derivable all intuitively plausible formulas and rules of inference that we can find in the classical (or intuitionistic) propositional calculus.

c) There are particular logical principles, as for example the rule of inference

$$\frac{\mathfrak{A};\ \mathfrak{B}}{\mathfrak{A}\ \&\ \mathfrak{B}},$$

which are valid in the classical (intuitionistic) propositional calculus and compatible with the nonderivability of (1) but are still not derivable in the calculi that are to be presented. Omission of such a principle must be justified by the presence of some desirable properties of the calculi we are constructing.

This paper is a result of an attempt to construct and investigate properties of exactly such propositional calculi.

We mention some results.

1. Essentially we retain all traditional rules and techniques of inference of the classical (intuitionistic) propositional calculi only if the hypotheses are actually used in the proofs of their consequences.(3) For example, the analogs of the ordinary deduction theorem, equivalent substitutions, etc. are valid.

2. In §3 the symbol $\vdash$ (simple deduction) is defined in a formal way. We give the following definition of $\vdash_{\overline{\text{rel}}}$. We shall say that $\mathfrak{A}_1; \ldots; \mathfrak{A}_n \vdash_{\overline{\text{rel}}} \mathfrak{B}$ if there is a finite sequence of formulas

$$\text{(A)} \qquad \mathfrak{C}_1; \ldots; \mathfrak{C}_k \quad (k \geqslant 3)$$

satisfying the following conditions:

a) $\mathfrak{C}_k$ is $\mathfrak{B}$ and $\mathfrak{C}_k$ follows by an analog of *modus ponens* from two preceding formulas in (A).

b) Every hypothesis $\mathfrak{A}_i$ $(i = 1, \ldots, n)$ in $\vdash_{\overline{\text{rel}}}$ is required in (A).

c) Every formula $\mathfrak{C}_j$ of the sequence (A) is either different from all hypotheses $\mathfrak{A}_i$ $(i = 1, \ldots, n)$ in $\vdash_{\overline{\text{rel}}}$ or is provable in a relevant calculus, or is obtained by an analog of *modus ponens* from two preceding formulas $\mathfrak{C}_q$ and $\mathfrak{C}_p$ in (A). In the last case, to the formula $\mathfrak{C}_j$ we assign an index set $\{q, p\}$ whose elements are obtained from indices q and p of those formulas $\mathfrak{C}_q$ and $\mathfrak{C}_p$ from which $\mathfrak{C}_j$ is obtained by the analog of *modus ponens*. Obviously, $q < j$ and $p < j$.

d) Every indexed formula in (A) has only one index set.

e) The index j of every formula $\mathfrak{C}_j$ in (A) for $j < k$ necessarily is contained in an index set that corresponds by c) to a formula from (A).

---

(3) A formal definition of "actually used in the proofs" is given in §4 of Chapter III.

It can be shown that $\mathfrak{A}_1; \ldots ; \mathfrak{A}_n \vdash_{\overline{\text{rel}}} \mathfrak{B}$ if and only if $\mathfrak{A}_1; \ldots ; \mathfrak{A}_n \vdash \mathfrak{B}$ is a simple proof.

3. The logical connective conjunction receives a new sense: $\mathfrak{A} \,\&\, \mathfrak{B}$ must now be interpreted, according to circumstances, either as a fact of interaction between $\mathfrak{A}$ and $\mathfrak{B}$, or as a form of interconnection between $\mathfrak{A}$ and $\mathfrak{B}$, or as a fact of a particular "intersection", "the unity of" $\mathfrak{A}$ and $\mathfrak{B}$ or in an analogous way. Here the formulas $\mathfrak{A} \longrightarrow (\mathfrak{B} \longrightarrow \mathfrak{C})$ and $(\mathfrak{A} \,\&\, \mathfrak{B} \longrightarrow \mathfrak{C})$ have to be interpreted as if they refer to essentially different logical situations. It is obvious that in particular such an interpretation of conjunction is in accordance with the consistency of the list of formulas $\{ \neg(\mathfrak{A} \,\&\, \mathfrak{B}); \mathfrak{A}; \mathfrak{B}\}$.

4. Analogous remarks hold with respect to disjunction.

5. The implication has distinctive and interesting properties of this kind. An extensive account of this is given in the present paper, in the section on the relevance theorem.

6. The law of the excluded middle, $\mathfrak{A} \vee \neg \mathfrak{A}$, is now not equivalent to the deletion of double negation $\neg\neg \mathfrak{A} \longrightarrow \mathfrak{A}$, since the latter is not derivable from the former.

7. In the calculi to be investigated the rule of inference

$$\frac{\mathfrak{A}; \mathfrak{B}}{\mathfrak{A} \,\&\, \mathfrak{B}}$$

is invalid, but the rule of inference

$$\frac{\mathfrak{A} \to \mathfrak{B}; \mathfrak{A} \to \mathfrak{C}}{\mathfrak{A} \to \mathfrak{B} \,\&\, \mathfrak{C}}$$

guarantees that in the construction of a conjunction of two propositions $\mathfrak{B}$ and $\mathfrak{C}$ it is required that both $\mathfrak{B}$ and $\mathfrak{C}$ are deductively derivable from the same hypotheses; that is, $\mathfrak{B}$ and $\mathfrak{C}$, so to speak, are covered by a situation described by the same hypotheses.

The basic results were published in a complete and very condensed form (without proofs) in [7].

Finally, let us note that the formulas and rules of inference that are not valid in our systems, but are valid in the classical (intuitionistic) propositional calculus, testify rather convincingly to the fundamental importance of the principle $A \longrightarrow (B \longrightarrow A)$ in the classical (intuitionistic) propositional calculus. This is shown in Chapter II of this paper.

On the other hand, the formulas and rules of inference which continue to hold in the formalisms we are considering, and properties inherent in such formalisms make manifest the richest possibilities of similar propositional calculi, in spite of the absence of $A \longrightarrow (B \longrightarrow A)$.

In this situation everything that can be formalized in the classical (intuitionistic) propositional calculus always has its image, its analogy, its formalization in the propositional calculi we are presenting here. To this fact Chapter III is devoted.

CHAPTER I

§1. Basic definitions and terminology

In order to avoid ambiguities in the sequel, let us agree to call "traditional" the classical and intuitionistic propositional calculi presented as for instance in Kleene's book *Introduction to Metamathematics* [8]. Consequently we shall also call "traditional" other formal systems which are equivalent or isomorphic to the traditional classical and traditional intuitionistic formal systems of Kleene.

Let us also agree to call axioms of the aforementioned formal logical systems "postulates", leaving the term "axiom" to designate the basic suppositions of concrete systems (like axioms of arithmetic). From axioms we obtain formal deductive consequences by means of a logical formalism. The logical formalism is described by an appropriate list of postulates, inference rules and conditions applying to them.(4)

By *propositional letters* we shall understand letters from the beginning of the Latin alphabet, with or without indices from the set of natural numbers; for instance, $A, B, D, F, A_2, B_k, D_i$ and F_m are propositional letters.

Implication we shall denote by an arrow $\longrightarrow$, conjunction by the symbol &, and disjunction by the symbol $\vee$. Negation in the systems we are presenting is reduced to implication and consequently is not primitive.

Every propositional letter is a formula. The symbol "abs" we shall also consider as being a formula. Let us note that a, b and s are neither propositional letters nor formulas. We do not call the formula abs a propositional letter. The symbol abs is short for the word *absurdum* and in the relevant propositional calculi plays the role of a contradiction.

In the sequel the definitions of formulas are as usual. If $\mathfrak{A}$ and $\mathfrak{B}$ are formulas, then $(\mathfrak{A} \,\&\, \mathfrak{B})$, $(\mathfrak{A} \vee \mathfrak{B})$ and $(\mathfrak{A} \longrightarrow \mathfrak{B})$ are formulas. We shall use the usual conventions for omitting parentheses. For convenience we shall at times replace parentheses by brackets, etc.

In propositional calculi of relevance, formulas of the form $(\mathfrak{A} \longrightarrow \text{abs})$ are called "negatives". There formulas we shall, as usual, denote by using the symbol $\neg$, namely $\neg\, \mathfrak{A} =_{\text{df}} (\mathfrak{A} \longrightarrow \text{abs})$.

As usual, we introduce the notion of identity and difference between formulas, the notion of a subformula, the notion of an occurrence in a formula, the notion of antecedents and consequents in implications, etc.

The consequent in the implication $(\mathfrak{A} \longrightarrow \mathfrak{B})$ which is not itself an implication we shall call the *terminal consequent* in the implication $(\mathfrak{A} \longrightarrow \mathfrak{B})$.

In this paper we investigate four logical formalisms called by the common term "relevant propositional calculi."

Those formalisms differ from each other by the presence or absence of one or two postulates. The smallest number of postulates is contained in the following

(4) The calculi we are considering are free from restrictions on postulates and rules of inference. In what follows we call them "relevant", the sense of this being explained in §2 of Chapter II.

minimally-relevant propositional calculus. This calculus is formulated as follows.

We shall call each of the following formulas a postulate of the minimally-relevant propositional calculus.

I. $(\rightarrow)$-*postulates*

1. $(A \rightarrow B) \rightarrow [(B \rightarrow C) \rightarrow (A \rightarrow C)]$,
2. $[A \rightarrow (B \rightarrow C)] \rightarrow [B \rightarrow (A \rightarrow C)]$,
3. $[A \rightarrow (A \rightarrow B)] \rightarrow (A \rightarrow B)$.

II. $(\&)$-*postulates*

1. $A \,\&\, B \rightarrow A$,
2. $A \,\&\, B \rightarrow B$,
3. $(A \rightarrow B) \,\&\, (A \rightarrow C) \rightarrow (A \rightarrow B \,\&\, C)$.

III. $(\vee)$-*postulates*

1. $A \rightarrow A \vee B$,
2. $B \rightarrow A \vee B$.

IV. ((&) *and* $(\vee)$-*postulate*

1. $A \,\&\, (B \vee C) \rightarrow (A \,\&\, B) \vee (A \,\&\, C)$.

A formula is derivable in the minimally-relevant propositional calculus if and only if it can be effectively obtained by an application or applications of the following rules.

a) Every postulate of the minimally-relevant propositional calculus is derivable.

b) If formulas $\mathfrak{A}$ and $(\mathfrak{A} \rightarrow \mathfrak{B})$ are derivable in the minimally-relevant propositional calculus, then the formula $\mathfrak{B}$ is also derivable in this calculus.

c) If the formulas $(\mathfrak{A} \rightarrow \mathfrak{B})$ and $(\mathfrak{A} \rightarrow \mathfrak{C})$ are derivable in the minimally-relevant propositional calculus, then in the same calculus the formula $(\mathfrak{A} \rightarrow \mathfrak{B} \,\&\, \mathfrak{C})$ is derivable.

d) If the formulas $(\mathfrak{A} \rightarrow \mathfrak{C})$ and $(\mathfrak{B} \rightarrow \mathfrak{C})$ are derivable in the minimally-relevant propositional calculus, then in the same propositional calculus the formula $(\mathfrak{A} \vee \mathfrak{B} \rightarrow \mathfrak{C})$ is also derivable.

e) If in a formula derivable in the minimally-relevant propositional calculus, every occurrence of a propositional letter is replaced by the same arbitrary formula, then the formula obtained as a result of such a substitution is derivable in the minimally-relevant propositional calculus.

In particular we note that one may not substitute for abs, since abs is not a propositional letter.

Now it is worth noting that the most important formulas (logical principles) which are derivable in the traditional propositional calculi are also derivable in the minimally-relevant propositional calculus. This is true of formulas describing the properties of negation as well (for instance the principle of contraposition and others). We note that in the given propositional calculus the formula $(A \,\&\, \neg A \rightarrow \text{abs})$ is derivable. This is why a list of formulas is said to be inconsistent if these formulas deductively yield just abs.

If to the postulates of group (I) of the minimally-relevant propositional calculus we adjoin the formula $(\text{abs} \rightarrow A)$ as a fourth postulate, retaining all postulates and inference rules of the minimally-relevant propositional calculus, then the new formalism is called the *intuitionistically-nonrelevant propositional calculus.* If to either of these two formalisms we adjoin the formula $A \vee (A \longrightarrow \text{abs})$ as a new postulate in the group (III), then we obtain either the *minimally-completed propositional calculus* or the *intuitionistically-completed propositional calculus,* respectively.

§2. Connections with the traditional propositional calculi

First of all, it is worth noting that adjoining the formula $[A \longrightarrow (B \longrightarrow A)]$ as a new postulate to the intuitionistically-nonrelevant propositional calculus transforms it into the traditional intuitionistic propositional calculus. Adjoining $[A \longrightarrow (B \longrightarrow A)]$ as a new postulate to the intuitionistically-completed propositional calculus transforms it into the traditional classical propositional calculus. Since the formula $[A \longrightarrow (B \longrightarrow A)]$ (we shall establish this fact later) is not deducible in any of the four systems so far considered, the properties of relevant propositional calculi make manifest the fundamental role which is played by the postulate $[A \longrightarrow (B \longrightarrow A)]$ in the traditional propositional calculi. It follows that relevant propositional calculi are subsystems of the corresponding traditional formalisms and consequently incomplete. We must pay attention to the fact that in relevant propositional calculi the law of double negation is unprovable. By this very fact the law of the excluded middle is not equivalent to the law of double negation, in distinction to the traditional formalisms. This means that this equivalence in the traditional propositional calculi was guaranteed just by the postulate $[A \longrightarrow (B \longrightarrow A)]$.

Analogously, a study of all the other properties of relevant propositional calculi essentially turns repeatedly to the establishment of some aspect of the fundamental role played by $[A \longrightarrow (B \longrightarrow A)]$ in traditional propositional calculus. But this, it is understood, does not depreciate the value of relevant propositional calculi in their own right.

CHAPTER II. PROPERTIES OF THE RELEVANT CALCULI THAT FOLLOW FROM THE UNPROVABILITY OF CERTAIN FORMULAS

§1. The proof of unprovability of some formulas

First of all we note that the methods we are going to describe are insufficient to establish the fact that the formula

(1) $$(A \to B) \to [A \to (A \to B)].$$

is not provable. Since the proof of its unprovability is very cumbersome we shall omit it. But it does not follow that the methods we are going to develop here are not effecttive and fruitful in establishing the unprovability of other formulas. In any case much greater effort is needed in order to find formula (1), whose unprovability is not established by the means developed below.

If we can prove that a formula is not provable in the intuitionistically-completed propositional calculus (it has the greatest number of postulates compared with other relevant calculi), by this very fact we establish the unprovability of these formulas in all relevant formalisms. First of all we give a complete definition of the intuitionistically-completed propositional calculus.

Intuitionistically-completed propositional calculus.

I. ($\longrightarrow$)-*postulates*

1. $(A \longrightarrow B) \longrightarrow [(B \longrightarrow C) \longrightarrow (A \longrightarrow C)]$.
2. $[A \longrightarrow (B \longrightarrow C)] \longrightarrow [B \longrightarrow (A \longrightarrow C)]$.
3. $[A \longrightarrow (A \longrightarrow B)] \longrightarrow (A \longrightarrow B)$.
4. $(\text{abs} \longrightarrow A)$.

II. (&)-*postulates*

1. $A \mathbin{\&} B \longrightarrow A$,
2. $A \mathbin{\&} B \longrightarrow B$,
3. $(A \longrightarrow B) \mathbin{\&} (A \longrightarrow C) \longrightarrow (A \longrightarrow B \mathbin{\&} C)$.

III. ($\vee$)-*postulates*

1. $A \longrightarrow A \vee B$,
2. $B \longrightarrow A \vee B$,
3. $A \vee (A \longrightarrow \text{abs})$.

IV. ((&) *and* ($\vee$))-*postulate*

$$A \mathbin{\&} (B \vee C) \rightarrow (A \mathbin{\&} B) \vee (A \mathbin{\&} C).$$

Rules of inference.

1. *Modus ponens*:

$$\frac{\mathfrak{A};\ (\mathfrak{A} \rightarrow \mathfrak{B})}{\mathfrak{B}}.$$

2. *The rule of conjunction introduction*:

$$\frac{(\mathfrak{A} \rightarrow \mathfrak{B});\ (\mathfrak{A} \rightarrow \mathfrak{C})}{\mathfrak{A} \rightarrow \mathfrak{B} \mathbin{\&} \mathfrak{C}}.$$

3. *The rule of disjunction introduction*:

$$\frac{(\mathfrak{A} \rightarrow \mathfrak{C});\ (\mathfrak{B} \rightarrow \mathfrak{C})}{\mathfrak{A} \vee \mathfrak{B} \rightarrow \mathfrak{C}}.$$

4. *The rule of substitution.*

Now we shall give the proof.

Let Γ denote a distributive lattice having a maximal element ∞ and a minimal 0. Let us denote the order relation in Γ by $>$ and $<$. The symbols $>$ and $<$ denote strict ordering, i.e. equality is excluded.

Thus Γ has the following properties:

$$0 \in \Gamma, \tag{2}$$

$$\infty \in \Gamma, \tag{3}$$

$$\text{if } \alpha \in \Gamma, \quad \text{then } 0 \leqq \alpha \leqq \infty. \tag{4}$$

We shall give interpretations of propositional connectives. If $\alpha \in \Gamma$ and $\beta \in \Gamma$, then

$$\alpha \mathbin{\&} \beta = \alpha \cap \beta, \tag{5}$$

$$\alpha \vee \beta = \alpha \cup \beta, \tag{6}$$

$$(7) \qquad (\alpha \to \beta) = \begin{cases} \infty, & \text{if } \alpha = 0, \\ 0, & \text{if } \alpha > \beta \text{ or } \alpha \parallel \beta \\ \beta, & \text{if } \alpha < \beta \text{ or } \alpha = \beta \end{cases} \text{ where } \alpha \neq 0.$$

[Here $\alpha \parallel \beta$ means that α and β are incomparable in the lattice.] In this interpretation we have the following two theorems, which we shall prove simultaneously.

THEOREM I. *Any formula provable in the intuitionistically-completed propositional calculus takes a value distinct from* 0 *for every system of values from* Γ *of its propositional letters, where* abs $\equiv 0$.

THEOREM II. *Any formula provable in the minimally-completed propositional calculus takes a value distinct from* 0 *for every system of values from* Γ *of its propositional letters and* abs.

The following evident propositions play an essential role in the proof of these theorems.

LEMMA I. *If any consequent in an implication has the value* ∞, *then the implication itself has the value* ∞.

LEMMA II. *If in an implication any of the antecedent formulas equals* 0, *then the implication has the value* ∞.

LEMMA III. *For any formula* $(\mathfrak{A} \longrightarrow \mathfrak{B})$, *if under a given interpretation in* Γ *the formula* $\mathfrak{A}$ *takes the same value as* $\mathfrak{B}$ *for all valuations, i.e. if* $\mathfrak{A} \equiv \mathfrak{B}$, *then the implication* $(\mathfrak{A} \longrightarrow \mathfrak{B})$ *is never equal to* 0.

We shall prove both theorems by induction on the length of proof. The proofs will be different from each other only in the case of postulates (I–4). We shall pay attention to this fact when needed. Otherwise both proofs coincide. We first show that every postulate satisfies the conditions of the theorems.

Postulate (I–1). $(A \longrightarrow B) \longrightarrow [(B \longrightarrow C) \longrightarrow (A \longrightarrow C)]$. If in (I–1) any of the antecedent formulas equals 0 then (I–1) $= \infty$, i.e. (I–1) $\neq 0$. Hence we have to consider only the case

$$(8) \qquad \begin{cases} \text{a) } (A \to B) \neq 0, \\ \text{b) } (B \to C) \neq 0, \\ \text{c) } A \neq 0, \text{ i.e. } 0 < A. \end{cases}$$

Since $A \neq 0$, it follows from (8a) that $A \leqslant B$, i.e. $0 < A \leqslant B$; thus

$$(9) \qquad B \neq 0, \text{ i.e. } 0 < B \quad \text{(under the conditions (8)).}$$

But then, analogously, (8b) and (9) imply that $B \leqslant C$; thus

$$(10) \qquad C \neq 0 \quad \text{(under the conditions (8)).}$$

Since (8)–(10) hold, and since $A \leqslant B \leqslant C$ when (8) holds, we can easily see that (I–1) never equals 0.

Postulate (I–2). $[A \longrightarrow (B \longrightarrow C)] \longrightarrow [B \longrightarrow (A \longrightarrow C)]$. If in (I–2) any of the antecedent formulas equals 0, then (I–2) $= \infty$, i.e. (I–2) $\neq 0$. Therefore we have to consider only the case

$$(11)\qquad \begin{cases} \text{a) } [A \to (B \to C)] \neq 0, \\ \text{b) } B \neq 0, \ \text{ i.e. } \ 0 < B, \\ \text{c) } A \neq 0, \ \text{ i.e. } \ 0 < A. \end{cases}$$

From (11a) and (11c) we obtain $0 < A \leqslant (B \longrightarrow C)$, i.e.,

$$(12)\qquad (B \longrightarrow C) \neq 0 \ \text{ (under the conditions (11)).}$$

By (11b) we have

$$(13)\qquad B \leqslant C \ \text{ (under the conditions (11)).}$$

Thus from (12) and (13) we have $(B \longrightarrow C) = C$; hence we can write (11a) as

$$(14)\qquad (A \longrightarrow C) \neq 0 \ \text{ (under the conditions (11)).}$$

Therefore from (11c) it follows that

$$(15)\qquad A \leqslant C \ \text{ (under the conditions (11)).}$$

By (13) and (15), we see that, under the conditions (11), (I–2) never equals 0. Therefore (I–2) never equals 0.

Postulate (I–3) $[A \longrightarrow (A \longrightarrow B)] \longrightarrow (A \longrightarrow B)$. If any of the antecedent formulas in (I–3) equals 0, then (I–3) $\neq 0$ as before. We have to consider the case

$$(16)\qquad \begin{cases} \text{a) } [A \to (A \to B)] \neq 0, \\ \text{b) } A \neq 0, \ \text{ i.e. } \ 0 < A. \end{cases}$$

Consequently, $A \leqslant (A \longrightarrow B)$, i.e., since $0 < A$, $0 < (A \longrightarrow B)$, i.e. $(A \longrightarrow B) \neq 0$, and so $A \leqslant B$. This means that (I–3) $\neq 0$ under the conditions (16). Hence (I–3) never equals 0.

Postulate (I–4). (abs $\longrightarrow A$). This postulate satisfies only the conditions of Theorem I, when abs $\equiv 0$. Theorem II can be extended to propositional calculi containing no postulate of the form (abs $\longrightarrow A$). Obviously, if abs $\equiv 0$ then (abs $\longrightarrow A) \neq 0$.

Postulates (II–1) *and* (II–2). $(A \ \& \ B \longrightarrow A)$ and $(A \ \& \ B \longrightarrow B)$ obviously never equal 0, i.e. by the definition of the interpretation in the lattice Γ we always have $A \ \& \ B \leqslant A$ and $A \ \& \ B \leqslant B$.

Postulate (II–3). $(A \longrightarrow B) \ \& \ (A \longrightarrow C) \longrightarrow (A \longrightarrow B \ \& \ C)$. If either of the antecedent formulas (II–3) equals 0, then (II–3) $\neq 0$. Hence we have to consider only the case

$$(17)\qquad \begin{cases} \text{a) } (A \to B) \,\&\, (A \to C) \neq 0, \\ \text{b) } A \neq 0, \quad \text{i.e. } 0 < A. \end{cases}$$

But if $(A \longrightarrow B) = 0$ or if $(A \longrightarrow C) = 0$, then $(A \longrightarrow B) \,\&\, (A \longrightarrow C) = 0$. Therefore under the conditions (17) we have

$$(18)\qquad \begin{cases} \text{a) } (A \to B) \neq 0 \\ \text{b) } (A \to C) \neq 0 \end{cases} \text{(under the conditions (17)).}$$

From (17b) we have

$$(19)\qquad \begin{cases} \text{a) } 0 < A \leqq B \\ \text{b) } 0 < A \leqq C \end{cases} \text{(under the conditions (17)).}$$

as well as

$$(20)\qquad 0 < A \leqslant B \,\&\, C, \text{ i.e. } B \,\&\, C \neq 0 \text{ (under the conditions (17)).}$$

(19) and (20) mean that, under the conditions (17), (II–3) never equals 0. Therefore, always (II–3) $\neq 0$.

Postulates (III–1) *and* (III–2). $(A \longrightarrow A \vee B)$ and $(B \longrightarrow A \vee B)$ obviously never equal 0, since under the given interpretation $A \leqslant A \vee B$ and $B \leqslant A \vee B$.

Postulate (III–3). $(A \vee (A \longrightarrow \text{abs}))$. If $A = 0$, then $(A \longrightarrow \text{abs}) = \infty$ for any value of abs, i.e. (III–3) $\neq 0$. If $A \neq 0$, then $A \vee (A \longrightarrow \text{abs}) \neq 0$ for any value of abs.

Postulate (IV). $A \,\&\, (B \vee C) \longrightarrow (A \,\&\, B) \vee (A \,\&\, C)$ never equals 0, since the lattice Γ is distributive by hypothesis.

Let us consider the inductive steps.

The rule of modus ponens:

$$\frac{\mathfrak{A};\ (\mathfrak{A} \to \mathfrak{B})}{\mathfrak{B}}.$$

Suppose that

$$(21)\qquad \begin{array}{l} \text{a) } \mathfrak{A} \neq 0, \quad \text{i.e. } 0 < \mathfrak{A} \text{ always,} \\ \text{b) } (\mathfrak{A} \to \mathfrak{B}) \neq 0 \quad \text{always.} \end{array}$$

Hence we have $0 < \mathfrak{A} \leqslant \mathfrak{B}$, i.e. $\mathfrak{B} \neq 0$ under the conditions (21).

The rule of conjunction introduction:

$$\frac{(\mathfrak{A} \to \mathfrak{B});\ (\mathfrak{A} \to \mathfrak{C})}{\mathfrak{A} \to \mathfrak{B} \,\&\, \mathfrak{C}}.$$

Suppose that

$$(22)\qquad \begin{array}{l} \text{a) } (\mathfrak{A} \to \mathfrak{B}) \neq 0 \quad \text{always,} \\ \text{b) } (\mathfrak{A} \to \mathfrak{C}) \neq 0 \quad \text{always.} \end{array}$$

If $\mathfrak{A} = 0$, then $(\mathfrak{A} \to \mathfrak{B} \,\&\, \mathfrak{C}) \neq 0$. Thus we have to consider only the case when

$$\mathfrak{A} \neq 0, \text{ i.e. } 0 < \mathfrak{A}. \tag{23}$$

But under the condition (23), from (22) we have $0 < \mathfrak{A} \leqslant \mathfrak{B}$ and $0 < \mathfrak{A} \leqslant \mathfrak{C}$, so that $0 < \mathfrak{A} \leqslant \mathfrak{B} \,\&\, \mathfrak{C}$; hence $(\mathfrak{A} \to \mathfrak{B} \,\&\, \mathfrak{C}) \neq 0$ under the conditions of (23) and (22). This shows that under the conditions of (22) we always have $(\mathfrak{A} \to \mathfrak{B} \,\&\, \mathfrak{C}) \neq 0$.

The rule of disjunction introduction:

$$\frac{(\mathfrak{A} \to \mathfrak{C});\ (\mathfrak{B} \to \mathfrak{C})}{(\mathfrak{A} \vee \mathfrak{B} \to \mathfrak{C})}.$$

Suppose that

$$\begin{aligned} &\text{a)}\ (\mathfrak{A} \to \mathfrak{C}) \neq 0 \quad \text{always,} \\ &\text{b)}\ (\mathfrak{B} \to \mathfrak{C}) \neq 0 \quad \text{always.} \end{aligned} \tag{24}$$

If $\mathfrak{A} = 0$, then $\mathfrak{A} \vee \mathfrak{B} = 0 \vee \mathfrak{B} = \mathfrak{B}$, i.e. $(\mathfrak{A} \vee \mathfrak{B} \to \mathfrak{C})$ has the same value as $(\mathfrak{B} \to \mathfrak{C})$, and hence by (24b) we have $(\mathfrak{A} \vee \mathfrak{B} \to \mathfrak{C}) \neq 0$. Analogously we consider the case when $\mathfrak{B} = 0$.

Therefore we have to consider the case

$$\begin{aligned} &\text{a)}\ \mathfrak{A} \neq 0, \quad \text{i.e. } 0 < \mathfrak{A}, \\ &\text{b)}\ \mathfrak{B} \neq 0, \quad \text{i.e. } 0 < \mathfrak{B}. \end{aligned} \tag{25}$$

Hence by (24) we have $0 < \mathfrak{A} \leqslant \mathfrak{C}$ and $0 < \mathfrak{B} \leqslant \mathfrak{C}$, i.e. $0 < \mathfrak{A} \vee \mathfrak{B} \leqslant \mathfrak{C}$ under the conditions (24) and (25). This means that under the same conditions (24) and (25) $(\mathfrak{A} \vee \mathfrak{B} \to \mathfrak{C}) \neq 0$. Therefore under the conditions (24) only we will have $(\mathfrak{A} \vee \mathfrak{B} \to \mathfrak{C}) \neq 0$.

The rule of substitution is considered in a trivial manner, as usual.

This proves both theorems.

Now it is easy to establish the unprovability of many formulas in the relevant calculi.

Obviously the set of all subsets of the set $\{1, 2, 3\}$ is a distributive lattice, with the order treated in a natural way as a relation of inclusion. Let us introduce the following symbols.

$$\begin{cases} 0 = \Lambda \quad \text{(the empty set)} \\ \alpha = \{1\} \\ \beta = \{2\} \\ \gamma = \{3\} \\ \delta = \{1; 2\} \\ \varrho = \{1; 3\} \\ \lambda = \{2; 3\} \\ \infty = \{1; 2; 3\}. \end{cases} \tag{26}$$

As usual, the distributive lattice is presented diagrammatically as follows:

(26a)

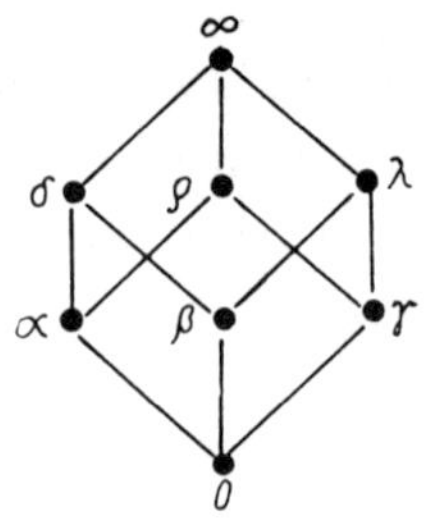

From the following table it is easy to see what values we have to give in accordance with (26) and (26a) to the propositional letters and to abs in order that the corresponding formula takes 0 as a value. In this way we can establish the unprovability of the formulas listed in the table in relevant propositional calculi, using the theorems just proved.

Unprovable formula	The values			
	A	B	C	abs
$A \to (B \to A)$	α	δ	—	—
$A \to (B \to A \& B)$	α	β	—	—
$(A \& B \to C) \to [A \to (B \to C)]$	α	β	0	—
abs	—	—	—	0
$[(A \to \text{abs}) \to \text{abs}] \to A$	α	—	—	0
$(A \& B \to \text{abs}) \to (A \to \text{abs}) \vee (B \to \text{abs})$	α	β	—	0
$[(A \to B) \to \text{abs}] \to A \& (B \to \text{abs})$	α	0	—	0
$(A \to \text{abs}) \vee B \to (A \to B)$	δ	α	—	0
$A \vee B \to [(A \to \text{abs}) \to B]$	0	α	—	0
$(A \& B \to C) \to (A \to C) \vee (B \to C) \vee [A \to (B \to C)]$	α	β	0	—
$(C \to A \vee B) \to (C \to A) \vee (C \to B)$	δ	ϱ	λ	—
$(A \to B) \to [(A \to C) \to (A \to B \& C)]$	α	∞	α	—
$(A \to C) \to [(B \to C) \to (A \vee B \to C)]$	0	α	α	—
$(A \to B) \to (A \to A \& B)$	α	δ	—	—
$(A \to B) \to (A \to A)$	α	δ	—	—
$(A \& B \to \text{abs}) \to [A \to (B \to \text{abs})]$	α	β	—	0
$(A \to \text{abs}) \to (B \to B)$	0	β	—	0
$(A \to A) \& (B \to B)$	α	β	—	—

The unprovability of the last formula $(A \longrightarrow A)$ & $(B \longrightarrow B)$ despite the provability in relevant propositional calculi of both formulas $A \longrightarrow A$ and $B \longrightarrow B$ shows that the rule of inference

$$\frac{\mathfrak{A};\ \mathfrak{B}}{\mathfrak{A}\ \&\ \mathfrak{B}}$$

does not hold in such systems.

§2. Properties of the formalisms considered so far with respect to the implication $\longrightarrow$. Theorems of relevance

1. We introduce some new concepts. If the implication

$$\mathfrak{A}_1 \to (\ldots \to (\mathfrak{A}_n \to \mathfrak{B})\ldots) \tag{27}$$

is such that the formula $\mathfrak{B}$ contains no implication, then $\mathfrak{B}$ is called the *last consequent* of the implication (27), and the list $\{\mathfrak{A}_1; \ldots; \mathfrak{A}_n\}$ will be called the *full enumeration of the hypotheses* of the implication (27), whereas the enumeration of formulas $\{\mathfrak{A}_1; \ldots; \mathfrak{A}_n;\ \mathfrak{B}\}$ is the *full enumeration of the hypotheses and the last consequent* of (27).

We shall call two formulas $\mathfrak{A}$ and $\mathfrak{B}$ *directly relevant to each other* if there is a propositional letter α that is a subformula of both $\mathfrak{A}$ and $\mathfrak{B}$. In this case we shall call α the *factor of relevance* for $\mathfrak{A}$ and $\mathfrak{B}$. Since by definition abs is not a propositional letter, abs cannot be a factor of relevance of two formulas: the relevance between formulas is always effected by propositional letters, never by abs.

Suppose we are given enumeration of formulas

$$\{\mathfrak{A}_1; \ldots; \mathfrak{A}_n\}. \tag{28}$$

We shall call the formulas $\mathfrak{A}_i$ and $\mathfrak{A}_j$ of this enumeration *relevant* (*indirectly relevant*) *with respect to the enumeration* (28) if there is a sequence (chain) of formulas

$$\mathfrak{A}_i;\ \mathfrak{A}_{k_1}; \ldots; \mathfrak{A}_{k_r};\ \mathfrak{A}_j \tag{29}$$

constructed from the formulas of the original enumeration (28) such that each pair of adjacent formulas of the chain (29) are directly relevant to each other.

We shall call the sequence of formulas (28) *relevant* if each pair of formulas is relevant with respect to this sequence. We shall call the implication $\mathfrak{A}$ *relevant* if the full enumeration of its hypotheses and the last consequent is relevant.

Put in another way, in this sense of relevant implication, the hypothesis and the last consequent always refer to things that are, generally speaking, connected in a single situation.

2. In the minimally relevant propositional calculus implication has the simplest properties of relevance. Here we have

THEOREM III. *If in an implicational formula provable in the minimally relevant propositional calculus we replace every occurrence of* abs *by the same propositional letter, we obtain a formula provable in the same minimally-relevant propositional calculus.*

This theorem essentially claims that in the provable formulas of the minimally relevant calculus abs can occur only as a result of substitution of propositional letters. In other words, it is always possible to find a formula provable in the minimally relevant calculus having no occurrence of abs, from which by substituting abs for propositional letters we can obtain any required provable formula having occurrences of abs. This was why in Theorem II abs ranged over the set of elements of the lattice Γ, like a propositional letter. Consequently, in the minimally relevant propositional calculus, when establishing the relevance of implication abs plays the same role of a link (factor of relevance) between the hypotheses and the last consequent as propositional letters. Theorem III reveals that in fact the relevance between hypotheses and the last consequent in an implication provable in the minimally relevant calculus is effected by propositional letters rather than by abs, which can be substituted for propositional letters later on. It can be shown that, under the conditions mentioned, in minimally relevant calculi every provable implication is relevant. This is the reason for calling this calculus *minimally* relevant. This is the meaning and content of Theorem III.

The proof of Theorem III is divided into a sequence of steps stated as lemmas.

LEMMA I. *If in a formula provable in the minimally relevant propositional calculus we replace every occurrence of* abs *by the same propositional letter, then the formula obtained is provable in the same minimally-relevant propositional calculus.*

The lemma is proved by induction on the length of proof.

In order to state Lemma II it is necessary to introduce a

DEFINITION. The enumerations of formulas $\{\mathfrak{A}_1, \ldots, \mathfrak{A}_n\}$ and $\{\mathfrak{B}_1, \ldots, \mathfrak{B}_r\}$ will be called *relevant to each other* if at least one formula of one sequence is directly relevant to at least one formula of the other sequence (in other words, the intersection of the sets of propositional letters of the first and second sequence is nonempty).

LEMMA II. *An enumeration of formulas is relevant if and only if every decomposition into two disjoint nonempty subsequences is a decomposition into subsequences relevant to each other.*

LEMMA III. *If a formula* $\mathfrak{A}$ *contains no occurrence of* abs, *and if in its interpretation in* Γ *as described above the propositional letters of* $\mathfrak{A}$ *assume as values one and the same nonzero element* $\alpha \in \Gamma$, *then* $\mathfrak{A}$ *assumes the same value* $\alpha \in \Gamma$.

The lemma is proved by induction on the number of logical connectives in $\mathfrak{A}$.

Given the preceding three lemmas, it is easy to complete the proof of Theorem III. In fact, by Lemma I, by substituting an arbitrary propositional letter for all occurrences of abs in the provable implication we obtain the provable formula

$$\widetilde{\mathfrak{A}} \rightleftharpoons [\mathfrak{A}_1 \to (\mathfrak{A}_2 \to \cdots (\mathfrak{A}_n \to \mathfrak{B}) \ldots)], \tag{1.5}$$

where $\mathfrak{B}$ is the last consequent of $\widetilde{\mathfrak{A}}$ and where there is now no occurrence of abs. The formula $\widetilde{\mathfrak{A}}$ has the following full enumeration of hypotheses and the last consequent:

$$\{\mathfrak{A}_1; \mathfrak{A}_2; \ldots; \mathfrak{A}_n; \mathfrak{B}\}. \tag{2.5}$$

Let us decompose (2.5) in an arbitrary way into two disjoint nonempty enumerations which together come to (2.5); thus

$$\{\mathfrak{A}_{i_1}; \ldots; \mathfrak{A}_{i_k}\} \tag{3.5}$$

and

$$\{\mathfrak{A}_{j_1}; \ldots; \mathfrak{A}_{j_r}; \mathfrak{B}\}. \tag{4.5}$$

It is easy to see that (3.5) and (4.5) are relevant, since otherwise if we take the elements $0 < \beta < \lambda < \infty$ from Γ and if we suppose that in all formulas of (3.5) all propositional letters take the value λ, and if we suppose that in all formulas of (3.5) all propositional letters take the value β, then by Lemma III the implication (1.5) assumes the value zero, which is impossible since (1.5) is derivable. Then by Lemma II the enumeration (2.5) is relevant, which proves Theorem 3.

3. In the intuitionistically-completed propositional calculi the properties of relevance of implication are more complicated.

Let us consider the lattices Γ containing a greatest element ∞ and a least element 0 such that the set

$$B = \Gamma \setminus \{0; \infty\} \tag{30}$$

obtained by omitting zero and ∞ from Γ is a sublattice of Γ, i.e. having the property that

$$\text{if } \alpha \in B \text{ and } \beta \in B, \text{ then } (\alpha \cup \beta) \in B \text{ and } (\alpha \cap \beta) \in B. \tag{31}$$

It is easy to see that if we adjoin the greatest element ∞ and the least element 0 to an arbitrary distributive lattice B, the resulting lattice is distributive as well, and B is a sublattice of Γ.

In this way we easily construct the distributive lattices Γ that will be required in the sequel.

Suppose that Γ and its sublattice B, obtained by omitting the greatest and the least element from Γ, are such that we can define (at least) one function $f(\alpha, \beta)$ in Γ, for all $\alpha, \beta \in B$, and having the following properties for all $\alpha \in B$:

$$\begin{aligned} &\text{a)} \quad \text{if } \beta \in B, \quad \text{then } f(\alpha, \beta) \geqq \alpha \text{ and } f(\alpha, \beta) \in B, \\ &\text{b)} \quad \text{if } \beta < \gamma, \quad \text{then } f(\alpha, \beta) < f(\alpha, \gamma), \\ &\text{c)} \quad \text{if } \beta \parallel \gamma, \quad \text{TO } f(\alpha, \beta) \parallel f(\alpha, \gamma), \\ &\text{d)} \ f(\alpha; \beta \cup \gamma) = f(\alpha, \beta) \cup f(\alpha, \gamma), \\ &\text{e)} \ f(\alpha; \beta \cap \gamma) = f(\alpha, \beta) \cap f(\alpha, \gamma). \end{aligned} \tag{32}$$

Any lattice Γ having these properties will be called a *lattice admitting an isomorphic binary operation.*

An example of such a lattice is the set of natural numbers where the ordering relation $n < k$ denotes that k is divisible by n, where l.u.b. is the union, g.l.b. is the intersection and $f(n, k) = n \cdot k$, etc.

The theorem of relevance in the intuitionistically-completed calculi([5]) can be stated in the following way:

THEOREM IV. *Let Γ be a distributive lattice having the following properties.*

1. *Γ contains a greatest element ∞ and a least element* 0.

2. *The set B obtained from Γ by omitting zero and ∞ is a sublattice of Γ (i.e. B is a lattice with the same ordering as Γ).*

3. *Γ is a lattice admitting an isomorphic binary operation.*

4. *If $\alpha \in B$, then there is a $\beta \in B$ such that $\alpha < \beta$ (i.e. for every α from B there is a greater element in B).*

The formulas of intuitionistically-completed propositional calculi will be interpreted as before (see (5)–(7)), supposing that abs $\equiv 0$.

Then if an implicational formula provable in the intuitionistically-completed propositional calculus takes a value different from ∞ (that is, takes at least one value from B), it follows that this formula is relevant, and takes infinitely many different values from B.

As we have seen, the set of natural numbers ordered according to divisibility is an example of a sublattice B admitting an isomorphic binary operation: $f(n, k) = n \cdot k$ or $f(n, k) = (2k)^n$, etc.

Thus, completing the natural numbers in order of magnitude by a minimal element 0 and by a maximal element ∞, with respect to the given ordering, we obtain a lattice Γ to which we can apply Theorem IV on the relevance of formulas provable in the intuitionistically-completed propositional calculi.

We shall derive some consequences of Theorem IV.

If in an implicational formula provable in the intuitionistically-completed propositional calculus abs has no occurrence, then, provided that the propositional letters of that formula take as values the same element $\alpha \in B$, the whole formula takes the value α. From Theorem IV follows

THEOREM V. *Any implicational formula provable in an intuitionistically completed propositional calculus and having no occurrence of* abs *is relevant.*

So, in the propositional calculi considered so far, irrelevance in provable implicational formulas occurs only when abs occurs.

We can understand this fact if we bear in mind that the only nonrelevant intuitionistic principle among the postulates of implication is (abs $\longrightarrow A$). This "obvious" fact was proved in Theorem V.

It is interesting to investigate the following contrast.

On the one hand, the provable nonrelevant implications of the intuitionistically completed propositional calculus, according to Theorem IV, always take the value ∞ in the interpretations in lattices Γ having properties given in Theorem IV.

([5]) A. V. Kuznecov, of Kishinev (USSR), in 1969 strengthened the present result, proving that derivable implications are relevant without the conditions 3 and 4.

On the other hand, every implication having contradictory antecedents is provable in the intuitionistically-completed propositional calculi, and in the interpretations of this kind always takes the value ∞.

We then raise the natural question: is this the only reason (the inconsistency of antecedents in some subformula of an implication) for the irrelevance of provable implications?

4. Let us investigate the relevance in provable implications. Given an implication

$$\mathfrak{A}_1 \to (\mathfrak{A}_2 \to \cdots (\mathfrak{A}_n \to \mathfrak{B}) \ldots), \tag{33}$$

let Γ be a lattice admitting an isomorphic binary operation. We interpret the formulas of intuitionistically-completed propositional calculi as before. Let the set G contain the last consequent of (33) as well as those antecedents which in the given interpretation have the value zero. For any implication (33), let us call the set G obtained in this way the *core* of the given implication (33) with respect to the given interpretation. One and the same implication in different interpretations (from the class of interpretations described) may have disjoint cores. The following important theorem holds.

THEOREM VI. *If a formula provable in one of the propositional calculi thus far considered, with a given interpretation in some lattice Γ admitting an isomorphic binary operation, takes a value different from ∞ then* 1) *the core of this implication is relevant, and* 2) *every antecedent of the implication not contained in the core is relevant to it.*

Since the selection of antecedents for the core of the implication (33) depends on the interpretation, we are interested in cores containing the smallest number of hypotheses. Obviously, by completing the sublattice B, and by adjoining new elements and obtaining a new sublattice playing the role of B, we have, generally speaking, additional ways of giving the value zero to those formulas which in a previous B never had the value zero. Hence, if we want to reduce the number of antecedents belonging to the cores of provable implications to a minimum, we have to use in the interpretations a denumerable linear ordering of the lattices B such that in the finite lattices condition 4) of Theorem IV does not hold.

The theorems of this section have the following interesting consequences.

In a particular class of provable implications (not identically equal to ∞) the relevance of all formulas is guaranteed by the relevance of the core. But in the core there are antecedents selected in such a way that they can assume the value zero (in some sense they can be false). If a formula assumes the value zero in the interpretations considered thus far, this means that such a formula is unprovable (Theorem I). In this way the relevance of the class of provable implications under consideration is guaranteed by the mutual relevance of nonprovable antecedents. In other words, in order to obtain a relevant provable implication it is first necessary to connect unprovable antecedents with each other, at least using the same terms only to build in such relevance. Usually such a formal link is guaranteed by the link in meaning between antecedents, i.e. by its content. If a formula not in the core is provable, it must be formulated in the language which corresponds to the terms of the core (at the expense of an application of the substitution rule).

This is the reason why in relevant propositional calculi adjoining new hypotheses in a provable formula is not allowed, regardless of whether the consequent is provable (see the theorems of relevance in this section). Moreover, it turns out that in provable implications, generally, we are not allowed to reiterate hypotheses, i.e. to adjoin as new hypotheses those we already have. For this reason it is useful to call those antecedents in provable implications *formulas occurring in the proof* of this implication. This is important in itself; and in addition it will be useful to have available the phrase, *formula occurring in the proof* (of this or that consequent) rather than the term "hypothesis" or "antecedent". Secondly, if in a particular class of provable implications the consequents are deducible only from *unprovable* antecedents, then the logical formalism of relevant calculi appears to be a set of rules for deducing conclusions from unprovable propositions—and nothing more. The formulas provable in relevant formalisms are as a rule not allowed to be adjoined as theorems or axioms to the axioms of any particular system (for example, to the axioms of formal arithmetic). In other words, as a result we have a splitting of the concept of "axiom", which is indivisible in traditional formalisms, to the essentially different concepts of a "postulate" (provable formula) and of an "axiom" which was exactly secured by our terminology. It follows that in axiomatic constructions of the sort of formal arithmetic, formal set theory, etc., the apparatus of relevant propositional calculi does not allow us to obtain consequences "*ex nihilo*" at the expense of adjoining provable formulas to the consequents obtained from the axioms of, say, arithmetic. In relevant calculi the consequents are deduced exclusively from the axioms, and not without them, as is the case in the traditional calculi. It follows that if the formalism of relevant propositional calculi contributes anything at all to the axiomatic constructions of the sort of formal arithmetic, set theory, analysis, etc., this is much less than is the case with traditional formalisms. As a rule, everything is restricted by the fact that the relevant formalisms play the role of an apparatus (rules) for transforming propositions—and that's all. By means of relevant calculi we develop consequences of the axioms, but we do not construct such consequences without axioms or without the axioms being actually and essentially used.

CHAPTER III. THE METHODS OF OBTAINING PROVABLE FORMULAS IN RELEVANT PROPOSITIONAL CALCULI AND PROPERTIES THAT FOLLOW THEREFROM

§1. General remarks

1. In §2 of Chapter I, we remarked that relevant propositional calculi are subsystems of the corresponding traditional propositional calculi. A natural question is raised: excluding the principle $(A \longrightarrow (B \longrightarrow A))$ from the provable formulas, do we at the same time sacrifice logical methods which it would be undesirable or unallowable to sacrifice? Is our narrowing of the class of provable formulas judicious? How good and flexible can relevant propositional calculi be in the formalization of intuitive mathematical reasoning, compared with the traditional calculi?

It turns out that the relevant propositional calculi considered here offer no fewer possibilities than the traditional ones. Moreover, the relevant propositional calculi contain analogs of almost all rules of inference and metatheorems of the corresponding traditional formalisms. This fact allows us to use, in order to prove formulas, habitual techniques of inference developed in the traditional calculi, so that no new training is needed. This chapter is devoted to the facts of just this kind.

2. In relevant propositional calculi we find analogs of the deduction theorem, of the equivalent substitution theorem, of Glivenko's theorem, and of almost all traditional inference rules and theorems. We state these and other analogs in the sequel. At the moment we shall only note that the richness and variety of the possibilities for obtaining provable formulas show that the class of provable formulas has been narrowed in a reasonable way: the class of formulas provable in relevant propositional calculi is sufficiently extensive for the formalization of anything formalizable in the traditional calculi. Moreover, the process of formalization is no less natural in the relevant formalisms than in the traditional ones. This fact should be kept in mind in reading what follows.

§2. Examples of provable formulas

Since in relevant propositional calculi there is an analog of the traditional logical connective "equivalence" (this will later on be shown to be useful), for the sake of economy we take advantage of this fact in the present section.

For example, in the minimally-relevant propositional calculus the following formulas are provable:

(1) $A \sim A,$

(2) $A \,\&\, B \sim B \,\&\, A,$

(3) $A \vee B \sim B \vee A,$

(4) $A \vee (A \,\&\, B) \sim A,$

(5) $A \,\&\, (A \vee B) \sim A,$

(6) $A \vee A \sim A,$

(7) $A \sim A \,\&\, A,$

(8) $A \,\&\, (B \,\&\, C) \sim (A \,\&\, B) \,\&\, C,$

(9) $A \vee (B \vee C) \sim (A \vee B) \vee C.$

In other words, if we interpret the symbols & and $\vee$ as operations on elements of a set $\mathfrak{N}$, then the $\mathfrak{N}$ must be a lattice in which intersection coincides with & and union coincides with $\vee$. We already took advantage of this fact when stating the interpretation of implication.

(10) $A \,\&\, (B \vee C) \sim (A \,\&\, B) \vee (A \,\&\, C),$

(11) $$A \vee (B \& C) \sim (A \vee B) \& (A \vee C),$$

(12) $$(A \to B) \to (\neg B \to \neg A),$$

This is the principle of contraposition. It is obtained by the substitution of abs for C in the postulate (I–1).

(13) $$(A \vee B \to C) \sim (A \to C) \& (B \to C).$$

Substituting everywhere abs for C, we obtain

(13a) $$\neg(A \vee B) \sim \neg A \& \neg B,$$

(14) $$(C \to A \& B) \sim (C \to A) \& (C \to B),$$

(15) $$(A \to C) \vee (B \to C) \to (A \& B \to C).$$

Substituting everywhere abs for C, we obtain

(15a) $$\neg A \vee \neg B \to \neg(A \& B).$$

As we have seen, the contrapositive of this formula does not hold in relevant propositional calculi.

(16) $$[A \vee (A \to C) \to C] \to C.$$

Substituting abs for C, we obtain

(16a) $$\neg\neg(A \vee \neg A).$$

We note that this formula is provable in the minimally relevant propositional calculus where the law of excluded middle $A \vee \neg A$ is not provable.

(17) $$A \to [(A \to C) \to C]$$

is obtained from $[(A \to C) \to (A \to C)]$ by permutation of antecedents. Substituting abs for C in (17), we obtain

(17a) $$A \to \neg\neg A.$$

(18) $$\{[(A \to C) \to C] \to C\} \to (A \to C)$$

is easily obtained in the following way. Adjoining (17) as an additional and first hypothesis to (18), we construct a formula obtainable from the postulate (I–1) by appropriate substitutions. Since (17) is provable, by *modus ponens* we obtain (18).

Substituting abs for C in (18), we have

(18a) $$\neg\neg\neg A \to \neg A,$$

(19) $$(A \vee \neg A) \& (B \vee \neg B) \to (A \& B) \vee \neg(A \& B).$$

Attention should be paid to the fact that in the minimally-relevant propositional calculi a considerable number of formulas are provable that are analogs of the theorems for negation in the traditional propositional calculi. Also, all these analogs are obtained by substitution of abs for a propositional letter in some formulas. This is an illustration of Theorem III in §5 of the preceding chapter. In this way it turns out that many properties of negation can be eliminated by the following definition: $\neg\mathfrak{A} \rightleftarrows (\mathfrak{A} \longrightarrow \text{abs})$, i.e. it is already contained in the relations between conjunction, disjunction and implication, and their properties.

§3. Simple deduction. Deduction theorem(6)

Consider an alphabet L; let us adjoin each formula from the formalisms so far considered, and also the symbol $\vdash$, to L. The formulas having occurrences to the left of $\vdash$ (if any) in those words of L having only one occurence of $\vdash$ will be called *hypotheses* of this word; the formulas occurring to the right of $\vdash$ (if any) will be called *consequents* of this word. In the words in the alphabet L we separate the formulas by commas.

Definition of simple deduction. By a *simple deduction* (in any of the given propositional calculi) we shall understand words in the alphabet L having only one occurrence of $\vdash$ and which can be obtained by the following rules.

1) If $\mathfrak{A}$ is provable, then the word $\vdash \mathfrak{A}$ in the alphabet L is a simple deduction.

2) $\mathfrak{B} \vdash \mathfrak{B}$ is a simple deduction for any formula $\mathfrak{B}$.

3) If $\mathfrak{A}_1; \ldots; \mathfrak{A}_n \vdash \mathfrak{B}$ and $\mathfrak{B} \vdash \mathfrak{C}$ are simple deductions, then so is $\mathfrak{A}_1; \ldots; \mathfrak{A}_n \vdash \mathfrak{C}$.

4) If $\mathfrak{A}_1; \ldots; \mathfrak{A}_n \vdash (\mathfrak{B} \longrightarrow \mathfrak{C})$ and $\vdash \mathfrak{B}$ are simple deductions, then so is $\mathfrak{A}_1; \ldots; \mathfrak{A}_n \vdash \mathfrak{C}$.

5) If $\mathfrak{A}_1; \ldots; \mathfrak{A}_n \vdash \mathfrak{B}$ and $\mathfrak{B}_1; \ldots; \mathfrak{B}_k \vdash (\mathfrak{B} \longrightarrow \mathfrak{C})$ are simple deductions, then so is $\mathfrak{A}_1; \ldots; \mathfrak{A}_n; \mathfrak{B}_1; \ldots; \mathfrak{B}_k \vdash \mathfrak{C}$.

6) A simple deduction may be obtained by permutations of hypotheses from a given simple deduction.

Obviously, a simple deduction always contains a consequent. The consequent is a single formula. This is a characteristic feature of a simple deduction, which distinguishes it from other types of deduction that we shall consider later. Obviously, every simple deduction with no hypotheses has a provable formula as its consequent.

The following proposition holds for simple deductions.

THEOREM VII (Deduction theorem). *In the propositional calculi considered here, if* $\mathfrak{A}_1; \ldots; \mathfrak{A}_{i-1}; \mathfrak{A}_i; \mathfrak{A}_{i+1}; \ldots; \mathfrak{A}_n \vdash \mathfrak{B}$ *is a simple deduction, then* $\mathfrak{A}_1; \ldots; \mathfrak{A}_{i-1}; \mathfrak{A}_{i+1}; \ldots; \mathfrak{A}_n \vdash (\mathfrak{A}_i \longrightarrow \mathfrak{B})$ *is also a simple deduction (for any i).*

We first prove a lemma:

The formula $(A \longrightarrow A)$ *is provable in the minimally relevant propositional calculus.*

From

$$\frac{(A \to A \vee A);\ (A \to A \vee A)}{A \to (A \vee A)\ \&\ (A \vee A)}$$

we obtain the provable formula $A \longrightarrow (A \vee A)\ \&\ (A \vee A)$. By distributivity we have

$$A \to [(A \vee A)\ \&\ A] \vee [(A \vee A)\ \&\ A]. \tag{1.1}$$

(6) For the symbol $\vdash$ defined here formally as a variant of the notion of derivability of $\mathfrak{B}$ from $\mathfrak{A}_1; \ldots; \mathfrak{A}_n$, see the paragraph numbered 2 on p. 2.

Hence from

$$\frac{[(A \vee A) \,\&\, A \rightarrow A];\ [(A \vee A) \,\&\, A \rightarrow A]}{[(A \vee A) \,\&\, A] \vee [(A \vee A) \,\&\, A] \rightarrow A},$$

using (1.1), we obtain $(A \longrightarrow A)$. This proves the lemma.

We shall prove the deduction theorem by induction on the length of simple deductions. In the definition of simple deduction 1) and 2) are the basis and 3)–6) are inductive steps.

The condition of the theorem (the enumeration of hypotheses is nonempty) excludes 1). Part 2) of the definition requires that for any formula $\mathfrak{B}$ the formula $(\mathfrak{B} \longrightarrow \mathfrak{B})$ is provable. But this has been proved.

3) Supposing that the deduction theorem holds for the simple deduction $\mathfrak{A}_1; \ldots; \mathfrak{A}_n \vdash \mathfrak{B}$, we have to prove that if

$$\vdash (\mathfrak{B} \rightarrow \mathfrak{C}) \tag{2.1}$$

the deduction theorem holds for the simple deduction $\mathfrak{A}_1; \ldots; \mathfrak{A}_n \vdash \mathfrak{C}$ as well.

By the inductive hypothesis we have

$$\mathfrak{A}_1; \ldots; \mathfrak{A}_{i-1}; \mathfrak{A}_{i+1}; \ldots; \mathfrak{A}_n \vdash (\mathfrak{A}_i \rightarrow \mathfrak{B}). \tag{3.1}$$

We also have

$$\vdash [(\mathfrak{A}_i \rightarrow \mathfrak{B}) \rightarrow ((\mathfrak{B} \rightarrow \mathfrak{C}) \rightarrow (\mathfrak{A}_i \rightarrow \mathfrak{C}))]. \tag{4.1}$$

By 3), from (4.1), (3.1) and (2.1) we obtain the simple deduction $\mathfrak{A}_1; \ldots; \mathfrak{A}_{i-1}; \mathfrak{A}_{i+1}; \ldots; \mathfrak{A}_n \vdash (\mathfrak{A}_i \longrightarrow \mathfrak{C})$.

4) Supposing that the deduction theorem holds for the simple deduction $\mathfrak{A}_1; \ldots; \mathfrak{A}_n \vdash (\mathfrak{B} \longrightarrow \mathfrak{C})$, we have to prove that if

$$\vdash \mathfrak{B} \tag{5.1}$$

then the deduction theorem holds for the simple deduction $\mathfrak{A}_1; \ldots; \mathfrak{A}_n \vdash \mathfrak{C}$ as well. By the inductive hypothesis we have

$$\mathfrak{A}_1; \ldots \mathfrak{A}_{i-1}; \mathfrak{A}_{i+1}; \ldots; \mathfrak{A}_n \vdash [\mathfrak{A}_i \rightarrow (\mathfrak{B} \rightarrow \mathfrak{C})]. \tag{6.1}$$

Permuting the hypotheses $\mathfrak{A}_i$ and $\mathfrak{B}$ in the consequent of (6.1), and then using (5.1), by 4) we obtain the simple deduction $\mathfrak{A}_1; \ldots; \mathfrak{A}_{i-1}; \mathfrak{A}_{i+1}; \ldots; \mathfrak{A}_n \vdash (\mathfrak{A}_i \longrightarrow \mathfrak{C})$.

5) We must prove that if the theorem holds for the simple deductions

$$\mathfrak{A}_1; \ldots; \mathfrak{A}_n \vdash \mathfrak{B}, \tag{7.1}$$

$$\mathfrak{B}_1; \ldots; \mathfrak{B}_k \vdash (\mathfrak{B} \rightarrow \mathfrak{C}), \tag{8.1}$$

then it holds for the simple deduction

$$\mathfrak{A}_1; \ldots; \mathfrak{A}_n; \mathfrak{B}_1; \ldots; \mathfrak{B}_k \vdash \mathfrak{C}. \tag{9.1}$$

By the inductive hypothesis, from (7.1) and (8.1) we obtain

$$\mathfrak{A}_1; \ldots; \mathfrak{A}_{i-1}; \mathfrak{A}_{i+1}; \ldots; \mathfrak{A}_n \vdash (\mathfrak{A}_i \rightarrow \mathfrak{B}), \tag{10.1}$$

(11.1) $\mathfrak{B}_1; \ldots; \mathfrak{B}_{j-1}; \mathfrak{B}_{j+1}; \ldots; \mathfrak{B}_k \vdash [\mathfrak{B}_j \to (\mathfrak{B} \to \mathfrak{C})]$.

From (11.1) it follows trivially that

(12.1) $\mathfrak{B}_1; \ldots; \mathfrak{B}_{j-1}; \mathfrak{B}_{j+1}; \ldots; \mathfrak{B}_k \vdash [\mathfrak{B} \to (\mathfrak{B}_j \to \mathfrak{C})]$.

We also have

(13.1) $\vdash (\mathfrak{A}_i \to \mathfrak{B}) \to \{[\mathfrak{B} \to (\mathfrak{B}_j \to \mathfrak{C})] \to [\mathfrak{A}_i \to (\mathfrak{B}_j \to \mathfrak{C})]\}$.

(10.1) and (13.1), by 3), and then (12.1), by 5), give

(14.1) $\mathfrak{A}_1;\ldots;\mathfrak{A}_{i-1};\mathfrak{A}_{i+1};\ldots;\mathfrak{A}_n;\mathfrak{B}_1;\ldots;\mathfrak{B}_{j-1};\mathfrak{B}_{j+1};\ldots;\mathfrak{B}_k \vdash [\mathfrak{A}_i \to (\mathfrak{B}_j \to \mathfrak{C})]$.

Using the simple deductions $\mathfrak{A}_i \vdash \mathfrak{A}_i$ or $\mathfrak{B}_j \vdash \mathfrak{B}_j$, 5) and 6), as well as permutation of $\mathfrak{A}_i$ and $\mathfrak{B}_j$ in (14.1), we obtain both

$$\mathfrak{A}_1; \ldots; \mathfrak{A}_n; \mathfrak{B}_1; \ldots; \mathfrak{B}_{j-1}; \mathfrak{B}_{j+1}; \ldots; \mathfrak{B}_k \vdash (\mathfrak{B}_j \to \mathfrak{C})$$

and

$$\mathfrak{A}_1; \ldots; \mathfrak{A}_{i-1}; \mathfrak{A}_{i+1}; \ldots; \mathfrak{A}_n; \mathfrak{B}_1; \ldots; \mathfrak{B}_k \vdash (\mathfrak{A}_i \to \mathfrak{C}).$$

6) Trivial.

The deduction theorem is proved.

THEOREM VIII. *In the propositional calculi considered here, if in a simple deduction we omit an occurrence of a hypothesis which occurs more than once, we obtain a simple proof.*

THEOREM IX (The converse deduction theorem). *In the propositional calculi considered here, if* $\mathfrak{A}_1; \ldots; \mathfrak{A}_n \vdash (\mathfrak{B} \longrightarrow \mathfrak{C})$ *is a simple deduction, then so is* $\mathfrak{A}_1; \ldots; \mathfrak{A}_n; \mathfrak{B} \vdash \mathfrak{C}$.

Concluding the present section, we give some examples of simple deductions in the minimally relevant propositional calculus.

Examples of simple deductions in the minimally-relevant propositional calculus.

(20) $\mathfrak{B}; (\mathfrak{B} \to \mathfrak{C}) \vdash \mathfrak{C}$,

(21) $(\mathfrak{A} \to \mathfrak{B}); (\mathfrak{B} \to \mathfrak{C}) \vdash (\mathfrak{A} \to \mathfrak{C})$,

(22) $\mathfrak{A} \,\&\, \mathfrak{B} \vdash \mathfrak{A}$,

(23) $\mathfrak{A} \,\&\, \mathfrak{B} \vdash \mathfrak{B}$,

(24) $(\mathfrak{A} \to \mathfrak{B}) \,\&\, (\mathfrak{A} \to \mathfrak{C}) \vdash (\mathfrak{A} \to \mathfrak{B} \,\&\, \mathfrak{C})$,

(25) $\mathfrak{A} \vdash \mathfrak{A} \vee \mathfrak{B}$,

(26) $\mathfrak{B} \vdash \mathfrak{A} \vee \mathfrak{B}$,

(27) $\mathfrak{A} \,\&\, (\mathfrak{B} \vee \mathfrak{C}) \vdash (\mathfrak{A} \,\&\, \mathfrak{B}) \vee (\mathfrak{A} \,\&\, \mathfrak{C})$,

(28) $(\mathfrak{A} \to \mathfrak{C}) \,\&\, (\mathfrak{B} \to \mathfrak{C}) \vdash (\mathfrak{A} \vee \mathfrak{B} \to \mathfrak{C})$,

(29) $(\mathfrak{A} \to \mathfrak{B}); (\mathfrak{C} \to \mathfrak{R}) \vdash (\mathfrak{A} \vee \mathfrak{C} \to \mathfrak{B} \vee \mathfrak{R})$,

(30) $(\mathfrak{A} \to \mathfrak{B}); (\mathfrak{B} \to \mathfrak{A}) \vdash (\mathfrak{A} \to \mathfrak{B}) \,\&\, (\mathfrak{B} \to \mathfrak{A})$.

The simple deduction (20) has a special role in the sequel as well as in the methods of obtaining provable formulas. It is obtained by application of Theorem IX to the simple deduction $(\mathfrak{B} \longrightarrow \mathfrak{C}) \vdash (\mathfrak{B} \longrightarrow \mathfrak{C})$. The simple deduction (30) expresses the basic property of the logical connective equivalence, and, of course, it ought to be provable.

§4. Complex deduction. Its transitivity. Chain deduction

1. Using the deduction theorem, we can obtain the corresponding provable formula from a simple deduction in a relevant propositional calculus. We also have, by the converse theorem, that from any provable implication we can easily obtain the corresponding simple deduction. Therefore, having the simple deduction, we also have a useful inference rule, and we easily obtain the corresponding provable formula. If the methods of construction of simple deductions from other simple deductions are more uniform and flexible, i.e. if they are closer to logical acts of intuitive mathematical thought(7) than complex methods of obtaining provable formulas from other provable formulas, then operations on simple deductions are of primary interest. The situation is similar in traditional propositional calculi.

From one of those same hypotheses (for example axioms) it is sometimes possible to deduce consequences (for example theorems); then from the consequences obtained (theorems) we can deduce new consequences, and so on, until we eventually derive the next theorem. Later on we always suppose that the last consequence (theorem) is deduced from the original enumeration of hypotheses (i.e. axioms).

The scheme of deduction described is typical, and we very often find it in intuitive mathematical thinking. As we know, this scheme of reasoning can be found (and proved) in traditional formal propositional calculi as well.

We show that the same useful scheme of reasoning can be formalized in the relevant propositional calculi. This section is devoted to the formalization of this scheme, and for this purpose the concepts of *complex deduction* and *chain deduction* will be introduced.

2. We shall introduce deductions (complex deductions) having several properties. It will be convenient to use the following

DEFINITION (occurrence in a deduction). We shall say that a formula which is a hypothesis in a simple deduction *occurs in this simple deduction* (of its consequent).

We shall say that any formula not coinciding with one of the hypotheses of a simple deduction *does not occur in this simple deduction* (of its consequent).

Using this terminology, a *complex deduction*

$$\mathfrak{A}_1; \ldots; \mathfrak{A}_n \vdash \mathfrak{B}_1; \ldots; \mathfrak{B}_k \tag{31}$$

is a word in the alphabet L which has only one occurrence of $\vdash$, and each hypothesis

(7) Reflecting the provability of one formula from others.

of which occurs in a simple deduction of at least one consequent of the complex deduction (31). By this we mean that every consequent in (31) must be deduced (in a simple deduction) from some hypothesis of the complex deduction (31), or else must be provable.

This is, so to speak, a descriptive definition of complex deduction.

We now give a formal definition of the same concept.

DEFINITION (Complex deduction). Suppose we have several simple deductions

$$(32)\qquad \begin{cases} \mathfrak{A}_{11}; \mathfrak{A}_{12}; \ldots; \mathfrak{A}_{1n_1} \vdash \mathfrak{B}_1, \\ \mathfrak{A}_{21}; \mathfrak{A}_{22}; \ldots; \mathfrak{A}_{2n_2} \vdash \mathfrak{B}_2, \\ \ldots\ldots\ldots\ldots \\ \mathfrak{A}_{k1}; \mathfrak{A}_{k2}; \ldots; \mathfrak{A}_{kn_k} \vdash \mathfrak{B}_k. \end{cases}$$

Starting with these simple deductions we construct a special string by the following rules (recursive definition).

By a *complex deduction* corresponding to the list of simple deductions (32), we shall mean words in the alphabet L which can be obtained by one of the following rules.

a) Write down all hypotheses of all simple proofs (32) in alphabetical order, then write down the symbol $\vdash$, and then write down in alphabetical order the consequents of the simple deductions (32). Thus we obtain

$$(33)\quad \mathfrak{A}_{11}; \mathfrak{A}_{12}; \ldots; \mathfrak{A}_{1n}; \mathfrak{A}_{21}; \mathfrak{A}_{22}; \ldots; \mathfrak{A}_{2n_2}; \ldots; \mathfrak{A}_{k1}; \mathfrak{A}_{k2}; \ldots; \mathfrak{A}_{kn_k} \vdash \mathfrak{B}_1; \mathfrak{B}_2; \ldots; \mathfrak{B}_k.$$

The string (word) (33) is a complex deduction corresponding to the list of simple deductions (32).

b) If (33) is a complex deduction corresponding to the list of simple deductions (32), then after an arbitrary permutation of all formulas in (33) such that each formula occurring in (33) to the left of $\vdash$ remains to the left of $\vdash$ and each formula occurring in (33) to the right of $\vdash$ remains to the right of $\vdash$, we obtain a new complex deduction corresponding to the same list of simple deductions.

c) If in the complex deduction corresponding to the list of simple deductions (32) we omit any occurrence of a formula having more than one occurrence to the left of $\vdash$, we obtain a complex deduction corresponding to the same list of simple deductions (32).

In a complex deduction we shall call the full enumeration of formulas to the left of $\vdash$ the *complex hypothesis* of this complex deduction, and we shall call any formula of this enumeration a *hypothesis* of this complex deduction.

In a complex deduction we shall call the full enumeration of formulas to the right of $\vdash$ the *complex consequent* of this complex deduction, and we shall call any formula of this enumeration a *consequent* of this complex deduction.

Obviously, every complex deduction with no hypotheses has as consequents only provable formulas. It is also easy to see that every simple deduction is a complex deduction and every complex deduction with a single consequent is a simple deduction.

Therefore, if we are able to transform complex deductions in such a way that the number of consequents is decreased, and if we eventually obtain a single consequent, then we have a method of obtaining a simple deduction. This is done by chain deductions.

3. Before we introduce chain deductions, it is necessary to pay attention to the following important facts.

Complex deductions are essentially a very condensed and compact expression for a list of simple deductions corresponding to it. Therefore, whenever in the sequel we deal with complex deductions, we must not forget the list of simple deductions which is summarized by the complex deduction. In other words, we must be able to deduce every consequent of the complex deduction from some hypotheses, using the simple deduction built up from these hypotheses and this consequent. Every hypothesis of a complex deduction must occur in at least one simple deduction of at least one consequent, and every consequent of the complex deduction must be derivable from some hypotheses or must be a theorem. In this way we can always construct a possible list of simple deductions corresponding to the given complex deduction.

In the sequel we shall not explicitly mention the original simple deductions the list of which corresponds to the given complex deduction. In such cases the reader should establish for himself at least one list of simple deductions corresponding to the given complex deduction. This is not difficult.

Using complex deductions instead of a list of simple deductions, we can arrive at condensed forms for expressing formal transformations which are nevertheless easily understandable.

In writing complex deductions we shall omit simple deductions in order to avoid obvious, trivial and unimportant points in our formalisms.

4. We must pay attention to the fact that, in general, we are not allowed to omit occurrences of consequents having more than one occurrence in a complex deduction (we can omit formulas only on the left of $\vdash$). If we omit consequents having more than one occurrence, we could obtain strings that are not complex deductions.

5. The following proposition holds for complex deductions.

THEOREM X (The "fusion" of complex deductions). *In the propositional calculi considered here, let* $\mathfrak{A}_1; \ldots; \mathfrak{A}_n \vdash \mathfrak{B}_1; \ldots; \mathfrak{B}_k$ *and* $\mathfrak{C}_1; \ldots; \mathfrak{C}_r \vdash \mathfrak{K}_1; \ldots; \mathfrak{K}_l$ *be complex deductions. Then* $\mathfrak{A}_1; \ldots; \mathfrak{A}_n; \mathfrak{C}_1; \ldots; \mathfrak{C}_r \vdash \mathfrak{B}_1; \ldots; \mathfrak{B}_k; \mathfrak{K}_1; \ldots; \mathfrak{K}_l$ *is also a complex deduction.*

This proposition can be written in the following form:

$$\frac{\begin{cases} \mathfrak{A}_1; \ldots; \mathfrak{A}_n \vdash \mathfrak{B}_1; \ldots; \mathfrak{B}_k \\ \mathfrak{C}_1; \ldots; \mathfrak{C}_r \vdash \mathfrak{R}_1; \ldots; \mathfrak{R}_l \end{cases}}{\mathfrak{A}_1; \ldots; \mathfrak{A}_n; \mathfrak{C}_1; \ldots; \mathfrak{C}_r \vdash \mathfrak{B}_1; \ldots; \mathfrak{B}_k; \mathfrak{R}_1; \ldots; \mathfrak{R}_l.}$$

Obviously, any finite set of complex deductions has a fusion.

THEOREM XI (Transitivity of complex deductions). *In the propositional calculi considered here, let* $\mathfrak{A}_1; \ldots; \mathfrak{A}_n \vdash \mathfrak{B}_1; \ldots; \mathfrak{B}_k$ *and* $\mathfrak{B}_1; \ldots; \mathfrak{B}_k \vdash \mathfrak{C}_1; \ldots; \mathfrak{C}_r$ *be two complex deductions. Then* $\mathfrak{A}_1; \ldots; \mathfrak{A}_n \vdash \mathfrak{C}_1; \ldots; \mathfrak{C}_r$ *is also a complex deduction.*

6. The last fact allows a partial ordering of the enumeration of formulas $\mathfrak{N}$, i.e. the set $\mathfrak{N}$ can be transformed into a lattice $\mathfrak{N}$. Linear sublattices of $\mathfrak{N}$ will be called *chain deductions.*

In more detail, this concept can be defined (this will be more useful for our purpose) as follows.

DEFINITION (chain deduction). Let

$$(34)\qquad \begin{cases} \mathfrak{A}_1; \ldots; \mathfrak{A}_n \vdash \mathfrak{B}_1; \ldots; \mathfrak{B}_k, \\ \mathfrak{B}_1; \ldots; \mathfrak{B}_k \vdash \mathfrak{C}_1; \ldots; \mathfrak{C}_r, \\ \mathfrak{C}_1; \ldots; \mathfrak{C}_r \vdash \mathfrak{R}_1; \ldots; \mathfrak{R}_l, \\ \cdots\cdots\cdots\cdots\cdots \end{cases}$$

be a finite ordered sequence of complex deductions, where the enumeration of consequents of each complex deduction is the enumeration of hypotheses of the complex deduction that immediately follows it in the sequence (34). Let us represent the sequence of complex deductions (34) by the string (word in the alphabet L; §3)

$$(35)\qquad \mathfrak{A}_1; \ldots; \mathfrak{A}_n \vdash \mathfrak{B}_1; \ldots; \mathfrak{B}_k \vdash \mathfrak{C}_1; \ldots; \mathfrak{C}_r \vdash \mathfrak{R}_1; \ldots; \mathfrak{R}_l \vdash \cdots.$$

We shall call this string a *chain deduction.*

Every sequence of formulas between two occurrences of the sign $\vdash$ in (35) we shall call a *link* of the chain deduction (35).

Let us agree to write indices, numerals, words, sentences and other signs above the symbol $\vdash$ in order to point out what postulates, theorems, axioms, schemata, rules of inference, operations on deductions, etc. have been applied at the given step of deduction of complex consequents. In this connection we shall always require that such information be sufficient for finding out what details have been omitted from formal transformations on deductions.

Obviously, complex deduction and simple deduction are both special cases of chain deduction. Moreover, applying the theorem of transitivity of complex deductions, we obtain the following important proposition.

THEOREM XII. (The basic property of chain deduction). *In the propositional calculi considered here, in any chain deduction any link implies each link to the right of it.*

The fact that we have the simple deduction $\mathfrak{B}; (\mathfrak{B} \longrightarrow \mathfrak{C}) \vdash \mathfrak{C}$ allows us in constructing (extending) a chain deduction to decrease the number of consequents in its last link, obtaining eventually a single consequent, i.e. a simple deduction.

§5. Operations on complex deductions

In the propositional calculi considered here the following rules for obtaining new complex deductions hold. Any of these operations on complex deductions can be considered as an analog of the corresponding transformation valid in the traditional propositional calculi. The difference consists in the fact that the corresponding rules of transformation in the traditional propositional calculi allow more freedom, due to the fact that we have the principle $(A \longrightarrow (B \longrightarrow A))$. Almost always this greater freedom can be reduced to an unlimited possibility of adjoining arbitrary hypothesis to deductions. The propositional calculi considered here exclude such a possibility. This fact has a bearing on the formulation of all derived rules of deduction. But we still have unlimited possibilities of obtaining new complex deductions, and this evidently shows that the class of provable formulas has been restricted in a reasonable way, compared with the traditional propositional calculi.

I. *The fusion of complex deductions.* In the propositional calculi considered here several complex deductions can be put together in one. The hypotheses of the new complex deduction are given by the enumeration of *distinct* hypotheses of all complex deductions, and the consequents by the enumeration of all consequents (including repetitions, if any).

II. In the propositional calculi considered here an arbitrary formula can be adjoined to both parts of a complex deduction. To all links of a chain deduction we can adjoin the same arbitrary formulas:

$$\frac{\mathfrak{A}_1; \ldots; \mathfrak{A}_n \vdash \mathfrak{B}_1; \ldots; \mathfrak{B}_r}{\mathfrak{A}_1; \ldots; \mathfrak{A}_n; \mathfrak{C}_1; \ldots; \mathfrak{C}_g \vdash \mathfrak{B}_1; \ldots; \mathfrak{B}_r; \mathfrak{C}_1; \ldots; \mathfrak{C}_g.}$$

We emphasize that the converse proposition does not hold. One is not allowed to omit a formula from both parts of a complex deduction.

III. In the propositional calculi considered here the following scheme for obtaining a new complex deduction from two others holds (generalized transitivity of complex deductions):

$$\frac{\mathfrak{R}_1; \ldots; \mathfrak{R}_l \vdash \mathfrak{A}_1; \ldots; \mathfrak{A}_n \qquad \mathfrak{A}_1; \ldots; \mathfrak{A}_n; \mathfrak{B}_1; \ldots; \mathfrak{B}_r \vdash \mathfrak{C}_1; \ldots; \mathfrak{C}_k}{\mathfrak{R}_1; \ldots; \mathfrak{R}_l; \mathfrak{B}_1; \ldots; \mathfrak{B}_r \vdash \mathfrak{C}_1; \ldots; \mathfrak{C}_k.}$$

In other words, instead of hypotheses $\mathfrak{A}_1; \ldots; \mathfrak{A}_n$ of one complex deduction, we take hypotheses $\mathfrak{R}_1; \ldots; \mathfrak{R}_l$ of another given complex deduction (compare Gentzen's cut).

IV. If $\mathfrak{R}$ is a provable formula, then $\mathfrak{C}_1; \ldots; \mathfrak{C}_r \vdash \mathfrak{C}_1; \ldots; \mathfrak{C}_r; \mathfrak{R}$ is a complex deduction.

This fact is often used to extend chain deductions. If we adjoin a provable formula as a new consequent to a simple deduction, the result will not be a simple deduction, but a complex deduction. We have to bear in mind that a simple deduction is a complex deduction with a single consequent.

V. In the propositional calculi considered here we have the following rule for obtaining new complex deductions from a given one:

$$\frac{\mathfrak{B}_1; \ldots; \mathfrak{B}_k; \mathfrak{R} \vdash \mathfrak{C}_1; \ldots; \mathfrak{C}_r}{\mathfrak{B}_1; \ldots; \mathfrak{B}_k; \mathfrak{R} \vdash \mathfrak{C}_1; \ldots; \mathfrak{C}_r; \mathfrak{R}}.$$

This rule is used together with the preceding one. In extending a chain deduction we often have to adjoin new formulas to the last link in order to restrict the consequent in the last link, using the deduction $\mathfrak{A}; (\mathfrak{A} \longrightarrow \mathfrak{B}) \vdash \mathfrak{B}$. Using rule V, we substitute an enumeration of formulas $\{\mathfrak{C}_1; \ldots; \mathfrak{C}_r\}$ for the complex consequent

$\{\mathfrak{C}_1; \ldots ; \mathfrak{C}_r; \mathfrak{R}\}$: we have a new consequent $\mathfrak{R}$. On the other hand, in rule IV the provable formula $\mathfrak{R}$ is adjoined as a consequent to the link $\{\mathfrak{C}_1; \ldots ; \mathfrak{C}_r\}$.

In extending a chain deduction it is often useful to apply the generalized transitivity of complex deduction (rule III) in the following form.

VI. In the propositional calculi considered here, if the last line $\mathfrak{C}_1; \ldots ; \mathfrak{C}_r; \mathfrak{A}_1; \ldots ; \mathfrak{A}_n$ in a chain deduction is such that some subsequence $\{\mathfrak{A}_1; \ldots ; \mathfrak{A}_n\}$ of its formulas implies its consequents $\{\mathfrak{B}_1; \ldots ; \mathfrak{B}_k\}$, then, adjoining to this sequence $\{\mathfrak{B}_1; \ldots ; \mathfrak{B}_k\}$ of consequents those formulas from the last link $\{\mathfrak{C}_1; \ldots ; \mathfrak{C}_r; \mathfrak{A}_1; \ldots ; \mathfrak{A}_n\}$ which have not been used in the deduction of consequents $\{\mathfrak{B}_1; \ldots ; \mathfrak{B}_k\}$, we obtain the new link $\{\mathfrak{C}_1; \ldots ; \mathfrak{C}_r; \mathfrak{B}_1; \ldots ; \mathfrak{B}_k\}$ of the chain deduction.

VII. *The conjunctive fusion of simple deductions.* In the propositional calculi considered here the consequents of simple deductions having the same nonempty enumerations of hypotheses can be conjoined by conjunction; the result will be a simple deduction.

In other words, we have the operation

$$\frac{(\&)\begin{cases}\mathfrak{A}_1; \ldots ;\mathfrak{A}_n \vdash \mathfrak{B}\\ \mathfrak{A}_1; \ldots ;\mathfrak{A}_n \vdash \mathfrak{C}\end{cases}}{\mathfrak{A}_1; \ldots ;\mathfrak{A}_n \vdash \mathfrak{B} \,\&\, \mathfrak{C}.}$$

We note that, in this rule, 1) the number of hypotheses in both given simple deductions must be the same, and 2) the hypotheses in both simple deductions must be the same. If either of these conditions is not fulfilled we cannot conjoin the consequents of these deductions, since this could yield strings which are not simple deductions.

There is an easy inductive proof of this rule. If $n = 1$, the rule has the following form:

$$\frac{(\mathfrak{A} \to \mathfrak{B}); (\mathfrak{A} \to \mathfrak{C})}{\mathfrak{A} \to \mathfrak{B} \,\&\, \mathfrak{C}.}$$

For $n + 1$ we have

$$\mathfrak{A}_1; \ldots ;\mathfrak{A}_n \vdash (\mathfrak{A}_{n+1} \to \mathfrak{B}), \tag{1.2}$$

obtained by the deduction theorem from $\mathfrak{A}_1; \ldots ; \mathfrak{A}_n; \mathfrak{A}_{n+1} \vdash \mathfrak{B}$. Analogously,

$$\mathfrak{A}_1; \ldots ;\mathfrak{A}_n \vdash (\mathfrak{A}_{n+1} \to \mathfrak{C}) \tag{2.2}$$

is obtained from $\mathfrak{A}_1; \ldots ; \mathfrak{A}_n; \mathfrak{A}_{n+1} \vdash \mathfrak{C}$. By the inductive hypothesis we can apply fusion to (1.2) and (2.2). We have

$$\mathfrak{A}_1; \ldots ;\mathfrak{A}_n \vdash (\mathfrak{A}_{n+1} \to \mathfrak{B}) \,\&\, (\mathfrak{A}_{n+1} \to \mathfrak{C}), \tag{3.2}$$

and hence by (III–3)

$$\mathfrak{A}_1; \ldots ;\mathfrak{A}_n \vdash (\mathfrak{A}_{n+1} \to \mathfrak{B} \,\&\, \mathfrak{C}), \tag{4.2}$$

i.e.

(5.2) $$\mathfrak{A}_1; \ldots; \mathfrak{A}_n; \mathfrak{A}_{n+1} \vdash \mathfrak{B} \,\&\, \mathfrak{C}.$$

VIII. In the propositional calculi considered here, consequents that are nonprovable formulas can be conjoined arbitrarily with a single hypothesis in a complex deduction.

IX. *The converse of the deduction theorem for complex deductions.* In the propositional calculi considered here, in complex deductions the antecedents of the consequents can be transformed into hypotheses of complex deductions, i.e.

$$\frac{\mathfrak{A}_1; \ldots; \mathfrak{A}_n \vdash (\mathfrak{B} \to \mathfrak{C}); \mathfrak{R}_1; \ldots; \mathfrak{R}_l}{\mathfrak{A}_1; \ldots; \mathfrak{A}_n; \mathfrak{B} \vdash \mathfrak{C}; \mathfrak{R}_1; \ldots; \mathfrak{R}_l.}$$

X. *The conjunctive fusion of simple deductions.* In the propositional calculi considered here, we can obtain a new simple deduction, given two simple deductions which differ from each other in a single hypothesis. In other words, we have

$$(\vee)\frac{\begin{cases}\mathfrak{A}_1; \ldots; \mathfrak{A}_n; \mathfrak{B} \vdash \mathfrak{R} \\ \mathfrak{A}_1; \ldots; \mathfrak{A}_n; \mathfrak{C} \vdash \mathfrak{R}\end{cases}}{\mathfrak{A}_1; \ldots; \mathfrak{A}_n; \mathfrak{B} \vee \mathfrak{C} \vdash \mathfrak{R}.}$$

PROOF. The given simple deductions can be transformed by the deduction theorem into provable formulas

(1.3) $$\mathfrak{B} \longrightarrow \mathfrak{M},$$

(2.3) $$\mathfrak{C} \longrightarrow \mathfrak{M},$$

where

(3.3) $$\mathfrak{M} \rightleftharpoons (\mathfrak{A}_1 \to (\mathfrak{A}_2 \to \cdots (\mathfrak{A}_n \to \mathfrak{R}) \ldots)).$$

If (1.3) and (2.3) are provable, then the formula

(4.3) $$\mathfrak{B} \vee \mathfrak{C} \longrightarrow \mathfrak{M}$$

is also provable, i.e. we have the simple deduction $\mathfrak{B} \vee \mathfrak{C} \vdash \mathfrak{M}$, and hence

(5.3) $$\mathfrak{A}_1; \ldots; \mathfrak{A}_n; \mathfrak{B} \vee \mathfrak{C} \vdash \mathfrak{R}.$$

XI. *The disjunctive fusion of deductions in the intuitionistically-nonrelevant or intuitionistically-completed propositional calculi.* Suppose that $\{\mathfrak{A}_{i_1}; \ldots; \mathfrak{A}_{i_k}\}$ is a subsequence (possibly empty) of the sequence $\{\mathfrak{A}_1; \ldots; \mathfrak{A}_n\}$. Then in the intuitionistically-nonrelevant or intuitionistically-completed propositional calculi we have the rule

$$\frac{\left\{(\vee)\begin{matrix}\mathfrak{A}_1; \ldots; \mathfrak{A}_n; \mathfrak{B} \vdash \mathfrak{R} \\ \mathfrak{A}_{i_1}; \ldots; \mathfrak{A}_{i_k}; \mathfrak{C} \vdash \text{abs}\end{matrix}\right.}{\mathfrak{A}_1; \ldots; \mathfrak{A}_n; \mathfrak{B} \vee \mathfrak{C} \vdash \mathfrak{R}.}$$

PROOF. Suppose that $\mathfrak{A}_{j_1}; \ldots; \mathfrak{A}_{j_r}$ are all formulas from the sequence $\{\mathfrak{A}_1; \ldots; \mathfrak{A}_n\}$ which are not in the sequence $\{\mathfrak{A}_{i_1}; \ldots; \mathfrak{A}_{i_k}\}$. Then from the simple deduction

(1.4) $$\mathfrak{A}_{i_1};\ldots;\mathfrak{A}_{i_k};\mathfrak{C}\vdash \text{abs}$$

and the provable (only in the above-mentioned calculi) formula

(2.4) $$\text{abs}\to[\mathfrak{A}_{j_1}\to(\mathfrak{A}_{j_2}\to\cdots(\mathfrak{A}_{j_r}\to\mathfrak{R})\ldots)]$$

we obtain the simple deduction

(4.4) $$\mathfrak{A}_{i_1};\ldots;\mathfrak{A}_{i_k};\mathfrak{C}\vdash[\mathfrak{A}_{j_1}\to(\mathfrak{A}_{j_2}\to\cdots(\mathfrak{A}_{j_r}\to\mathfrak{R})\ldots)].$$

Hence

(4.5) $$\mathfrak{A}_1;\ldots;\mathfrak{A}_n;\mathfrak{C}\vdash\mathfrak{R}.$$

The problem is thus reduced to the preceding theorem.

DEFINITION. We shall say that the formula $\mathfrak{R}$ occurring in the sequence $\{\mathfrak{R};\mathfrak{A}_1;\ldots;\mathfrak{A}_n\}$ *depends* (*deductively*) on this sequence of formulas, if it can be deduced from some (possibly empty) subsequence $\{\mathfrak{A}_{i_1};\ldots;\mathfrak{A}_{i_k}\}$ of the original sequence $\{\mathfrak{R};\mathfrak{A}_1;\ldots;\mathfrak{A}_n\}$ and does not occur in the subsequence $\{\mathfrak{A}_{i_1};\ldots;\mathfrak{A}_{i_k}\}$.

Every formula provable in the propositional calculi considered here is dependent (deductively) on a sequence of formulas, according to our definition, only if it occurs in this sequence of formulas.

XII. *Omission of dependent* (*redundant*) *hypotheses from complex deductions.* In the propositional calculi considered here we can omit hypotheses which depend (deductively) on the sequence of hypotheses, i.e. if $\mathfrak{R}$ depends on the sequence of formulas $\{\mathfrak{R};\mathfrak{A}_1;\ldots;\mathfrak{A}_n\}$, then we have the rule:

$$\frac{\mathfrak{R};\mathfrak{A}_1;\ldots;\mathfrak{A}_n\vdash\mathfrak{B}_1;\ldots;\mathfrak{B}_r}{\mathfrak{A}_1;\ldots;\mathfrak{A}_n\vdash\mathfrak{B}_1;\ldots;\mathfrak{B}_r.}$$

As a result we obtain the following rule:

XIIa. In the propositional calculi considered here, any provable or dependent formula can be omitted from the set of hypotheses as well as adjoined as a consequent in the given complex deduction.

We give an illustration of the use of the operations presented above and of chain deduction in the construction of the following simple deduction:

(36) $$(A\ \&\ B)\vee(A\ \&\ C)\vdash A\ \&\ (B\vee C),$$

i.e. we prove that (36) is a simple deduction.

The proof is represented in the following condensed form:

$$\frac{(\vee)\begin{cases}A\ \&\ B\vdash A;B\ \big|\!\underline{\ \text{postulate (III-1)}\ }A;(B\vee C)\ \big|\!\underline{\ \text{VII}\ }\ A\ \&\ (B\vee C)\\ A\ \&\ C\vdash A;C\ \big|\!\underline{\ \text{postulate (III-2)}\ }A;(B\vee C)\ \big|\!\underline{\ \text{VII}\ }\ A\ \&\ (B\vee C)\end{cases}}{(A\ \&\ B)\vee(A\ \&\ C)\vdash A\ \&\ (B\vee C)}.$$

The reader can without difficulty restore the details omitted in the formal proof: the necessary information is contained in the given representation.

§6. The rules of substitution

1. As another illustration of the use of rules for complex deduction, we prove the following important theorem.

THEOREM XIII. *In the propositional calculi considered here we have the following simple deduction*:

$$(A \to B); (B \to A) \vdash (A \to B) \,\&\, (B \to A).$$

PROOF. We have $A; (A \to B); (B \to A) \vdash B; (B \to A) \vdash A$, i.e.

$$\frac{\begin{cases} A; (A \to B); (B \to A) \vdash A \\ \qquad (A \to B) \vdash (A \to B) \end{cases}}{A; (A \to B); (B \to A) \vdash A; (A \to B) \vdash B,}$$

or, in other words, $A; (A \to B); (B \to A) \vdash B$. Hence, by the deduction theorem,

$$(A \to B); (B \to A) \vdash (A \to B). \tag{37}$$

Interchanging A and B in (37), we obtain

$$(A \to B); (B \to A) \vdash (B \to A). \tag{38}$$

From (37) and (38) we obtain

$$(A \to B); (B \to A) \vdash (A \to B) \,\&\, (B \to A).$$

The theorem just proved allows us to introduce equivalence $\sim$, namely:

$$(\mathfrak{A} \sim \mathfrak{B}) =_{\text{df}} (\mathfrak{A} \to \mathfrak{B}) \,\&\, (\mathfrak{B} \to \mathfrak{A}). \tag{39}$$

As usual, we shall say that two formulas $\mathfrak{A}$ and $\mathfrak{B}$ are *equivalent* if $\vdash (\mathfrak{A} \sim \mathfrak{B})$.

In relevant propositional calculi we have the following theorems of substitution.(8)

THEOREM XIV (equivalent substitution). *In the propositional calculi considered here, if* $(\mathfrak{A} \sim \mathfrak{B})$ *is provable, then for every formula* $\mathfrak{C}$ *and every propositional letter* A *the formula* $S_A^{\mathfrak{A}}[\mathfrak{C}] \sim S_A^{\mathfrak{B}}[\mathfrak{C}]$ *is also provable.*

If we define

$$\begin{cases} \mathfrak{C}(\mathfrak{A}) =_{\text{df}} S_A^{\mathfrak{A}}[\mathfrak{C}], \\ \mathfrak{C}(\mathfrak{B}) =_{\text{df}} S_A^{\mathfrak{B}}[\mathfrak{C}], \end{cases} \tag{40}$$

then Theorem XIV can be written in the form of the following rule of inference:

$$\frac{\mathfrak{A} \sim \mathfrak{B}}{\mathfrak{C}(\mathfrak{A}) \sim \mathfrak{C}(\mathfrak{B})}. \tag{41}$$

(8) $S_A^{\mathfrak{A}}[\mathfrak{C}]$ denotes the result of substituting $\mathfrak{A}$ for every occurrence of the propositional letter A in $\mathfrak{C}$. If $\mathfrak{C}$ contains no occurrence of A, then $S_A^{\mathfrak{A}}[\mathfrak{C}] =_{\text{df}} \mathfrak{C}$.

Since both the formula $\mathfrak{C}$ and the propositional letter A are arbitrary, A may not occur in $\mathfrak{C}$, or may occur several times. If the latter is the case, it is always possible to substitute a propositional letter having no occurrence in $\mathfrak{C}$ for A at some (not necessarily all) places in $\mathfrak{C}$. Let this new variable be P. Substituting P for A at some arbitrarily chosen places in $\mathfrak{C}$, we obtain a new formula $\widetilde{\mathfrak{C}}$ such that $S_P^{\mathfrak{A}}[\widetilde{\mathfrak{C}}]$ and $S_P^{\mathfrak{B}}[\widetilde{\mathfrak{C}}]$ are the result of replacing not necessarily all occurrences of A in $\mathfrak{C}$ by either $\mathfrak{A}$ or $\mathfrak{B}$ respectively; in fact, the result of substituting either $\mathfrak{A}$ or $\mathfrak{B}$ for A in some particular occurrences in $\mathfrak{C}$. This is why $\widetilde{\mathfrak{C}}$ is called a *pointing* formula. The introduction of pointing formulas allows us to replace the propositional letter A by formulas $\mathfrak{A}$ or $\mathfrak{B}$ only in some particular, chosen occurrences in the application of Theorem XIV.

Let us note that in the relevant calculi the formula $(\mathfrak{A} \sim \mathfrak{B}) \longrightarrow [\mathfrak{C}(\mathfrak{A}) \sim \mathfrak{C}(\mathfrak{B})]$ is not provable, for the antecedent may be irrelevant to the consequent.

THEOREM XV (The principle of conjunctive substitution). *In the propositional calculi considered here the following substitution rule holds*:

$$\frac{\mathfrak{A} \rightarrow \mathfrak{B}}{\mathfrak{C}(\mathfrak{A}) \sim \mathfrak{C}(\mathfrak{A} \,\&\, \mathfrak{B})}. \tag{42}$$

Here we used the following definitions:

$$\begin{cases} \mathfrak{C}(\mathfrak{A}) =_{\mathrm{df}} S_A^{\mathfrak{A}}[\mathfrak{C}], \\ \mathfrak{C}(\mathfrak{A} \,\&\, \mathfrak{B}) =_{\mathrm{df}} S_A^{\mathfrak{A} \& \mathfrak{B}}[\mathfrak{C}]. \end{cases} \tag{43}$$

To prove this let us note that if $\mathfrak{A} \vdash \mathfrak{B}$, then by conjunctive fusion we obtain $\mathfrak{A} \vdash \mathfrak{A} \,\&\, \mathfrak{B}$, i.e. then $(\mathfrak{A} \sim \mathfrak{A} \,\&\, \mathfrak{B})$ is provable.

THEOREM XVI (The principle of abs-disjunctive substitution). *In the intuitionistic-nonrelevant and in the intuitionistically-completed propositional calculus the following rule of substitution holds*:

$$\frac{\mathfrak{A} \rightarrow \mathrm{abs}}{\mathfrak{C}(\mathfrak{B}) \sim \mathfrak{C}(\mathfrak{A} \vee \mathfrak{B})}. \tag{44}$$

Here we used the following definitions:

$$\begin{cases} \mathfrak{C}(\mathfrak{B}) =_{\mathrm{df}} S_A^{\mathfrak{B}}[\mathfrak{C}], \\ \mathfrak{C}(\mathfrak{A} \vee \mathfrak{B}) =_{\mathrm{df}} S_A^{\mathfrak{A} \vee \mathfrak{B}}[\mathfrak{C}]. \end{cases} \tag{45}$$

To prove this, let us note that if $\mathfrak{A} \vdash \mathrm{abs}$, then by the intuitionistic principle $(\mathrm{abs} \longrightarrow \mathfrak{B})$ we have $\mathfrak{A} \vdash \mathfrak{B}$. By disjunctive fusion we obtain $\mathfrak{A} \vee \mathfrak{B} \vdash \mathfrak{B}$, i.e. $\mathfrak{B} \sim \mathfrak{A} \vee \mathfrak{B}$.

2. In the propositional calculi considered here we have the following analogue of the theorem of Glivenko.

THEOREM XVII. *For any formula provable in the minimally-completed (intuitionistically-completed) propositional calculus, its double negation is provable in the minimally relevant (intuitionistically-nonrelevant) propositional calculus.*

3. Formalizing the methods of mathematical induction in traditional formalisms, we often use the following inference schema (in the predicate calculus):

$$(46)\qquad \frac{\begin{cases} \mathfrak{A}_1;\ldots;\mathfrak{A}_k \vdash \mathfrak{B}(0) \\ \mathfrak{A}_1;\ldots;\mathfrak{A}_k;\mathfrak{A}_{k+1};\ldots;\mathfrak{A}_n \vdash \forall n[\mathfrak{B}(n) \to \mathfrak{B}(n+1)] \end{cases}}{\mathfrak{A}_1;\ldots;\mathfrak{A}_n \vdash \mathfrak{B}(0) \,\&\, \forall n[\mathfrak{B}(n) \to \mathfrak{B}(n+1)].}$$

When using induction to prove an implication $\forall n[\mathfrak{B}(n) \longrightarrow \mathfrak{B}(n)]$, we often use the inference schema

$$(47)\qquad \frac{\begin{cases} \mathfrak{A}_1;\ldots;\mathfrak{A}_k;\mathfrak{B}(0) \vdash \text{abs} \\ \mathfrak{A}_1;\ldots;\mathfrak{A}_k;\mathfrak{A}_{k+1};\ldots;\mathfrak{A}_n \vdash \forall n\{[\mathfrak{B}(n) \to \mathfrak{C}(n)] \to [\mathfrak{B}(n+1) \to \mathfrak{C}(n+1)]\} \end{cases}}{\mathfrak{A}_1;\ldots;\mathfrak{A}_n \vdash [\mathfrak{B}(0) \to \mathfrak{C}(0)] \,\&\, \forall n\{[\mathfrak{B}(n) \to \mathfrak{C}(n)] \to [\mathfrak{B}(n+1) \to \mathfrak{C}(n+1)]\}.}$$

The schema (47) holds in the intuitionistically-nonrelevant or in the intuitionistically completed propositional calculus, since there we have the intuitionistic principle $(\text{abs} \longrightarrow A)$.

On the other hand, the schema (46) holds only if the principle $[A \longrightarrow (B \longrightarrow A)]$ is valid. *Prima facie*, it seems that we must not sacrifice the principle $[A \longrightarrow (B \longrightarrow A)]$, since we then sacrifice the possibility of inductive proofs.

In fact, in the relevant propositional calculi it is possible to formalize inductive proofs. The fact that the schema (45) does not hold shows only that in the relevant propositional calculi we have to require that the corresponding hypotheses are connected by conjunction. Here we see the exceptional, essential role of conjunction.

Indeed, in schema (46) the sequence of hypotheses $\{\mathfrak{A}_1;\ldots;\mathfrak{A}_k\}$ of the first original simple deduction is a subsequence of hypotheses $\{\mathfrak{A}_1;\ldots;\mathfrak{A}_k;\mathfrak{A}_{k+1};\ldots;\mathfrak{A}_n\}$ of the second original simple deduction. In the relevant calculi, as we have seen, the scheme of the conjunctive fusion holds:

$$(46\text{a})\qquad \frac{(\&)\begin{cases} \widetilde{\mathfrak{A}}_1;\ldots;\widetilde{\mathfrak{A}}_k \vdash \mathfrak{B}(0) \\ \widetilde{\mathfrak{A}}_1;\ldots;\widetilde{\mathfrak{A}}_k \vdash \forall n[\mathfrak{B}(n) \to \mathfrak{B}(n+1)] \end{cases}}{\widetilde{\mathfrak{A}}_1;\ldots;\widetilde{\mathfrak{A}}_k \vdash \mathfrak{B}(0) \,\&\, \forall n[\mathfrak{B}(n) \to \mathfrak{B}(n+1)].}$$

A necessary condition here is that the sequences of both original simple deductions coincide. In this sense the hypotheses $\{\mathfrak{A}_{k+1};\ldots;\mathfrak{A}_n\}$ in the second original deduction of the schema (46) are "redundant." Obviously, if these "redundant" hypotheses were in any way conjunctively fused (distributed) with the general hypotheses $\{\mathfrak{A}_1;\ldots;\mathfrak{A}_k\}$; then, if we mark by $\sim$ the new hypotheses obtained in this way, we would have

$$(48)\qquad \widetilde{\mathfrak{A}}_1;\ldots;\widetilde{\mathfrak{A}}_k \vdash \mathfrak{A}_1;\ldots;\mathfrak{A}_k$$

i.e. $\widetilde{\mathfrak{A}}_i$ is either $\mathfrak{A}_i$ or a conjunction with $\mathfrak{A}_i$ as a member. If in (46) we have $\mathfrak{A}_1;\ldots;\mathfrak{A}_k \vdash \mathfrak{B}(0)$, then we have also $\widetilde{\mathfrak{A}}_1;\ldots;\widetilde{\mathfrak{A}}_k \vdash \mathfrak{B}(0)$. The second original deduction in

(46) is replaced by

(48a) $$\mathfrak{A}_1; \ldots; \mathfrak{A}_k \vdash \forall n[\mathfrak{B}(n) \to \mathfrak{B}(n+1)].$$

Applying the inference scheme (46a) to (48) and (48a), we obtain the scheme (under the conditions of (48))

(49) $$\frac{\begin{cases} \mathfrak{A}_1; \ldots; \mathfrak{A}_k \vdash \mathfrak{B}(0) \\ \mathfrak{A}_1; \ldots; \mathfrak{A}_k \vdash \forall n[\mathfrak{B} n) \to \mathfrak{B}(n+1)] \end{cases}}{\mathfrak{A}_1; \ldots; \mathfrak{A}_k \vdash \mathfrak{B}(0) \mathbin{\&} \forall n[\mathfrak{B}(n) \to \mathfrak{B}(n+1)].}$$

Thus, in order to formalize recursive considerations in relevant propositional calculi, it is necessary to connect the hypotheses by conjunction. We note that since the axioms in formalizations of arithmetic, set theory, etc., can always be given in the form of a single conjunction of these axioms, in formal set theory or in formal arithmetic the formalizability of all recursive considerations is provided, so to speak, a priori. In the general case we must, even against our will, investigate by what kind of conjunctions the hypotheses are conjoined in this or that conjunctive group. This additional investigation is justified essentially by the relevant propositional calculi as well as by the fact that the hypotheses cannot be given separately without describing the specific connections between them, i.e. without corresponding conjunctions.

There are many other facts which can be used to illustrate the following statements:

1) Every inference scheme of the traditional propositional calculi has, as a rule, its analog, its image in the relevant propositional calculi.

2) Every axiomatic theory formalizable in traditional propositional calculi is also formalizable by means of relevant propositional calculi (has its image in them).

All this shows that in obtaining the relevant logics from traditional propositional calculi the class of provable formulas has been restricted in a reasonable way.

BIBLIOGRAPHY

1. Wilhelm Ackermann, *Begründung einer strengen Implikation*, J. Symbolic Logic **21** (1956), 113–128.

2. Alan Ross Anderson and Nuel D. Belnap, Jr., *The pure calculus of entailment*, J. Symbolic Logic **27** (1962), 19–52.

3. ———, *First degree entailments*, Math. Ann. **149** (1962/63), 302–319.

4. Nuel D. Belnap, Jr., *Entailment and relevance*, J. Symbolic Logic **25** (1960), 144–146.

5. Storrs McCall, *Connexive implication*, J. Symbolic Logic **31** (1966), 415–433.

6. V. V. Dončenko, *Some questions connected with the decision problem for Ackermann's system of rigorous implication*, Problems of Logic, Izdat. Akad. Nauk SSSR, Moscow, 1963, pp. 18–24. (Russian)

7. M. I. Semenenko, *An example of the formalization of a semantic relation in propositional calculi*, Naučno-Tehn. Informacija (VINITI) Ser. 2: Informacionnye Processy i Sistemy **1967**, no. 10, 9–14. (Russian)

8. Stephen Cole Kleene, *Introduction to metamathematics*, Van Nostrand, Princeton, N. J., 1952.

Translated by A. R. ANDERSON and A. KRON

Amer. Math. Soc. Transl.
(2) Vol. 113, 1979

ON A THEOREM IN ADDITIVE NUMBER THEORY*

A. F. LAVRIK

The following conjecture was stated in connection with Linnik's theorem [1], [2], according to which every even number $N > N_0$ can be represented as a sum of two prime numbers and a bounded number of powers of a fixed integer $g \geqslant 2$, and in connection with Romanov's theorem [3] that the sequence consisting of sums of a prime number and a power of a fixed integer $g \geqslant 2$ is a sequence with positive density. The conjecture states that there exists a constant number k (which may be arbitrarily large, but fixed) such that "almost all" even (or odd) numbers can be expressed in the form

$$p + g^{x_1} + \dots + g^{x_k}, \tag{1}$$

where p is a prime, $g \geqslant 2$ is a fixed integer, and the $x_i \geqslant 1$ are integers. In other words, the asymptotic density of the sequence of all integers representable in the form (1) is equal to ½. Attempts to prove this proposition by analytic methods have not been successful.

The purpose of this note is to explain that this conjecture is always false, except perhaps for the case $g = 2$. Let $N > 0$ denote any large number; let $\epsilon > 0$ be an arbitrarily small constant.

For fixed $g > 2$ we define the integer $d > 1$ as follows: $d = 3$ if $g = 2^\gamma$ and γ is an even number; if $g = 2^\gamma$ and $\gamma > 1$ is odd, then we set d equal to the least prime divisor of $2^\gamma - 1$; finally, if $g = 2^\alpha t$, where $t > 1$ is odd and $\alpha \geqslant 0$ is an integer, then d denotes the least divisor of t. We further take an arbitrary sequence of positive odd numbers $a_1, a_2, \dots$ such that, for the given $g > 2$, among the terms of $\{a_i\}$ not exceeding N no more than $O(N \ln^{-k-\epsilon} N)$ of them are multiples of d. We then have the following proposition: for every fixed integer $k \geqslant 1$ the asymptotic density of the sequence of numbers not representable in the form

$$a_i + g^{x_1} + \dots + g^{x_k}, \tag{2}$$

where the $x_i \geqslant 1$ are integers, is no less than $(d + 1)/2d$.

Hence, in particular, for $a_i = p_i$, where p_i is the ith odd prime number, it follows that the asymptotic density of the sequence of numbers representable by (1) does not exceed $(d - 1)/2d$, and this just means that the above conjecture is false for all $g > 2$.

To fix ideas, suppose k is odd. We consider the case $g = 2^\gamma$, where γ is even. We show that "almost all" numbers in the progression $6n + k$ $(n = 1, 2, \dots)$ are not representable by (2). We have

---

AMS (MOS) subject classifications (1970). Primary 10J15.

*Translation of Uspehi Mat. Nauk **14** (1959), no. 1 (85), 197–198. MR **21** #31.

$$a_i+g^{x_1}+\dots+g^{x_k}=a_i+k+(2^{\frac{\gamma x_1}{2}}-1)(2^{\frac{\gamma x_1}{2}}+1)$$
$$+\dots+(2^{\frac{\gamma x_k}{2}}-1)(2^{\frac{\gamma x_k}{2}}+1)=a_i+3u+k,$$

where $u>0$ is an integer, since for every integer $x_i\geqslant 1$ in this case

$$(2^{\frac{\gamma x_i}{2}}+1)(2^{\frac{\gamma x_i}{2}}-1)\equiv 0 \pmod 3.$$

Consequently, if 3 does not divide a_j, then $6n+k\neq a_j+g^{x_1}+\cdots+g^{x_k}$. But for a_j divisible by 3 the number of numbers

$$6n+k=a_j+g^{x_1}+\dots+g^{x_k}\leqslant N \tag{3}$$

is estimated by $O(\ln^k N/\ln^k g)$, so that there are always no more than

$$O\left(\frac{N}{\ln^{k+\varepsilon}N}\frac{\ln^k N}{\ln^k g}\right)=O\left(\frac{N}{\ln^{\varepsilon}N}\right)=o(N)$$

numbers of the form $6n+k$ satisfying (3). Since for $g=2^\gamma$ all even numbers are also not representable, we find that the asymptotic density of the sequence of integers $m\neq a_i+2^{\gamma x_1}+\cdots+2^{\gamma x_k}$ is no less than $2/3$, i.e., $(d+1)/2d$. Now let $g=2^\gamma$, where $\gamma>1$ is odd; then

$$a_i+g^{x_1}+\dots+g^{x_k}=a_i+(2^{\gamma x_1}-1)+\dots+(2^{\gamma x_k}-1)+k=a_i+du+k, \tag{4}$$

where $u>0$ is an integer and $d>1$ is a divisor of $2^\gamma-1$, since for $\gamma>1$ and for any integer $x_i\geqslant 1$ we have $2^{\gamma x_i}\equiv 1 \pmod d$, because $2^{\gamma x_i}\equiv 1 \pmod{(2^\gamma-1)}$. Reasoning as above, we find that "almost all" numbers $2dn+k$ ($n=1,2,\dots$) cannot be represented in the form (4). This implies the assertion in the case $g=2^\gamma$, $\gamma>1$ odd. Finally, for $g=2^\alpha t$, where $t>1$ and $\alpha\geqslant 0$, we have

$$a_i+g^{x_1}+\dots+g^{x_k}=a_i+2^\alpha du,$$

where u is a positive integer.

Considering the cases $\alpha=0$ and $\alpha>0$ separately, in the same way as before we obtain our final proposition under the assumption that k is odd. The case of even k is analogous.

BIBLIOGRAPHY

1. Ju. V. Linnik, *Addition of prime numbers with powers of one and the same number*, Mat. Sb. **32** (74) (1953), 3–60. (Russian) MR **15**, 602.

2. A. I. Vinogradov, *On an "almost binary" problem*, Izv. Akad. Nauk SSSR Ser. Mat. **20** (1956), 713–750. (Russian) MR **19**, 393.

3. N. P. Romanoff [Romanov], *Uber einige Sätze der additiven Zahlentheorie*, Math. Ann. **109** (1934), 668–678.

Translated by N. KOBLITZ

Amer. Math. Soc. Transl.
(2) Vol. 113, 1979

ON THE THEORY OF DISTRIBUTION OF PRIMES BASED ON I. M. VINOGRADOV'S METHOD OF TRIGONOMETRIC SUMS*

A. F. LAVRIK

For integers $u_1, \dots, u_m$, $m \geqslant 1$, let $\pi(x; u_1, \dots, u_m)$ denote the number of primes p in the interval $(0, x)$ for which each of the numbers $p + u_1, \dots, p + u_m$ is also prime.

This function is of great importance in prime number theory. In particular, it expresses the twin prime problem, the problem of representing even numbers as the sum of two primes, and many others.

The study of $\pi(x; u_1, \dots, u_m)$ in the general case was carried out by the "sieve" method [15]–[18]. However, this method only yielded upper estimates.

Using the investigations of I. M. Vinogradov [1] relating to trigonometric sums involving primes, we obtain in this paper asymptotic equalities with remainder terms for the growth of $\pi(x; u_1, \dots, u_m)$, valid for each nontrivial system $u_1, \dots, u_m$ in an interval $(0, y(x))$, except for an insignificantly small number of such systems. Thus, we have

THEOREM 1. *Let $\gamma_m(p)$ be the number of classes to which $0, u_1, \dots, u_m$ belong modulo the prime p. Then*

$$\pi(x; u_1, \dots, u_m) = \prod_p \frac{p - \gamma_m(p)}{p\left(1 - \frac{1}{p}\right)^{m+1}} \left(\int_2^x \frac{dx}{\ln^{m+1} x} + O\left(\frac{x}{\ln^{m+c} x} \right) \right) + O(1)$$

for each system of distinct numbers $u_1, \dots, u_m$ in the interval $[1, X]$, $X = x(\ln x)^{-m-c}$, except for at most $\ll X^m (\ln x)^{-M}$ of them, where $M > 0$ and $c > 1$ are arbitrarily given constants; the symbols $\ll$ and O do not depend on x.

REMARK. Since the number of systems of distinct $u_1, \dots, u_m$ in $[1, X]$ is a quantity of order X^m, this establishes, in particular, the validity of the Hardy-Littlewood conjecture regarding $\pi(x; u_1, \dots, u_m)$[2] (with a correction in the term $O(1)$, without which the conjecture is false) for almost all systems $u_1, \dots, u_m$.

The preceding theorem admits a generalization to progressions. It turns out that for almost all systems $u_1, \dots, u_m$ the sets of primes $p, p + u_1, \dots, p + u_m$ are uniformly distributed with respect to all primitive arithmetic progressions whose difference D does not exceed any fixed power of the logarithm of the number of terms in the progression. Namely,

---

AMS (MOS) subject classifications (1970). Primary 10G05, 10H20, 10J15.
*Translation of Trudy Mat. Inst. Steklov. **64** (1961), 90–125. MR **28** #69.

THEOREM 2. *The number* $\pi(x; u_1, \ldots, u_m; D)$ *of sets of primes* $p, p + u_1, \ldots, p + u_m$ *in the interval* $(0, x)$ *belonging respectively to the progressions* $Dn + l$, $Dn_1 + l_1, \ldots, Dn_m + l_m$, *with* $0 < D \leqslant (\ln x)^A$, *where* $A > 0$ *is an arbitrary constant, and* $0 < l, l_1, \ldots, l_m \leqslant D$ *are relatively prime to D, for each system of integers* $u_1, \ldots, u_m$ *satisfying the natural conditions*

$1°.\ 1 \leqslant u_1, \ldots, u_m \leqslant X;\ X = x(\ln x)^{-m-c}$,

$2°.\ u_\nu \equiv l_\nu - l \pmod D, (\nu = 1, \ldots, m)$,

$3°.\ u_s \neq u_t \neq 0$ *if* $s \neq t$,

except for at most $\ll X^m D^{-m} (\ln X)^{-M}$ *of them, where* $M > 0$ *and* $c > 1$ *are arbitrarily given constants, can be expressed in the form*

$$\pi(x; u_1, \ldots, u_m; D) = \frac{1}{D} \prod_p \frac{p - \gamma_m(p, D)}{p\left(1 - \frac{1}{p}\right)^{m+1}} \left(\int_2^x \frac{dx}{\ln^{m+1} x} + O\left(\frac{x}{\ln^{m+c} x}\right)\right) + O(1), \tag{1}$$

where the product extends over all primes p; $\gamma_m(p, D)$ *is the number of solutions of the congruence*

$$(D\xi + l)(D\xi + l + u_1) \ldots (D\xi + l + u_m) \equiv 0 \pmod p;$$

and the symbols $\ll$ *and* O *do not depend on* x.

Note that in view of the inequality

$$\gamma_m(p, D) \leqslant \min(p, m+1)$$

the product in (1) is always nonnegative. If, however, this product is equal to zero, the corresponding system $u_1, \ldots, u_m$ is called trivial, and in this case

$$\pi(x; u_1, \ldots, u_m; D) \ll 1.$$

For each nontrivial system $u_1, \ldots, u_m$ we have

$$\frac{1}{D} \prod_p \frac{p - \gamma_m(p, D)}{p\left(1 - \frac{1}{p}\right)^{m+1}} > \frac{c_m}{\varphi(D)},$$

where $c_m > 0$ depends only on m, and φ is the Euler function. To this we add that the number of trivial and nontrivial systems have the same order of growth. The corresponding computations are not difficult, but rather tedious, and we will not consider them here.

We state a proposition corresponding to the case $m = 1$ of Theorem 2 with $x = |u_1|$, $u_1 < 0$, which proves the conjecture of Sylvester [3] on the number of representations of an even integer N as the sum of two primes, asymptotically for all N.

THEOREM 3. *For each even* $N \equiv l' + l'' \pmod D$ *in the interval* $(0, x)$, *except for at most* $\ll x/D(\ln x)^M$ *of them, where* $M > 0$ *is an arbitrarily given constant, the number* $I(N; D)$ *of solutions of the equation* $N = p' + p''$ *in prime numbers* $p' \equiv$

l' (mod D), $p'' \equiv l''$ (mod D) with $0 \leqslant l', l'' \leqslant D$, $(l', D) = (l'', D) = 1$, $D \leqslant (\ln N)^A$ and $A > 0$ arbitrary, can be expressed by the formula

$$I(N, D) = 2\prod_{p>2} \frac{p(p-2)}{(p-1)^2} \prod_{\substack{p\mid D\\ p>2}} \frac{p-1}{p-2} \frac{N}{\varphi(D)\ln^2 N} \prod_{\substack{p>2\\ p\mid N;\, p\nmid D}} \frac{p-1}{p-2}\left(1 + O\left(\frac{1}{\ln N}\right)\right),$$

where p ranges over the primes, φ is the Euler function, and $\ll$ and O depend only on A and M.

The proofs of the above theorems rely principally on I. M. Vinogradov's fundamental theorem (§1.3) on trigonometric sums involving primes, and also the Page-Siegel-Walfisz theorem on primes in progressions with a "growing difference" (§1.3, Lemma 1.2).

In its methodical aspect our paper is close to those of K. K. Mardžanišvili [4], N. G. Čudakov [5] and others dealing with the additive theory of primes.

Our results were announced earlier in [9]–[11], but in a less complete form and without proofs. Various applications are given in [12]–[14].

Below is a complete proof of Theorem 2.

CHAPTER I

§1.1. Notation

$M, A, c > 1$ are arbitrarily given positive constants; $m \geqslant 1$ is an integer; x is a real variable; $x > x_0(m, M, A, c)$; $L = \ln x$; $X = x \cdot L^{-m-c}$; (n, k) is the greatest common divisor of n and k; D is an integer such that $0 < D \leqslant L^A$; $l, l_1, \ldots, l_m$ are nonnegative integers less than D and relatively prime to D; $p, p_1, \ldots, p_m$ are primes; B is a function bounded by a quantity depending only on M, m, A and c; $\varphi(n)$ is the Euler function; $\mu(n)$ the Möbius function; $\tau_1(n)$ the number of divisors of n; $\vartheta(n)$ the number of prime divisors of n; χ is a character mod D; $\bar{\chi}$ is the character conjugate to χ; $\epsilon > 0$ an arbitrarily small constant, not always with the same value; $\gamma_1, \gamma_2, \ldots,$ are positive constants; and

$$\Lambda = L^{\alpha}, \quad \alpha = M + (m+2)\cdot(m+c), \quad \omega = L^{\beta},$$
$$\beta = 64\cdot(2^{2m} + M + (m+2)\cdot(m+c+A+1)), \quad \eta = \omega^{\frac{63}{64}},$$
$$\Delta = \omega^{0,1}, \quad \tau = x\cdot\omega^{-1}.$$

For $\nu = 1, \ldots, m$ we set

$$S_\nu(\alpha_\nu;\ x, D, l_\nu) = \sum_\chi \bar{\chi}(l_\nu) \sum_{3\leqslant p\leqslant x} \chi(p)\cdot\exp(2\pi i p\alpha_\nu), \tag{1.1}$$

and

$$S(\alpha_1, \ldots, \alpha_m;\ x, D, l) = \sum_\chi \bar{\chi}(l_\nu) \sum_{3\leqslant p\leqslant x} \chi(p)\cdot\exp\left(-2\pi i p\sum_{\nu=1}^m \alpha_\nu\right), \tag{1.2}$$

where Σ_χ denotes the sum over all characters mod D; and

$$H(\alpha_1, \dots, \alpha_m) = H(\alpha_1, \dots, \alpha_m; x, D, l, l_1, \dots, l_m)$$
$$= S(\alpha_1, \dots, \alpha_m; x, D, l) \cdot \prod_{\nu=1}^{m} S_\nu(\alpha_\nu; x, D, l_\nu). \tag{1.3}$$

We also introduce $(m+1)$ functions F_ν which, as we will show, approximate the corresponding functions S_ν defined by (1.1) and (1.2). Thus for $\nu = 1, \dots, m$ we let

$$F_{a_\nu q_\nu}(\alpha_\nu) = F_{a_\nu q_\nu}(\alpha_\nu; x, D, l_\nu)$$
$$= E(q_\nu) \frac{\mu(\delta_\nu)}{\varphi(\delta_\nu)} \cdot \exp\left(2\pi i \frac{a_\nu V_\nu l_\nu}{q_\nu}\right) \sum_{n=3}^{\leqslant x} \frac{1}{\ln n} \exp\left(2 \cdot \pi \cdot i \cdot n \left(\alpha_\nu - \frac{a_\nu}{q_\nu}\right)\right), \tag{1.4}$$

where a_ν and q_ν are nonnegative integers, $a_\nu \leqslant q_\nu$, and

$$(a_\nu, q_\nu) = 1, \quad \delta_\nu = q_\nu / d_\nu, \quad d_\nu = (q_\nu, D);$$
$$E(q_\nu) = \begin{cases} 1, & \text{if } (\delta_\nu, D) = 1, \\ 0, & \text{if } (\delta_\nu, D) > 1; \end{cases}$$

also, $V_\nu = \delta_\nu \cdot g_\nu$, where g_ν is the residue mod d_ν such that $\delta_\nu g_\nu \equiv 1 \pmod{d_\nu}$. Furthermore, suppose that a and g are defined by the relations

$$\frac{a}{q} = \sum_{\nu=1}^{m} \frac{a_\nu}{q_\nu}, \quad (a, q) = 1;$$

if d, δ, g, V and $E(q)$ are correspondingly defined, let

$$F_{a_1 q_1, \dots, a_m q_m}(\alpha_1, \dots, \alpha_m) = F_{a_1 q_1, \dots, a_m q_m}(\alpha_1, \dots, \alpha_m; x, D, l)$$
$$= E(q) \cdot \frac{\mu(\delta)}{\varphi(\delta)} \cdot \exp\left(-2 \cdot \pi \cdot i \frac{a \cdot V \cdot l}{q}\right) \sum_{n=3}^{\leqslant x} \frac{1}{\ln n}$$
$$\times \exp\left(-2 \cdot \pi \cdot i \cdot n \sum_{\nu=1}^{m} \left(\alpha_\nu - \frac{a_\nu}{q_\nu}\right)\right). \tag{1.5}$$

Finally, let

$$N(\alpha_1, \dots, \alpha_m) = N(\alpha_1, \dots, \alpha_m; x, D, l, l_1, \dots, l_m)$$
$$= \sum_{q_1, \dots, q_m}^{[1, \Delta]} \sum_{a_1, \dots, a_m}^{q_1, \dots, q_m} F_{a_1 q_1, \dots, a_m q_m}(\alpha_1, \dots, \alpha_m) \cdot \prod_{\nu=1}^{m} F_{a_\nu q_\nu}(\alpha_\nu), \tag{1.6}$$

where $q_1, \dots, q_m$ range independently over the integers in the segment $[1, \Delta]$, and $a_1, \dots, a_m$ range over reduced systems of residues mod $q_1, \dots,$ mod q_m respectively. We will use this sort of abbreviated notation throughout this paper.

Our first goal is to obtain an upper estimate for the expression

$$I_m(x) = \int_0^1 \cdots \int_0^1 |H(\alpha_1, \dots, \alpha_m) - N(\alpha_1, \dots, \alpha_m)|^2 \, d\alpha_1 \dots d\alpha_m, \tag{1.7}$$

which is the dispersion of the function $\pi(x; u_1, \ldots, u_m; D)$ under investigation.

§1.2. Preliminary estimates

When no confusion can arise, we expediently replace

$$N(\alpha_1, \ldots, \alpha_m), \quad H(\alpha_1, \ldots, \alpha_m), \quad F_{a_1 q_1, \ldots, a_m q_m}(\alpha_1, \ldots, \alpha_m),$$
$$F_{a_\nu q_\nu}(\alpha_\nu), \quad S_\nu(\alpha_\nu), \quad S(\alpha_1, \ldots, \alpha_m), \quad \Gamma_{a_\nu q_\nu}$$

by N_m, H_m, F, F_ν, S_ν, S and Γ_ν, respectively.

Also, replacing $|H_m - N_m|$ in (1.7) by the quantity

$$|H_m - N_m|$$

$$|H_m - F_1 \cdots F_m \cdot F| + |N_m - F_1 \cdots F_m \cdot F|,$$

we have

$$I_m(x) = B \cdot I'_m(x) + B \cdot I''_m(x), \tag{1.8}$$

where

$$I'_m(x) = \int_0^1 \cdots \int_0^1 |H_m - F_1 \ldots F_m \cdot F|^2 \, d\alpha_1 \ldots d\alpha_m \tag{1.9}$$

and

$$I''_m(x) = \int_0^1 \cdots \int_0^1 |N_m - F_1 \ldots F_m \cdot F|^2 \, d\alpha_1 \ldots d\alpha_m. \tag{1.10}$$

Consider $I'_m = I'_m(x)$.

As a consequence of the periodicity of the integrand in (1.9), the intervals of integration $[0, 1]$ may be replaced by $[-\tau^{-1}, 1 - \tau^{-1}]$. Furthermore, we decompose the domain of integration by means of Farey arcs.

As is well known, for any $\alpha_\nu \in [-\tau^{-1}, 1 - \tau^{-1}]$ there exist integers a_ν and q_ν such that

$$\alpha_\nu = \frac{a_\nu}{q_\nu} + \beta_\nu, \quad 1 \leqslant q_1 \leqslant \tau; \quad (a_\nu, q_\nu) = 1, \quad |\beta_\nu| \leqslant \frac{1}{\tau \cdot q_\nu}.$$

Thus for each rational point $a_\nu/q_\nu \in [-\tau^{-1}, 1 - \tau^{-1}]$ there is an associated interval $\Gamma_\nu = \Gamma_{a_\nu q_\nu}$ of length

$$\zeta_{a_\nu q_\nu} = B \cdot (\tau \cdot q_\nu)^{-1}. \tag{1.11}$$

The set of intervals Γ_ν such that

$$1 \leqslant q_\nu \leqslant \tau_1, \quad (a_\nu, q_\nu) = 1, \quad a_\nu \leqslant q_\nu$$

is denoted by $\mathfrak{M}_\nu$. All of these are contained in $[-\tau^{-1}, 1 - \tau^{-1}]$ and for sufficiently large x, henceforth assumed, do not intersect.

The set of intervals remaining in $[-\tau^{-1}, 1 - \tau^{-1}]$ after extracting all $\Gamma_\nu \subset \mathfrak{M}_\nu$ is denoted by $\mathfrak{N}_\nu$.

Thus each α_ν belonging to some Γ_ν in $\mathfrak{N}_\nu$ can be represented in the form

$$\alpha_\nu = \frac{a_\nu}{q_\nu} + \beta_\nu, \quad |\beta_\nu| \leqslant \frac{1}{\tau \cdot q_\nu}, \quad \eta < q_\nu \leqslant \tau.$$

Corresponding to the indicated decomposition of the domain of integration, I'_m can be represented in the following form:

$$I'_m = I'_{m0} + I'_{m1} + \cdots + I'_{mm},$$

where for $j = 0, 1, \ldots, m$

$$I'_{mj} = \int \underset{\Omega_j}{\cdots} \int |H_m - F_1 \cdots F_m \cdot F|^2 \, d\alpha_1 \cdots d\alpha_m$$

and Ω_j denotes the set of all domains composed of j Γ_ν-intervals of $\mathfrak{N}_\nu$ and $(m-j)$ Γ_ν-intervals of $\mathfrak{M}_\nu$.

Suppose $j = 0$. Then, in view of (1.3),

$$|H_m - F_1 \cdots F_m \cdot F| = |S_1 \cdots S_m \cdot S - F_1 \cdot \ldots \cdot F_m \cdot F|$$
$$\leqslant |(S_1 - F_1) S_2 \cdots S_m \cdot S| + |F_1 (S_2 \cdots S_m \cdot S - F_2 \cdots F_m \cdot F)|$$
$$\leqslant \sum_{\nu=0}^{m} |S_{\nu+1} - F_{\nu+1}| \cdot \prod_{n=0}^{\nu} |F_n| \cdot \prod_{t=\nu+2}^{m+2} |S_t|,$$

where $F_0 = S_{m+2} = 1$, $F_{m+1} = F$ and $S_{m+1} = S$; and, consequently,

$$I'_{m0} = \int_{\mathfrak{M}_1} \cdots \int_{\mathfrak{M}_m} |H_m - F_1 \cdots F_m \cdot F|^2 \, d\alpha_1 \cdots d\alpha_m$$
$$= \sum_{\nu=0}^{m} B \cdot \sum_{q_1, \ldots, q_m}^{[1, \eta]} \sum_{a_1, \ldots, a_m}^{q_1, \ldots, q_m} \int_{\Gamma_1} \cdots \int_{\Gamma_m} |S_{\nu+1} - F_{\nu+1}|^2$$
$$\times \prod_{n=0}^{\nu} |F_n|^2 \cdot \prod_{t=\nu+2}^{m+2} |S_t|^2 \, d\alpha_1 \cdots d\alpha_m.$$

If $1 \leqslant j \leqslant m$, we decompose I'_{mj} into two parts:

$$I'_{mj} = B \cdot I'_{mj'} + B \cdot I'_{mj''},$$

where

$$I'_{mj'} = \int \underset{\Omega_j}{\cdots} \int | S_1 \cdots S_m \cdot S|^2 \, d\alpha_1 \cdots d\alpha_m$$

and

$$I'_{mj''} = \int \underset{\Omega_j}{\cdots} \int |F_1 \cdots F_m \cdot F|^2 \, d\alpha_1 \cdots d\alpha_m.$$

This completes the transformations of I'_m. Each of the expressions defined in

(1.14), (1.16) and (1.17) admits the necessary upper estimate. We will arrive at these estimates after obtaining a series of preliminary lemmas.

Now consider $I''_m = I''_m(x)$. We have

$$I''_m = \int_0^1 \cdots \int_0^1 | N_m - F_1 \cdots F_m \cdot F |^2 \, d\alpha_1 \ldots d\alpha_m$$

$$= \int_{-\tau^{-1}}^{1-\tau^{-1}} \cdots \int_{-\tau^{-1}}^{1-\tau^{-1}} | N_m - F_1 \cdots F_m \cdot F |^2 \, d\alpha_1 \ldots d\alpha_m$$

$$= \sum_{q_1, \ldots, q_m}^{[1, \tau]} \sum_{a_1, \ldots, a_m}^{q_1, \ldots, q_m} \int_{\Gamma_{a_1 q_1}} \cdots \int_{\Gamma_{a_m q_m}} \Bigg| \sum_{q'_1, \ldots, q'_m}^{[1, \Delta]} \sum_{a'_1, \ldots, a'_m}^{q'_1, \ldots, q'_m} F_{a'_1 q'_1}(\alpha_1) \cdots F_{a'_m q'_m}(\alpha_m) \cdot F_{a'_1 q'_1, \ldots, a'_m q'_m}(\alpha_1, \ldots, \alpha_m) - F_{a_1 q_1}(\alpha_1) \cdots F_{a_m q_m}(\alpha_m) \cdot F_{a_1 q_1, \ldots, a_m q_m}(\alpha_1, \ldots, \alpha_m) \Bigg|^2 d\alpha_1 \cdots d\alpha_m.$$

For $\nu = 1, \ldots, m$ we split each sum over q_ν in (1.18) into two according to the conditions $1 \leqslant q_\nu \leqslant \Delta$ and $\Delta + 1 \leqslant q_\nu \leqslant \tau$.

From the 2^m (m-fold) sums thus obtained we form $(m + 1)$ sets of sums by grouping together those with the same number of variables which vary from 1 to Δ. Namely, to the first set we assign the sum where all of the numbers $q_1, \ldots, q_m$ vary from 1 to Δ; to the second we assign all sums in which $(m - 1)$ of the numbers $q_1, \ldots, q_m$ vary from 1 to Δ and one number varies from $\Delta + 1$ to τ, and so on.

In view of the independence of the variables $q_1, \ldots, q_m$, it will suffice to restrict ourselves to estimating only one representative of each indicated set of sums.

We will deal with the sums

$$\sum_{q_1, \ldots, q_m}^{[1, \Delta]}, \quad \sum_{q_1, \ldots, q_{m-1}}^{[1, \Delta]} \sum_{q_m}^{[\Delta+1, \tau]}, \quad \ldots, \quad \sum_{q_1}^{[1, \Delta]} \sum_{q_2, \ldots, q_m}^{[\Delta+1, \tau]}, \quad \sum_{q_1, \ldots, q_m}^{[\Delta+1, \tau]}.$$

Consider the integrand in (1.18) for each of the above sums, denoting it by Φ_m, $\Phi_{m-1}, \ldots, \Phi_0$. It is not difficult to see that

$$\Phi_m = \Bigg| \sum_{\substack{q'_1, \ldots, q'_m \\ a_\nu q'_\nu - a'_\nu q_\nu \neq 0 \\ 1 \leqslant \nu \leqslant m}}^{[1, \Delta]} \sum_{a'_1, \ldots, a'_m}^{q'_1, \ldots, q'_m} F_{a'_1 q'_1}(\alpha_1) \cdots F_{a'_m q'_m}(\alpha_m) F_{a'_1 q'_1, \ldots, a'_m q'_m}(\alpha_1, \ldots, \alpha_m) \Bigg|^2,$$

but Φ_j for $0 \leqslant j \leqslant m - 1$ satisfies the inequality

$$\Phi_j \leqslant \Phi_{j1} + \Phi_{j2},$$

where

$$\Phi_{j1}=\Bigl|\sum_{\substack{q_1,\ldots,q_m\\ j+1\leqslant\nu\leqslant m}}^{[1,\Delta]}\ \sum_{\substack{a_1',\ldots,a_m'\\ a_\nu q_\nu'-a_\nu' q_\nu\neq 0\\ q_1',\ldots,q_m'}} F_{a_1'q_1'}(\alpha_1)\cdots F_{a_m'q_m'}(\alpha_m)F_{a_1'q_1',\ldots,a_m'q_m'}(\alpha_1,\ldots,\alpha_m)\Bigr|, \tag{1.21}$$

and

$$\Phi_{j2}=|F_{a_1q_1}(\alpha_1)\cdots F_{a_mq_m}(\alpha_m)\cdot F_{a_1q_1,\ldots,a_mq_m}(\alpha_1,\ldots,\alpha_m)|^2. \tag{1.21'}$$

Consequently,

$$I_m''(x)=B\cdot I_{mm1}''+\ldots+B\cdot I_{m11}''+B\cdot I_{m01}'' \tag{1.21''}$$
$$+B\cdot I_{m(m-1)2}''+\ldots+B\cdot I_{m02}'',$$

where for $0\leqslant j\leqslant m$

$$I_{mj1}''=\sum_{q_1,\ldots,q_j}^{[1,\Delta]}\ \sum_{q_{j+1},\ldots,q_m}^{[\Delta+1,\tau]}\ \sum_{a_1,\ldots,a_m}^{q_1,\ldots,q_m}\int_{\Gamma_{a_1q_1}}\cdots\int_{\Gamma_{a_mq_m}}|\Phi_{j1}|\,d\alpha_1\ldots d\alpha_m, \tag{1.22}$$

and for $0\leqslant j\leqslant m-1$

$$I_{mj2}''=\sum_{q_1,\ldots,q_j}^{[1,\Delta]}\ \sum_{q_{j+1},\ldots,q_m}^{[\Delta+1,\tau]}\ \sum_{a_1,\ldots,a_m}^{q_1,\ldots,q_m}\int_{\Gamma_{a_1q_1}}\cdots\int_{\Gamma_{a_mq_m}}|\Phi_{j2}|\,d\alpha_1\ldots d\alpha_m, \tag{1.23}$$

where $\Phi_{m1}=\Phi_m$ by definition.

§1.3. The main lemmas

We state two fundamental propositions of prime number theory which form the basis of the proofs of our theorems. We have in mind the well-known results of I. M. Vinogradov on estimates of trigonometric sums involving primes and results on the distribution of primes in arithmetic progressions where the difference of the progression grows in a known way along with the number of terms.

I. M. VINOGRADOV'S THEOREM. *If* $0<Q\leqslant L^{A_1}$, $A_1>0$ *a constant*, $(a,q)=1$, $L^\gamma<q\leqslant x\cdot L^{-\gamma}$, *and*

$$S(x,Q)=\sum_{\substack{3\leqslant p\leqslant x\\ p\equiv l(\mathrm{mod}\ Q)}}\exp\left(2\pi i\,\frac{ap}{q}\right),$$

then for any preassigned λ *such that* $\gamma\geqslant 64(\lambda+A_1+1)$ *we have*

$$|S(x,Q)|=B\cdot x(QL^\lambda)^{-1}.$$

A detailed proof of this theorem can be found in Chapter VI of Hua's book [6].

LEMMA 1.1. *If* $q_\nu\leqslant\eta$ *for* $\nu=1,\ldots,m$, $\chi''(p)$ *is a character* mod Q, $Q\leqslant\eta^m$, *and*

$$S'_{m+1}(\alpha_1, \ldots, \alpha_m) = \sum_{3 \leqslant p \leqslant x} \chi''(p) \cdot \exp\left(-2\pi i p \sum_{\nu=1}^{m} \left(\alpha_\nu - \frac{a_\nu}{q_\nu}\right)\right),$$

then, uniformly in Q,

$$S'_{m+1}(\alpha_1, \ldots, \alpha_m) = \theta(\chi'') \cdot \sum_{n=3}^{\leqslant x} \frac{1}{\ln n} \exp\left(-2\pi i n \sum_{\nu=1}^{m} \left(\alpha_\nu - \frac{a_\nu}{q_\nu}\right)\right)$$
$$+ B \cdot x \cdot \exp(-\gamma_1 L^{0.5}),$$

where $\theta(\chi'') = 1$ *if* $\chi'' = \chi''_0$, *the principal character, and* $\theta(\chi'') = 0$ *for* $\chi'' \neq \chi''_0$.

LEMMA 1.2. *If* $0 < Q \leqslant L^A$ *and* $(l, Q) = 1$, *then*

$$\sum_{\substack{3 \leqslant p \leqslant x \\ p \equiv l (\mathrm{mod}\, Q)}} 1 = \frac{1}{\varphi(Q)} \operatorname{li} x + B \cdot x \cdot \exp(-\gamma_1 L^{0.5}).$$

Lemma 1.1 is equivalent to the law of primes in progressions (Lemma 1.2). The proof repreats the argument given, for example, in [7].

§ 1.4. General lemmas

LEMMA 1.3. *Suppose that* $q_1, \ldots, q_m$ *are positive integers such that* $F_1 \cdots F_m \neq 0$, $(a_\nu, q_\nu) = 1$, $a_\nu \leqslant q_\nu$, *and q is defined by the relations*

$$\frac{a}{q} = \frac{a_1}{q_1} + \ldots + \frac{a_m}{q_m}, \quad (a, q) = 1.$$

Then

$$\sum_{a_1, \ldots, a_m}^{q_1, \ldots, q_m} \frac{1}{q^2} = B \cdot D \cdot \frac{q_1 \cdots q_m}{q_j},$$

where j is any of the numbers $1, \ldots, m$.

PROOF. Putting $d_\nu = (q_\nu, D)$ and $\delta_\nu = q_\nu / d_\nu$, we have that $(d_j, \delta_\nu) = 1$ for $1 \leqslant j, \nu \leqslant m$, Otherwise, one of the numbers $E(q_1), \ldots, E(q_m)$ would be zero.

Consequently,

$$\sum_{a_1, \ldots, a_m}^{q_1, \ldots, q_m} \frac{1}{q^2} = \sum_{a'_1, \ldots, a'_m}^{d_1, \ldots, d_m} \frac{1}{q_{11}^2} \sum_{a''_1, \ldots, a''_m}^{\delta_1, \ldots, \delta_m} \frac{1}{q_{21}^2} \cdot (1) \qquad (1.24)$$

The identity (1.24) is a consequence of the well-known fact that if b_1 and b_2 range over reduced residue systems mod c_1 and mod c_2, then $b_1 c_2 + b_2 c_1$ ranges over a reduced residue system mod $c_1 \cdot c_2$, provided that $(c_1, c_2) = 1$. We will show, furthermore, that for an integer $n > 0$, $p^2 \nmid n$,

(1) The quantities $q_{11}, q_{21}, \ldots$ are defined in the same way as q.

$$\sum_{a_1,\ldots,a_m}^{n,\ldots,n} \frac{1}{q_{31}^2} = B \cdot n^{m-1}. \tag{1.25}$$

By (1.24) we have

$$\sum_{a_1,\ldots,a_m}^{n,\ldots,n} \frac{1}{q_{31}^2} = \prod_{p\mid n} \sum_{a_1,\ldots,a_m}^{p,\ldots,p} \frac{1}{q_{31}^2} = \prod_{p\mid n} \Bigg\{ \sum_{\substack{a_1+\ldots+a_m\equiv 0 \pmod p\\ 1\leqslant a_1,\ldots,a_m<p}} 1$$

$$+ \frac{1}{p^2} \sum_{\substack{a_1+\ldots+a_m\not\equiv 0 \pmod p\\ 1\leqslant a_1,\ldots,a_m<p}} 1 \Bigg\} = \prod_{p\mid n} \left\{ f_m(p) + \frac{\varphi^m(p) - f_m(p)}{p^2} \right\}$$

$$= \varphi^{m-1}(n) \cdot \prod_{p\mid n} \left\{ \frac{f_m(p)}{\varphi^{m-1}(p)} \left(1 - \frac{1}{p^2}\right) + \frac{p-1}{p^2} \right\},$$

where $f_m(p)$ denotes the number of solutions of the congruence $a_1 + \cdots a_m \equiv 0 \pmod p$ in positive integers $a_1, \ldots, a_m$ not greater than p. For $m = 1$ the estimate (1.25) is obvious. Suppose $m \geqslant 2$. In this case, for fixed $a_1, \ldots, a_{m-1}$ the number a_m is uniquely determined, and, therefore, $f_m(p) \leqslant \varphi^{m-1}(p)$.

From this and (1.26) it follows that (1.25) is valid for any integer $m \geqslant 1$.

Now set n equal to the greatest common divisor of $\delta_1, \ldots, \delta_m$, let $\delta'_\nu = \delta_\nu/n$, and note that under the hypotheses of the lemma the numbers $\delta_1, \ldots, \delta_m$ are square-free.

Then, by (1.24) and (1.25), we obtain

$$\sum_{1,\ldots,a_m}^{\delta_1,\ldots,\delta_m} \frac{1}{q_{21}^2} = \sum_{a'_1,\ldots,a'_m}^{n,\ldots,n} \frac{1}{q_{31}^2} \sum_{a_1,\ldots,a_m}^{\delta'_1,\ldots,\delta'_m} \frac{1}{q_{41}^2}$$

$$= B \cdot n^{m-1} \sum_{a_1,\ldots,a_{m-1}}^{\delta'_j \cdot \delta'_{j_1}, \delta'_{j_2}, \ldots, \delta'_{j_{m-1}}} \frac{1}{q_{51}^2},$$

where j is a given integer in the segment $[1, m]$, δ'_{j_1} is one of the numbers $\delta'_1, \ldots, \delta'_m$ not equal to δ'_j, and $j_2 < j_3 < \cdots < j_{m-1}$ are integers in $[1, m]$ different from j and j_1.

The incorporation of δ'_j and δ'_{j_1} into a product is possible if $(\delta'_j, \delta'_{j_1}) = 1$, which we are justified in assuming.

Thus the estimate of the m-fold sum in (1.27) is reduced to the estimate of the analogous $(m-1)$-fold sum. Since in the case of $m-1$, $m \geqslant 3$, we have

$$\sum_{a_1,\ldots,a_{m-1}}^{\delta_1,\ldots,\delta_{m-1}} \frac{1}{q_{61}^2} = B \cdot \frac{\delta_1 \ldots \delta_{m-1}}{\delta_j}, \quad 1 \leqslant j \leqslant m-1,$$

from (1.25) and (1.27) we deduce that

$$\sum_{a_1,\ldots,a_m}^{\delta'_1,\ldots,\delta'_m} \frac{1}{q_{21}^2} = B\cdot n^{m-1}\cdot\delta'_{j_2}\ldots\delta'_{j_{m-1}} = B\cdot\{\text{ g.c.d. } (\delta_1, \ldots, \delta_m)\} \tag{1.27'}$$

$$\times\,\delta_{j_2}\cdots\,\delta_{j_{m-1}} = B\cdot\delta_{j_1}\cdots\delta_{j_{m-1}} = B\cdot\frac{\delta_1\ldots\delta_m}{\delta_j}.$$

But when $m = 2$ we have

$$\sum_{a_1, a_2}^{\delta_1, \delta_2}\frac{1}{q_{71}^2} = B\cdot(\delta_1, \delta_2)\sum_{a}^{\delta''}\frac{(\delta_1, \delta_2)^4}{(\delta_1\cdot\delta_2)^2} = B\cdot\frac{(\delta_1, \delta_2)^3}{\delta_1\cdot\delta_2},$$

where $\delta'' = \delta_1\cdot\delta_2/(\delta_1, \delta_2)$. Also,

$$(\delta_1, \delta_2)^3 \leqslant \begin{cases}\delta_1^2\cdot\delta_2\\ \delta_1\cdot\delta_2^2.\end{cases}$$

Hence, (1.27′) is true for any integer $m \geqslant 1$.

Finally, from (1.24) and (1.27) we obtain

$$\sum_{a_1,\ldots,a_m}^{q_1,\ldots,q_m}\frac{1}{q^2} = B\cdot d_1\cdots d_m\cdot\frac{\delta_1\ldots\delta_m}{\delta_j} = B\cdot d_j\frac{q_1\ldots q_m}{q_j} = B\cdot D\cdot\frac{q_1\ldots q_m}{q_j}.$$

This proves Lemma 1.3.

LEMMA 1.4. *If $\chi_1(t)$ and $\chi_2(t)$ are characters* mod *Q and Q' respectively, then the product $\chi_1(t)\cdot\chi_2(t)$ is the principal character if and only if $\chi_1(t) = \overline{\chi}_2(t)$ for all $(t, Q'') = 1$ and χ_1 is a derived character generated by a character* mod *e, where $e = (Q, Q')$ and $Q'' = Q\cdot Q'/e$.*

LEMMA 1.5. *For $(j, k) = 1$, $e|k$ and $(r, e) = 1$ we have*

$$\sum_{\substack{1\leqslant l\leqslant k\\ l\equiv r\,(\mathrm{mod}\, e)\\ (l,k)=1}}\exp\left(-2\pi i\,\frac{jl}{k}\right) = \begin{cases}0, & \text{if } (e, k/e) > 1,\\ \mu\left(\frac{k}{e}\right)\exp\left(-2\pi i\,\frac{j\theta r}{e}\right), & \text{if } (e, k/e) = 1,\end{cases}$$

where θ is a solution of the congruence $(k/e)z \equiv 1$ (mod e).

The proof of Lemmas 1.4 and 1.5 can be found in Rademacher's paper [8].

LEMMA 1.6. $\varphi(q)\ln q \geqslant \gamma_2 q$.

The proof is common knowledge.

§1.5. Estimation of $I'_m(x)$

LEMMA 1.7. *If $1 \leqslant q_\nu \leqslant \eta$ for $\nu = 1, \ldots, m$, then uniformly in D*

$$|S - F| = B\cdot x\cdot\exp(-\gamma_3\cdot L^{0,5}).$$

PROOF. It follows from the definitions of q and S that

$$S=\sum_{\chi}\bar{\chi}(l)\sum_{\substack{3\leqslant p\leqslant x\\(p,q)=1}}\chi(p)\cdot\exp\left(-2\pi ip\left(\frac{a}{q}+\sum_{\nu=1}^{m}\beta_\nu\right)\right)+B\cdot D\cdot\vartheta(q).$$

By a well-known property of characters and an estimate for $\vartheta(q)$, the number of prime divisors of q, we see upon introducing the notation χ' for a character mod q that

$$S=\frac{1}{\varphi(q)}\sum_{\chi}\sum_{\chi'}\sum_{1\leqslant h\leqslant q}\bar{\chi}(l)\cdot\bar{\chi}'(h)\cdot\exp\left(-2\pi ih\frac{a}{q}\right)$$
$$\times\sum_{3\leqslant p\leqslant x}\chi''(p)\cdot\exp\left(-2\pi ip\sum_{\nu=1}^{m}\beta_\nu\right)+B\cdot L^{A+1}, \tag{1.28}$$

where $\chi''(p)=\chi(p)\chi'(p)$ is a character mod $Q=D\cdot q/d$, and $d=(D,q)$.

Furthermore, put

$$T\left(\frac{a}{q},\chi,\chi'\right)=\frac{1}{\varphi(q)}\sum_{\chi}\sum_{\chi'}\sum_{1\leqslant h\leqslant q}\bar{\chi}(l)\cdot\bar{\chi}(h)\cdot\exp\left(-2\pi ih\frac{a}{q}\right) \tag{1.29}$$

and consider the case $\chi\chi'=\chi_0''$, where χ_0'' is the principal character. When $Q=D$, $Q'=q$ and $Q''=D\cdot q/d$, the use of Lemma 1.4 converts the right-hand side of (1.29) to

$$\frac{1}{\varphi(q)}\sum_{\chi_d}\sum_{1\leqslant h\leqslant q}\bar{\chi}_d(l)\cdot\chi_d(h)\cdot\exp\left(-2\pi ih\frac{a}{q}\right).$$

Consequently,

$$T\left(\frac{a}{q},\ \chi\chi'=\chi_0''\right)=\frac{\varphi(d)}{\varphi(q)}\sum_{\substack{1\leqslant h\leqslant q\\h\equiv l\,(\mathrm{mod}\,d)\\(h,q)=1}}\exp\left(-2\pi ih\cdot\frac{a}{q}\right). \tag{1.30}$$

But this last sum, by Lemma 1.5, is equal in the notation of §1.1, to

$$E(q)\cdot\mu(\delta)\cdot\exp\left(-2\pi i\frac{aql}{d}\right). \tag{1.31}$$

Also, the sum over primes in (1.28) is equal, by Lemma 1.1, to

$$\theta(\chi'')\cdot\sum_{n=3}^{\leqslant x}\frac{1}{\ln n}\cdot\exp\left(-2\pi in\sum_{\nu=1}^{m}\left(\alpha_\nu-\frac{a_\nu}{q_\nu}\right)\right)+B\cdot x\cdot\exp\left(-\gamma_1\cdot L^{0.5}\right) \tag{1.32}$$

Combining (1.28)–(1.32), we obtain

$$S=E(q)\cdot\frac{\mu(\delta)}{\varphi(\delta)}\cdot\exp\left(-2\pi i\frac{aVl}{q}\right)$$
$$\times\sum_{n=3}^{\leqslant x}\frac{1}{\ln n}\exp\left(-2\pi in\sum_{\nu=1}^{m}\left(\alpha_\nu-\frac{a_\nu}{q_\nu}\right)\right)+B\cdot x\cdot\exp\left(-\gamma_4\cdot L^{0.5}\right).$$

This, in view of (1.5), implies the lemma.

LEMMA 1.8. *If* $1 \leqslant q_\nu \leqslant \eta$, *then, uniformly in D,*

$$|S_\nu - F_\nu| = B \cdot x \cdot \exp(-\gamma_1 \cdot L^{0,5}).$$

The proof is analogous to that of the previous lemma.

We first estimate the sum in (1.14) with $\nu = m$. Applying Lemma 1.7, we obtain

$$V_m(x) = B \cdot x^2 \cdot \exp(-2\gamma_4 \cdot L^{0,5}) \sum_{q_1, \ldots, q_m}^{[1, \eta]} \sum_{a_1, \ldots, a_m}^{q_1, \ldots, q_m} \int_{\Gamma_1} \cdots \int_{\Gamma_m} |F_1 \cdots F_m|^2 \, d\alpha_1 \ldots d\alpha_m$$

$$= B \cdot x^2 \cdot \exp(-2\gamma_4 \cdot L^{0,5}) \sum_{q_1, \ldots, q_m}^{[1, \eta]} \sum_{a_1, \ldots, a_m}^{q_1, \ldots, q_m} \int_0^1 \cdots \int_0^1 |F_1 \cdots F_m|^2 \, d\alpha_1 \ldots d\alpha_m.$$

But by Lemma 1.6

$$\int_0^1 |F_\nu|^2 \, d\alpha_\nu = B \cdot (\varphi(\delta_\nu))^{-2} \cdot \sum_{n=3}^{\leqslant x} \frac{1}{\ln^2 n} = B \cdot \frac{d_\nu^2 \cdot \ln^2 q_\nu \cdot x}{q_\nu^2 \cdot \ln^2 x}.$$

Therefore,

$$V_m(x) = B \cdot \frac{x^{m+2} \cdot D^{2(m+1)} \cdot \ln^{2m} L}{L^{2m} \cdot \exp(\gamma_5 \cdot L^{0,5})} \left(\sum_{q=1}^{\leqslant \eta} \frac{1}{q} \right)^m = B \cdot x^{m+2} \cdot \Lambda^{-1}. \tag{1.33}$$

Furthermore, since

$$\int_0^1 |S_\nu(\alpha_\nu)|^2 \, d\alpha_\nu = \varphi^2(D) \cdot \sum_{\substack{3 \leqslant p \leqslant x \\ p \equiv l_\nu (\mathrm{mod}\, D)}} 1, \tag{1.34}$$

it follows, on applying Lemmas 1.8, 1.2 and 1.6 as above, that from (1.14) for the expressions $V_\nu(x)$, $1 \leqslant \nu \leqslant m-1$, we have

$$V_\nu(x) = B \cdot x^{m+2} \cdot \Lambda^{-1}, \tag{1.35}$$

and it follows from (1.33) and (1.35) that

$$I'_{m0}(x) = B \cdot x^{m+2} \cdot \Lambda^{-1}. \tag{1.36}$$

We now turn to the expressions $I'_{mj'}$ defined for $1 \leqslant j' \leqslant m$ by (1.16). The necessary estimate for these expressions is achieved by means of Vinogradov's theorem, Namely,

$$\left| \sum_{\substack{3 \leqslant p \leqslant x \\ p \equiv l_\nu (\mathrm{mod}\, D)}} \exp(2\pi i p \alpha_\nu) - \sum_{\substack{3 \leqslant p \leqslant x \\ p \equiv l_\nu (\mathrm{mod}\, D)}} \exp\left(2\pi i \frac{a_\nu p}{q_\nu}\right) \right|$$

$$= B \cdot \sum_{\substack{3 \leqslant p \leqslant x \\ p \equiv l_\nu (\mathrm{mod}\, D)}} |\exp(2\pi i p \beta_\nu) - 1| = B \cdot |\beta_\nu| \cdot \frac{x^2}{\varphi(D) \cdot L} \tag{1.37}$$

$$= B \cdot \frac{x^2}{\eta \cdot \tau \cdot D} = B \cdot x \cdot (\Lambda \cdot D^m)^{-1},$$

if α_ν lies in some interval Γ_ν of the set $\mathfrak{N}_\nu$, since then

$$\alpha_\nu = \frac{a_\nu}{q_\nu} + \beta_\nu, \quad |\beta_\nu| \leqslant \frac{1}{\tau \cdot q_\nu} \leqslant \frac{1}{\eta \cdot \tau}.$$

Furthermore, putting $L^{\gamma} = \omega$, $l = l_{\nu}$ and $Q = D$ in Vinogradov's theorem, we have, under the above conditions,

$$\Big|\sum_{\substack{3\leqslant p\leqslant x\\ p\equiv l_\nu \pmod D}} \exp\Big(2\pi i\,\frac{a_\nu p}{q_\nu}\Big)\Big| = B\cdot x\cdot(\Lambda\cdot D^m)^{-1}, \tag{1.38}$$

and the conjunction of (1.37) and (1.38) yields

$$\max_{\alpha\subset\Gamma_\nu\subset\mathfrak{N}_\nu} |S_\nu(\alpha_\nu)| = \max\Big|\sum_{\substack{3\leqslant p\leqslant x\\ p\equiv l_\nu \pmod D}} \exp(2\pi i p\alpha_\nu)\Big| = B\cdot x\cdot(\Lambda\cdot D^m)^{-1}.$$

For $1 \leqslant j' \leqslant m$ we see from (1.16), (1.34) and (1.38) that

$$I'_{mj'} = B\cdot\max_{\alpha_1\subset\Gamma_1\subset\mathfrak{N}_1}|S_1(\alpha_\nu)|^2\cdot\int_{\mathfrak{N}_1}\cdots\int_{\mathfrak{N}_j}\int_{\mathfrak{M}_{j+1}}\cdots\int_{\mathfrak{M}_m}|S_2\cdots S_m\cdot S|^2\,d\alpha_1\ldots d\alpha_m$$

$$= Bx^2\cdot\Lambda^{-1}\cdot D^{-2m}\int_0^1\cdots\int_0^1|S_2\cdots S_m\cdot S|^2\,d\alpha_1\ldots d\alpha_m.$$

Integrating first with respect to α_1, then with respect to α_2, and so on, we finally obtain

$$I'_{mj'}(x) = B\cdot x^{m+2}\cdot\Lambda^{-1}. \tag{1.39}$$

Hence to estimate $I'_m(x)$ it remains to consider the expressions

$$I'_{mj''} = \int\cdots\int_{\mathfrak{L}_j}|F_1\cdots F_m\cdot F|^2\,d\alpha_1\ldots d\alpha_m$$

$$= \int_{\mathfrak{N}_1}\cdots\int_{\mathfrak{N}_j}\int_{\mathfrak{M}_{j+1}}\cdots\int_{\mathfrak{M}_m}|F_1\cdots F_m\cdot F|^2\,d\alpha_1\ldots d\alpha_m$$

$$= B\cdot x^2\cdot\sum_{q_1}^{[\eta+1,\tau]}\sum_{a_1}^{q_1}\int_{\Gamma_{a_1q_1}}|F_{a_1q_1}(\alpha_1)|^2\,d\alpha_1\int_0^1\cdots\int_0^1|F_2\cdots F_m|^2\,d\alpha_2\ldots d\alpha_m$$

$$= B\cdot x^{m+1}\sum_{q_1}^{[\eta+1,\tau]}\sum_{a_1}^{q_1}\frac{1}{q_1\tau}\,\frac{d_1^2\cdot\ln^2 q_1}{q_1^2}\,\frac{x^2}{L^2}$$

$$= B\cdot x^{m+3}\cdot\tau^{-1}\cdot D^2\sum_{q_1=[\eta]+1}^{\infty}\frac{1}{q_1^2} = B\cdot x^{m+3}\cdot(\tau\cdot\eta)^{-1}\cdot D^2.$$

Consequently, for any j'', $1 \leqslant j'' \leqslant m$,

$$I'_{mj''} = B\cdot x^{m+2}\cdot\Lambda^{-1}. \tag{1.40}$$

Now combining the estimates (1.33), (1.36), (1.40) and (1.12)–(1.17), we see ultimately that

$$I'_m(x) = \int_0^1\cdots\int_0^1|H_m - F_1\cdots F_m\cdot F|^2\,d\alpha_1\ldots d\alpha_m = B\cdot x^{m+2}\cdot\Lambda^{-1}. \tag{1.41}$$

§1.6. Estimation of $I''_m(x)$

Let us estimate I''_{mj2}. For $0 \leqslant j \leqslant m-1$ we have from (1.23) and (1.21) that

$$I''_{mj2} = \sum_{q_1,\ldots,q_j}^{[1,\Delta]} \sum_{\substack{q_{j+1},\ldots,q_m \\ F_1\ldots F_m\neq 0}}^{[\Delta+1,\tau]} \sum_{a_1,\ldots,a_m}^{q_1,\ldots,q_m} \int_{\Gamma_{a_1q_1}} \cdots \int_{\Gamma_{a_mq_m}} |F_{a_1q_1}(\alpha_1)$$

$$\cdots F_{a_mq_m}(\alpha_m) \cdot F_{a_1q_1,\ldots,a_mq_m}(\alpha_1,\ldots,\alpha_m)|^2 \, d\alpha_1 \ldots d\alpha_m$$

$$= B \cdot x^2 \cdot D^2 \sum_{q_1,\ldots,q_j}^{[1,\Delta]} \sum_{\substack{q_{j+1},\ldots,q_m \\ F_1\cdots F_m\neq 0}}^{[\Delta+1,\tau]} \sum_{a_1,\ldots,a_m}^{q_1,\ldots,q_m} \frac{1}{q^2} \int_0^1 |F_{a_1q_1}(\alpha_1)|^2 \, d\alpha_1 \tag{1.42}$$

$$\cdots \int_0^1 |F_{a_mq_m}(\alpha_m)|^2 \, d\alpha_m = B \cdot x^{m+2} \cdot D^2 \sum_{\substack{q_1,\ldots,q_{m-1} \\ F_1\ldots F_m\neq 0}}^{[1,\tau]} \frac{\mu^2(\delta_1)\ldots\mu^2(\delta_{m-1})}{\varphi^2(\delta_1)\ldots\varphi^2(\delta_{m-1})}$$

$$\times \sum_{q_m}^{[\Delta+1,\tau]} \frac{\mu^2(\delta_m)}{\varphi^2(\delta_m)} \sum_{a_1,\ldots,a_m}^{q_1,\ldots,q_m} \frac{1}{q^2}.$$

Applying Lemmas 1.3 and 1.6, we obtain

$$I''_{mj2} = B \cdot x^{m+2} \cdot D^{2(m+1)} \left(\sum_{n=1}^{\tau} \frac{1}{n} \right)^{m-1} \sum_{q_m=[\Delta]+1}^{\infty} \frac{1}{q_m^2}$$

$$= B \cdot x^{m+2} \cdot L^{2A(m+1)} \cdot \ln^{m-1}\tau \cdot \Delta^{-1} = B \cdot x^{m+2} \cdot \Lambda^{-1}.$$

Consequently,

$$B \cdot I''_{m(m-1)2} + \cdots + B \cdot I''_{m02} = B \cdot x^{m+2} \cdot \Lambda^{-1}. \tag{1.43}$$

Further estimates of $I''_m(x)$ are based on the following observation. If for some ν, $1 \leqslant \nu \leqslant m$, we have $a_\nu q'_\nu - a'_\nu q_\nu \neq 0$ and

$$\alpha_\nu \subset \Gamma_{a_\nu q_\nu}; \quad \alpha_\nu = \frac{a_\nu}{q_\nu} + \beta_\nu,$$

then, because of the periodicity of the function $\exp(\pm 2\pi i n(\alpha_\nu - a'_\nu/q'_\nu))$, to each a'_ν there corresponds a unique number a''_ν (and conversely) such that the following relations hold:

$$0 < \left| \frac{a_\nu}{q_\nu} - \frac{a''_\nu}{q'_\nu} \right| \leqslant \frac{1}{2}; \quad (a''_\nu, q'_\nu) = 1; \quad -\frac{1}{2} q'_\nu \leqslant a''_\nu \leqslant \frac{3}{2} q'_\nu;$$

$$F_{a'_\nu q'_\nu}(\alpha_\nu) = F_{a''_\nu q'_\nu}(\alpha_\nu);$$

moreover, the value of $F_{a'_1q'_1,\ldots,a'_mq'_m}(\alpha_1,\ldots,\alpha_m)$ does not change if we replace a'_ν by a''_ν.

In view of (1.4), Abel's transformation yields

$$\left|F_{a''_\nu q'_\nu}(\alpha_\nu)\right| \leqslant \left(\varphi\left(\frac{q'_\nu}{d'_\nu}\right)\right)^{-1} \cdot \max_{3\leqslant t\leqslant x}\left|\sum_{n=3}^{t}\exp\left(2\pi in\left(\alpha_\nu - \frac{a''_\nu}{q'_\nu}\right)\right)\right|$$
$$= B\cdot\frac{d'_\nu\cdot\ln L}{q'_\nu}\frac{1}{\left|\sin\pi\left(\alpha_\nu-\frac{a''_\nu}{q'_\nu}\right)\right|} = B\cdot\frac{D\cdot\ln L}{q'_\nu}\frac{1}{\left|\alpha_\nu-\frac{a''_\nu}{q'_\nu}\right|}, \tag{1.44}$$

since when $x > x_0$

$$\left|\alpha_\nu - \frac{a''_\nu}{q'_\nu}\right| \leqslant \left|\frac{a_\nu}{q_\nu} - \frac{a''_\nu}{q'_\nu}\right| + |\beta_\nu| \leqslant \frac{1}{2} + \frac{1}{\tau\cdot q_\nu} < \frac{3}{4}.$$

We also have

$$\left|\alpha_\nu - \frac{a''_\nu}{q'_\nu}\right| \geqslant \left|\frac{a_\nu}{q_\nu} - \frac{a''_\nu}{q'_\nu}\right| - |\beta_\nu| \geqslant \frac{1}{2}\left|\frac{a_\nu}{q_\nu} - \frac{a''_\nu}{q'_\nu}\right|, \tag{1.45}$$

inasmuch as

$$|\beta_\nu| \leqslant \frac{1}{\tau\cdot q_\nu} \leqslant \frac{1}{2\cdot\Delta\cdot q_\nu} \leqslant \frac{1}{2q_\nu\cdot q'_\nu} \leqslant \frac{1}{2}\left|\frac{a_\nu}{q_\nu} - \frac{a''_\nu}{q'_\nu}\right|.$$

It follows from (1.44) and (1.45) that

$$\left|F_{a''_\nu q'_\nu}(\alpha_\nu)\right|^2 = B\,\frac{D^2\cdot q_\nu^2\cdot\ln^2 L}{(a_\nu q'_\nu - a''_\nu q_\nu)^2}. \tag{1.46}$$

We first estimate I''_{m01} from (1.22). By the above remark, the Φ_{01} from (1.21) can be written in the form

$$\Phi_{01} = \left|\sum_{q'_1,\ldots,q'_m}^{[1,\Delta]}\ \sum_{\substack{0<\left|\frac{a_\nu}{q_\nu}-\frac{a''_\nu}{q'_\nu}\right|\leqslant\frac{1}{2}\\ 1\leqslant\nu\leqslant m}}\ \prod_{\nu=1}^{m} F_{a''_\nu q'_\nu}(\alpha_\nu)\cdot F_{a''_1 q'_1,\ldots,a''_m q'_m}(\alpha_1,\ldots,\alpha_m)\right|^2.$$

Then by the Cauchy inequality and (1.46) we obtain

$$I''_{m01} = \sum_{q_1,\ldots,q_m}^{[\Delta+1,\tau]}\ \sum_{a_1,\ldots,a_m}^{q_1,\ldots,q_m}\int_{\Gamma_{a_1q_1}}\cdots\int_{\Gamma_{a_mq_m}}|\Phi_{01}|\,d\alpha_1\ldots d\alpha_m$$
$$= Bx^2\cdot D^2\cdot\Delta^{2m}\left(\sum_{q}^{[\Delta+1,\tau]}\sum_{a}^{q}\frac{1}{q\tau}L^{2A}\cdot q^2\cdot\ln^2 L\sum_{q'}^{[1,\Delta]}\ \sum_{-0.5q'\leqslant a''\leqslant 1.5q'}\frac{1}{(aq'-a''q)^2}\right)^m \tag{1.47}$$
$$= B\cdot x^2\cdot L^{2A(m+1)}\cdot\Delta^{2m}\cdot\ln^{2m}L\left(\sum_{s=1}^{\infty}\frac{h(s)}{s^2}\right)^m,$$

where $h(s)$ denotes the number of solutions of the equation $s = aq' - a''q$, $s \neq 0$, in integers q, q' and a'' belonging to the intervals $1 \leqslant q \leqslant \tau$, $1 \leqslant q' \leqslant \Delta$ and $-0.5\cdot\Delta \leqslant a'' \leqslant 1.5\cdot\Delta$. But for given q, q' and a'' the number a is uniquely determined.

Hence, taking into account the limits of variation of q, q' and a'', we see that $h(s) = B \cdot \tau \cdot \Delta^2$. Then, by (1.47), we obtain

$$I''_{m01} = B \cdot x^{m+2} \cdot \Delta^{4m} \cdot L^{2A(m+1)} \cdot \ln^{2m} L \cdot \omega^{-m} = B \cdot x^{m+2} \cdot \Lambda^{-1}. \tag{1.48}$$

Similarly, we have

$$I''_{mm1} = B \cdot x^{m+2} \cdot \Lambda^{-1}. \tag{1.49}$$

Now suppose $j = 1, \ldots, m-1$ in (1.22). Since the arguments $b_1\xi_1, \ldots, b_m\xi_m$ of the function $F_{b_1\xi_1,\ldots,b_m\xi_m}(\alpha_1, \ldots, \alpha_m)$ are defined by the arguments $F_{b_\nu\xi_\nu}(\alpha_\nu)$, $\nu = 1, \ldots, m$, we will henceforth omit them.

We have

$$\Phi'_{j1} = \sum_{q'_1,\ldots,q'_m}^{[1,\Delta]} \sum_{\substack{a'_1,\ldots,a'_m \\ a_\nu q'_\nu - a'_\nu q_\nu \neq 0 \\ j+1 \leqslant \nu \leqslant m}}^{q'_1,\ldots,q'_m} F_{a'_1 q'_1}(\alpha_1) \cdots F_{a'_m q'_m}(\alpha_m) \cdot F(\alpha_1, \ldots, \alpha_m)$$

$$= \prod_{t=1}^{j} F_{a_t q_t}(\alpha_t) \sum_{q'_{j+1},\ldots,q'_m}^{[1,\Delta]} \sum_{\substack{a'_{j+1},\ldots,a'_m \\ a_\nu q'_\nu - a'_\nu q_\nu \neq 0 \\ j+1 \leqslant \nu \leqslant m}}^{q'_{j+1},\ldots,q'_m} F_{a'_{j+1} q'_{j+1}}(\alpha_{j+1}) \cdots F_{a'_m q'_m}(\alpha_m) \cdot F(\alpha_1, \ldots, \alpha_m)$$

$$+ \sum_{n_1}^{[1,j]} \prod_{\substack{t=1 \\ t \neq n_1}}^{j} F_{a_t q_t}(\alpha_t) \sum_{q'_{n_1}}^{[1,\Delta]} \sum_{\substack{a'_{n_1} \\ a_{n_1} q'_{n_1} - a'_{n_1} q_{n_1} \neq 0}}^{q'_{n_1}} \sum_{q'_{j+1},\ldots,q'_m}^{[1,\Delta]} \sum_{\substack{a'_{j+1},\ldots,a'_m \\ a_\nu q'_\nu - a'_\nu q_\nu \neq 0 \\ j+1 \leqslant \nu \leqslant m}}^{q'_{j+1},\ldots,q'_m} F_{a'_{n_1} q'_{n_1}}(\alpha_{n_1})$$

$$\times F_{a'_{j+1} q'_{j+1}}(\alpha_{j+1}) \cdots F_{a'_m q'_m}(\alpha_m) \cdot F(\alpha_1, \ldots, \alpha_m)$$

$$+ \cdots\cdots\cdots\cdots\cdots\cdots\cdots\cdots\cdots$$

$$+ \sum_{n_1,\ldots,n_k}^{[1,j]} \prod_{\substack{t=1 \\ t \neq n_1,\ldots,t \neq n_k}}^{j} F_{a_t q_t}(\alpha_t) \sum_{q'_{n_1},\ldots,q'_{n_k}}^{[1,\Delta]} \sum_{\substack{a'_{n_1},\ldots,a'_{n_k} \\ a_{n_i} q'_{n_i} - a'_{n_i} q_{n_i} \neq 0 \\ 1 \leqslant i \leqslant k}}^{q'_{n_1},\ldots,q'_{n_k}} \sum_{q'_{j+1},\ldots,q'_m}^{[1,\Delta]} \sum_{\substack{a'_{j+1},\ldots,a'_m \\ a_\nu q'_\nu - a'_\nu q_\nu \neq 0 \\ j+1 \leqslant \nu \leqslant m}}^{q'_{j+1},\ldots,q'_m}$$

$$\times F_{a'_{n_1} q'_{n_1}}(\alpha_{n_1}) \cdots F_{a'_{n_k} q'_{n_k}}(\alpha_{n_k}) \cdot F_{a'_{j+1} q'_{j+1}}(\alpha_{j+1}) \cdots F_{a'_m q'_m}(\alpha_m) \cdot F(\alpha_1, \ldots, \alpha_m)$$

$$+ \cdots\cdots\cdots\cdots\cdots\cdots\cdots\cdots\cdots$$

$$+ \sum_{q'_1,\ldots,q'_m}^{[1,\Delta]} \sum_{\substack{a'_1,\ldots,a'_m \\ a_\nu q'_1 - a'_\nu q_\nu \neq 0 \\ 1 \leqslant \nu \leqslant m}}^{q'_1,\ldots,q'_m} F_{a'_1 q'_1}(\alpha_1) \cdots F_{a'_m q'_m}(\alpha_m) \cdot F(\alpha_1, \ldots, \alpha_m)$$

$$= \prod_{t=1}^{j} F_{a_t q_t}(\alpha_t) \sum_{q'_{j+1},\ldots,q'_m}^{[1,\Delta]} \sum_{\substack{0 < \left| \frac{a_\nu}{q_\nu} - \frac{a''_\nu}{q'_\nu} \right| \leqslant 0.5 \\ j+1 \leqslant \nu \leqslant m}} F_{a''_{j+1} q'_{j+1}}(\alpha_{j+1}) \cdots F_{a''_m q'_m}(\alpha_m) \times$$

$$\times F(\alpha_1, \ldots, \alpha_m) + \sum_{n_1}^{[1,j]} \prod_{\substack{t=1\\ t\neq n_1}}^{j} F_{a_t q_t}(\alpha_t) \sum_{q'_{n_1}}^{[1,\Delta]} \sum_{0<\left|\frac{a_{n_1}}{q_{n_1}}-\frac{a''_{n_1}}{q'_{n_1}}\right|\leqslant 0.5} \sum_{q'_{j+1},\ldots,q'_m}^{[1,\Delta]}$$

$$\times \sum_{\substack{0<\left|\frac{a_\nu}{q_\nu}-\frac{a''_\nu}{q'_\nu}\right|\leqslant 0.5\\ j+1\leqslant\nu\leqslant m}} F_{a''_{n_1} q'_{n_1}}(\alpha_{n_1}) \cdot F_{a''_{j+1} q'_{j+1}}(\alpha_{j+1}) \cdots F_{a''_m q'_m}(\alpha_m) \cdot F(\alpha_1, \ldots, \alpha_m)$$

$$+ \cdots\cdots\cdots\cdots\cdots\cdots\cdots\cdots\cdots$$

$$+ \sum_{n_1,\ldots,n_k}^{[1,j]} \prod_{\substack{t=1\\ t\neq n_1,\ldots,t\neq n_k}}^{j} F_{a_t q_t}(\alpha_t) \sum_{q'_{n_1},\ldots,q'_{n_k}}^{[1,\Delta]} \sum_{\substack{0<\left|\frac{a_{n_i}}{q_{n_i}}-\frac{a''_{n_i}}{q'_{n_i}}\right|\leqslant 0.5\\ 1\leqslant i\leqslant k}}$$

$$\times \sum_{q'_{j+1},\ldots,q'_m}^{[1,\Delta]} \sum_{\substack{0<\left|\frac{a_\nu}{q_\nu}-\frac{a''_\nu}{q'_\nu}\right|\leqslant 0.5\\ j+1\leqslant\nu\leqslant m}} F_{a''_{n_1} q'_{n_1}}(\alpha_{n_1}) \cdots F_{a''_{n_k} q'_{n_k}}(\alpha_{n_k}) \tag{1.50}$$

$$\times F_{a''_{j+1} q'_{j+1}}(\alpha_{j+1}) \cdots + F_{a''_m q'_m}(\alpha_m) \cdot F(\alpha_1, \ldots, \alpha_m)$$

$$+ \cdots\cdots\cdots\cdots\cdots\cdots\cdots\cdots\cdots$$

$$+ \sum_{q'_1,\ldots,q'_m}^{[1,\Delta]} \sum_{\substack{0<\left|\frac{a_\nu}{q_\nu}-\frac{a''_\nu}{q'_\nu}\right|\leqslant 0.5\\ 1\leqslant\nu\leqslant m}} F_{a''_1 q'_1}(\alpha_1) \cdots F_{a''_m q'_m}(\alpha_m) \cdot F(\alpha_1, \ldots, \alpha_m)$$

$$= T_j(0) + \sum_{n_1}^{[1,j]} T_j(n_1) + \cdots + \sum_{n_1,\ldots,n_k}^{[1,j]} T_j(n_1, \ldots, n_k)$$

$$+ \cdots + T_j(1, \ldots, j).$$

We also note that $\Phi_{j1} = |\Phi'_{j1}|^2$. Consequently,

$$\Phi_{j1} = B \cdot |T_j(0)|^2 + B \sum_{n_1}^{[1,j]} |T_j(n_1)|^2 + \cdots$$
$$+ B \cdot \sum_{n_1,\ldots,n_k}^{[1,j]} |T_j(n_1, \ldots, n_k)|^2 + \cdots + B\,|T_j(1, \ldots, j)|^2. \tag{1.51}$$

We now turn to estimates for the expression I''_{mj1} in (1.22). It is evident that the numbers $n_1, \ldots, n_k$, where $1 \leqslant k \leqslant j$, are distinct. We take one combination of them, for example $n_1 = 1, \ldots, n_k = k$, and consider the expression

$$I''_{mj1k} = \sum_{q_1,\ldots,q_j}^{[1,\Delta]} \sum_{q_{j+1},\ldots,q_m}^{[\Delta+1,\tau]} \sum_{a_1,\ldots,a_m}^{q_1,\ldots,q_m\,'} \int_{\Gamma_{a_1q_1}} \cdots \int_{\Gamma_{a_mq_m}} |T_j(1,\ldots,k)|^2\, d\alpha_1 \ldots d\alpha_m$$

$$= \sum_{q_1,\ldots,q_j}^{[1,\Delta]} \sum_{q_{j+1},\ldots,q_m}^{[\Delta+1,\tau]} \sum_{a_1,\ldots,a_m}^{q_1,\ldots,q_m} \int_{\Gamma_{a_1q_1}} \cdots \int_{\Gamma_{a_mq_m}} |F_{a_1q_1}(\alpha_1) \cdots F_{a_kq_k}(\alpha_k)$$

$$\times \sum_{q'_{k+1},\ldots,q'_m}^{[1,\Delta]} \sum_{0<\left|\frac{a_\nu}{q_\nu}-\frac{a''_\nu}{q'_\nu}\right|\leqslant 0.5} F_{a''_{k+1}q'_{k+1}}(\alpha_{k+1}) \cdots F_{a''_mq'_m}(\alpha_m)$$

$$\times F_{a_1q_1,\ldots,a_kq_k,a''_{k+1}q'_{k+1},\ldots,a''_mq'_m}(\alpha_1,\ldots,\alpha_m)|^2\, d\alpha_1 \ldots d\alpha_m.$$

Using (1.46), the Cauchy inequality and the fact that the intervals $\Gamma_{a_\nu q_\nu}$ with $q_\nu \leqslant \Delta$ do not intersect, we obtain

$$I''_{mj1k} = B \cdot x^2 \cdot D^2 \cdot \Delta^{2(m-k-1)} \int_0^1 \cdots \int_0^1 |F_{a_1q_1}(\alpha_1) \cdots F_{a_kq_k}(\alpha_k)|^2\, d\alpha_1 \ldots d\alpha_k$$

$$\times \sum_{q_{k+1},\ldots,q_m}^{[1,\Delta]} \sum_{q_{j+1},\ldots,q_m}^{[\Delta+1,\tau]} \sum_{a_{k+1},\ldots,a_m}^{q_{k+1},\ldots,q_m} \frac{1}{q_{k+1}\tau \ldots q_m\tau} L^{2A(m-k-1)} \cdot q_{k+1}^2 \cdots q_m^2 \cdot \ln^{2(m-k-1)} L$$

$$\times \sum_{q'_{k+1},\ldots,q'_m}^{[1,\Delta]} \sum_{\substack{0<\left|\frac{a_\nu}{q_\nu}-\frac{a''_\nu}{q'_\nu}\right|\leqslant 0.5 \\ k+1\leqslant\nu\leqslant m}} \frac{1}{(a_{k+1}q'_{k+1}-a''_{k+1}q_{k+1})^2} \cdots \frac{1}{(a_mq'_m-a''_mq_m)^2}$$

$$= Bx^2L^{2A(m-k)} \cdot \Delta^{2(m-k-1)} \cdot \ln^{2m} L \cdot \left(\sum_q^{[1,\Delta]} \sum_a^{q} \sum_{q'}^{[1,\Delta]} \sum_{-0.5\cdot q'\leqslant a''\leqslant 1.5\cdot q'} \frac{1}{(a\cdot q'-a''q)^2} \right)^{j-k}$$

$$\times \left(\sum_q^{[1,\tau]} \sum_a^{q} \sum_{q'}^{[1,\Delta]} \sum_{-0.5\cdot q'\leqslant a''\leqslant 1.5\cdot q'} \frac{1}{(a\cdot q'-a''q)^2} \right)^{m-j}.$$

Repeating the computations involved in estimating I'_{m01}, we find that

$$I''_{mj1k} = B \cdot x^2 \cdot L^{2A(m-k)} \cdot \Delta^{2(m-k-1)} \cdot \ln^{2m} L \cdot \Delta^{3(j-k)} \cdot \tau^{m-j} \cdot \Delta^{2(m-j)}.$$

Hence for $1 \leqslant j \leqslant m-1$ and any k, $1 \leqslant k \leqslant j$, we have

$$I''_{mj1k} = B \cdot x^{m+2} \cdot \Lambda^{-1}.$$

Summing all such estimates over k, then over j, and using (1.21), (1.22), (1.43) and (1.51), we ultimately obtain

$$\int_0^1 \cdots \int_0^1 |N_m - F_1 \cdots F_m \cdot F|^2\, d\alpha_1 \ldots d\alpha_m = B \cdot x^{m+2} \cdot \Lambda^{-1}. \quad (1.52)$$

CHAPTER II

§2.1. A critical series of the problem

We will investigate the expression

$$\sigma_\Delta(u_1, \ldots, u_m; D) = \sigma_\Delta(u_1, \ldots, u_m; D, l, l_1, \ldots, l_m) = \sum_{q_1, \ldots, q_m}^{[1, \Delta]} G_{u_1, \ldots, u_m}(q_1, \ldots, q_m; D), \tag{2.1}$$

where

$$G_{u_1, \ldots, u_m}(q_1, \ldots, q_m; D) = G_{u_1, \ldots, u_m}(q_1, \ldots, q_m; D, l, l_1, \ldots, l_m) = E(q_1 \ldots q_m) \cdot \frac{\mu(\delta_1)}{\varphi(\delta_1)} \cdots \frac{\mu(\delta_m)}{\varphi(\delta_m)} \times \sum_{a_1, \ldots, a_m}^{q_1, \ldots, q_m} \frac{\mu(\delta)}{\varphi(\delta)} \prod_{\nu=1}^{m} \exp\left(2\pi i \frac{a_\nu}{q_\nu}(V_\nu l_\nu - Vl - u_\nu)\right). \tag{2.2}$$

Here for $\nu = 1, \ldots, m$ we have $\delta_\nu = q_\nu/d_\nu$, $d_\nu = (q_\nu, D)$, $\delta = q/d$ and $d = (q, D)$, where q is defined by the relations

$$\frac{a}{q} = \frac{a_1}{q_1} + \ldots + \frac{a_m}{q_m}, \quad (a, q) = 1;$$

$$E(q_1 \ldots q_m) = \begin{cases} 1, & \text{if } (\delta \cdot \delta_1 \cdots \delta_m, D) = 1, \\ 0, & \text{if } (\delta \cdot \delta_1 \cdots \delta_m, D) > 1; \end{cases}$$

$V_\nu = \delta_\nu \cdot g_\nu$ and $V = \delta \cdot g$, where g_ν and g are solutions of the congruences $\delta_\nu \cdot z \equiv 1 \pmod{d_\nu}$ and $\delta \cdot z \equiv 1 \pmod{d}$ respectively.

§2.2. Special lemmas

LEMMA 2.1. *If for any $1 \leqslant \nu, t \leqslant m$ we have $(q_\nu, q'_t) = 1$, then for given integers $u_1, \ldots, u_m$ the functions $G(q_1, \ldots, q_m)$ defined in* (2.2) *satisfy the relation*

$$G(q_1, \ldots, q_m) \cdot G(q'_1, \ldots, q'_m) = G(q_1 \cdot q'_1, \ldots, q_m q'_m).$$

PROOF. Under the hypotheses of the lemma we have

$$\begin{gathered} E(q_1 \cdots q_m) \cdot E(q'_1 \cdots q'_m) = E(q_1 \cdot q'_1 \cdots q_m q'_m), \\ V_\nu \equiv V''_\nu \pmod{d_\nu}, \quad V'_\nu \equiv V''_\nu \pmod{d'_\nu}, \quad V \equiv V'' \pmod{d}, \end{gathered} \tag{2.3}$$

where $V''_\nu = V_\nu V'_\nu$ and $V'' = VV'$. Therefore,

$$\exp\left(2\pi i l_\nu \left(\frac{a_\nu}{q_\nu} V_\nu + \frac{a'_\nu}{q'_\nu} V'_\nu\right)\right) = \exp\left(2\pi i \left(\frac{a_\nu}{q_\nu} + \frac{a'_\nu}{q'_\nu}\right) V''_\nu l_\nu\right). \tag{2.4}$$

Also, setting $\delta'' = \delta \cdot \delta'$, we see that

$$\exp\left(-2\pi i l \left(\frac{a_\nu}{q_\nu} V + \frac{a'_\nu}{q'_\nu} V'\right)\right) = \exp\left(2\pi i \left(\frac{a_\nu}{q_\nu} + \frac{a'_\nu}{q'_\nu}\right) V'' l\right). \tag{2.5}$$

From (2.3)–(2.5), taking into account the multiplicative nature of the Möbius and Euler functions, we obtain under the hypothesis $(q_\nu, q'_t) = 1$ that

$$G(q_1, \ldots, q_m) \cdot G(q'_1, \ldots, q'_m) = E(q''_1 \ldots q''_m) \frac{\mu(\delta''_1)}{\varphi(\delta''_1)} \cdots \frac{\mu(\delta''_m)}{\varphi(\delta''_m)}$$

$$\times \sum_{a_1, \ldots, a_m}^{q_1, \ldots, q_m} \frac{\mu(\delta)}{\varphi(\delta)} \sum_{a'_1, \ldots, a'_m}^{q'_1, \ldots, q'_m} \frac{\mu(\delta')}{\varphi(\delta')} \prod_{\nu=1}^{m} \exp\left(2\pi i \left(\frac{a_\nu}{q_\nu} + \frac{a'_\nu}{q'_\nu}\right)(V''_\nu l_\nu - V'' l - u_\nu)\right)$$

$$= E(q''_1 \cdots q''_m) \cdot \prod_{t=1}^{m} \frac{\mu(\delta''_t)}{\varphi(\delta''_t)} \sum_{a''_1, \ldots, a''_m}^{q''_1, \ldots, q''_m} \frac{\mu(\delta'')}{\varphi(\delta'')} \prod_{\nu=1}^{m} \exp\left(2\pi i \frac{a''_\nu}{q''_\nu}(V''_\nu l_\nu - V'' l - u_\nu)\right),$$

where $\delta''_\nu = \delta_\nu \cdot \delta'_\nu$, $\delta'' = \delta \cdot \delta'$, $q''_\nu = q_\nu \cdot q'_\nu$ and $a''_\nu = a_\nu \cdot q'_\nu + a'_\nu \cdot q_\nu$, since if a_ν and a'_ν range over reduced residue systems mod q_ν and mod q'_ν respectively, then a''_ν ranges over a reduced residue system mod q''_ν if $(q_\nu, q'_\nu) = 1$.

LEMMA 2.2. *If the system of numbers $u_1, \ldots, u_m$ satisfies the condition*

$$u_\nu \equiv l_\nu - l \pmod D \quad (\nu = 1, \ldots, m),$$

then

$$\sigma_\Delta(u_1, \ldots, u_m) = \sum_{\lambda_1 \mid D} \varphi(\lambda_1) \cdots \sum_{\lambda_m \mid D} \varphi(\lambda_m) \cdot \sigma'_{\frac{\Delta}{\lambda_1}, \ldots, \frac{\Delta}{\lambda_m}}(u_1, \ldots, u_m),$$

where

$$\sigma'_{\frac{\Delta}{\lambda_1}, \ldots, \frac{\Delta}{\lambda_m}}(u_1, \ldots, u_m) = \sum_{\substack{1 \leqslant \xi_1 \leqslant \Delta/\lambda_1 \\ (\xi_1, D) = 1}} \cdots \sum_{\substack{1 \leqslant \xi_m \leqslant \Delta/\lambda_m \\ (\xi_m, D) = 1}} G_{u_1, \ldots, u_m}(\xi_1, \ldots, \xi_m).$$

PROOF. Suppose for a given combination of $q_1, \ldots, q_m$ the numbers $d_1, \ldots, d_m$ have the values indicated in §2.1, and suppose that $E(q_1 \cdots q_m) \neq 0$. Then the hypotheses of Lemma 2.1 are satisfied with $q'_\nu = q_\nu / d_\nu$ and $q_\nu = d_\nu$, and hence we have

$$G_{u_1, \ldots, u_m}(q_1, \ldots, q_m) = G_{u_1, \ldots, u_m}\left(\frac{q_1}{d_1}, \ldots, \frac{q_m}{d_m}\right) \cdot G_{u_1, \ldots, u_m}(d_1, \ldots, d_m). \quad (2.6)$$

But if

$$u_\nu \equiv l_\nu - l \pmod D, \quad (2.7)$$

then the second factor in (2.6) is equal to

$$\varphi(d_1) \ldots \varphi(d_m). \quad (2.8)$$

From (2.6)–(2.8) we conclude that

$$\sigma_\Delta(u_1, \ldots, u_m) = \sum_{\substack{q_1, \ldots, q_m \\ E(q_1 \cdots q_m) \neq 0}}^{[1, \Delta]} G_{u_1, \ldots, u_m}\left(\frac{q_1}{d_1}, \ldots, \frac{q_m}{d_m}\right) \cdot \prod_{\nu=1}^{m} \varphi(d_\nu),$$

and the right-hand side of the last equality can then be transformed into

$$\sum_{\lambda_1 \mid D} \cdots \sum_{\lambda_m \mid D} \varphi(\lambda_1) \cdots \varphi(\lambda_m) \sum_{\substack{1 \leqslant \xi_1 \leqslant \Delta/\lambda_1 \\ (\xi_1, D)=1}} \cdots \sum_{\substack{1 \leqslant \xi_m \leqslant \Delta/\lambda_m \\ (\xi_m, D)=1}} G_{u_1, \ldots, u_m}(\xi_1, \ldots, \xi_m),$$

which proves Lemma 2.2.

We will now determine conditions on the system of numbers $u_1, \ldots, u_m$ under which the series

$$\sigma'(u_1, \ldots, u_m; D) = \sum_{\substack{\xi_1, \ldots, \xi_m \\ (\xi_1 \ldots \xi_m, D)=1}}^{[1, \infty]} G_{u_1, \ldots, u_m}(\xi_1, \ldots, \xi_m),$$

where $\xi_1, \ldots, \xi_m$ range independently over the natural numbers relatively prime to D, is absolutely convergent.

Consider the expression $G_{u_1,\ldots,u_m}(\xi_1, \ldots, \xi_m)$, where the numbers $\xi_1, \ldots, \xi_m$ have only one of two values: either p, a prime not dividing D $(p \nmid D)$, or unity.

Assume that among the numbers $\xi_1, \ldots, \xi_m$ there are j equal to p, $p \nmid D$, and $(m-j)$ equal to unity, and let $\nu_1 < \cdots < \nu_j$ be the indices of those ξ_ν equal to p.

Since when $\xi_\nu = 1$ we have that a_ν assumes only the two values 0, 1, we may for $G_{u_1,\ldots,u_m}(\xi_1, \ldots, \xi_m)$ write instead $G_{u_{\nu_1},\ldots,u_{\nu_j}}(p; p \nmid D)$. Also, we will assign to the empty sum the value zero, without mentioning this explicitly each time.

For an integer j in $[1, m]$ and $p \nmid D$, we have $d_\nu = 1$ for $\nu = 1, \ldots, m$; $d = 1$, $E(q_1 \cdots q_m) = 1$; $\delta_\nu = p$ for $\nu = \nu_1, \ldots, \nu_j$ and $\delta_\nu = 1$ for the remaining indices; and $V_\nu = 0$ for all ν; $V = 0$.

First, if $j = 1$, then $\delta = p$ and, therefore,

$$G_{u_{\nu_1}}(p;\ p \nmid D) = \frac{1}{\varphi^2(p)} \varphi_1(u_{\nu_1}, p), \tag{2.9}$$

where

$$\varphi_1(u, p) = \sum_a^p \exp\left(-2\pi i \frac{a \cdot u}{p}\right). \tag{2.10}$$

We next turn to the general case: $2 \leqslant j \leqslant m$. From (2.2) we obtain

$$\begin{aligned} G_{u_{\nu_1}, \ldots, u_{\nu_j}}(p;\ p \nmid D) &= \frac{\mu^j(p)}{\varphi^j(p)} \sum_{a_{\nu_1}, \ldots, a_{\nu_j}}^{p, \ldots, p} \frac{\mu(\delta)}{\varphi(\delta)} \cdot \exp\left(-2\pi i \sum_{t=1}^{j} \frac{a_{\nu_t} \cdot u_{\nu_t}}{p}\right) \\ &= \frac{\mu^j(p)}{\varphi^j(p)} \left\{ \sum_{\substack{a_{\nu_1}, \ldots, a_{\nu_j} \\ a_{\nu_1} + \ldots + a_{\nu_j} \equiv 0 \pmod p}}^{p, \ldots, p} + \frac{\mu(p)}{\varphi(p)} \sum_{\substack{a_{\nu_1}, \ldots, a_{\nu_j} \\ a_{\nu_1} + \ldots + a_{\nu_j} \not\equiv 0 \pmod p}}^{p, \ldots, p} \right\} \\ &\quad \times \exp\left(-2\pi i \sum_{t=1}^{j} \frac{a_{\nu_t} \cdot u_{\nu_t}}{p}\right), \end{aligned} \tag{2.11}$$

since

$$\delta=\begin{cases}1, & \text{if } a_{\nu_1}+\dots+a_{\nu_j}\equiv 0 \pmod p;\\ p, & \text{if } a_{\nu_1}+\dots+a_{\nu_j}\not\equiv 0 \pmod p).\end{cases}$$

If we now introduce

$$W_{u_{\nu_1},\dots,u_{\nu_j}}(p)=\sum_{a_{\nu_1},\dots,a_{\nu_j}}^{p,\dots,p}\exp\left(-2\pi i\sum_{t=1}^{j}\frac{a_{\nu_t}\cdot u_{\nu_t}}{p}\right), \tag{2.12}$$

then (2.11) can be written in the form

$$\begin{aligned}G_{u_{\nu_1},\dots,u_{\nu_j}}(p;\, p+D)&=\frac{\mu^{j+1}(p)}{\varphi^{j+1}(p)}\sum_{a_{\nu_1},\dots,a_{\nu_j}}^{p,\dots,p}\exp\left(-2\pi i\sum_{t=1}^{j}\frac{a_{\nu_t}\cdot u_{\nu_t}}{p}\right)\\ &+\left(\frac{\mu^{j}(p)}{\varphi^{j}(p)}-\frac{\mu^{j+1}(p)}{\varphi^{j+1}(p)}\right)\sum_{\substack{a_{\nu_1},\dots,a_{\nu_j}\\ a_{\nu_1}+\dots+a_{\nu_j}\equiv 0 \pmod p}}^{p,\dots,p}\exp\left(-2\pi i\sum_{t=1}^{j}\frac{a_{\nu_t}\cdot v_{\nu_t}}{p}\right)\\ &=\frac{\mu^{j+1}(p)}{\varphi^{j+1}(p)}\left\{\prod_{t=1}^{j}\left(\sum_{a_{\nu_t}}^{p}\exp\left(-2\pi i\,\frac{a_{\nu_t}\cdot u_{\nu_t}}{p}\right)\right)-p\cdot W_{u_{\nu_1},\dots,u_{\nu_j}}(p)\right\}\\ &=\frac{\mu^{j+1}(p)}{\varphi^{j+1}(p)}\left\{\prod_{t=1}^{j}\varphi_1(u_{\nu_t},\, p)-p\cdot W_{u_{\nu_1},\dots,u_{\nu_j}}(p)\right\}.\end{aligned} \tag{2.13}$$

We consider transformations of the sum W in (2.12). We have

$$\begin{aligned}W_{u_{\nu_1},\dots,u_{\nu_j}}(p)&=\sum_{\substack{a_{\nu_1},\dots,a_{\nu_j}\\ a_{\nu_1}+\dots+a_{\nu_j}\equiv 0 \pmod p}}^{p,\dots,p}\exp\left(-2\pi i\left(\sum_{t=1}^{j}\frac{a_{\nu_t}\cdot u_{\nu_t}}{p}-u_{\nu_1}\sum_{t=1}^{j}\frac{a_{\nu_t}}{p}\right)\right)\\ &=\sum_{\substack{a_{\nu_2},\dots,a_{\nu_j}\\ a_{\nu_2}+\dots+a_{\nu_j}\not\equiv 0 \pmod p}}^{p,\dots,p}\exp\left(-2\pi i\left(\sum_{t=2}^{j}\frac{a_{\nu_t}\cdot u_{\nu_t}}{p}-u_{\nu_1}\sum_{t=2}^{j}\frac{a_{\nu_t}}{p}\right)\right)\\ &=\left\{\sum_{a_{\nu_2},\dots,a_{\nu_j}}^{p,\dots,p}-\sum_{\substack{a_{\nu_2},\dots,a_{\nu_j}\\ a_{\nu_2}+\dots+a_{\nu_j}\equiv 0 \pmod p}}^{p,\dots,p}\right\}\exp\left(-2\pi i\sum_{t=2}^{j}\frac{a_{\nu_t}(u_{\nu_t}-u_{\nu_1})}{p}\right)\\ &=\sum_{a_{\nu_2},\dots,a_{\nu_j}}^{p,\dots,p}\prod_{t=2}^{j}\exp\left(-2\pi i\,\frac{a_{\nu_t}}{p}(u_{\nu_t}-u_{\nu_1})\right)-W_{u_{\nu_2},\dots,u_{\nu_j}}(p).\end{aligned} \tag{2.14}$$

If $j>2$, then, continuing the above reasoning by replacing u_{ν_1} by u_{ν_2}, we obtain

$$W_{u_{\nu_2},\dots,u_{\nu_j}}(p)=\sum_{a_{\nu_3},\dots,a_{\nu_j}}^{p,\dots,p}\prod_{t=3}^{j}\exp\left(-2\pi i\,\frac{a_{\nu_t}}{p}(u_{\nu_t}-u_{\nu_2})\right)-W_{u_{\nu_3},\dots,u_{\nu_j}}(p).$$

Hence after $(j-1)$ steps we have

$$W_{u_{\nu_1},\ldots,u_{\nu_j}}(p)=\prod_{t=2}^{j}\left(\sum_{a_{\nu_t}}^{p}\exp\left(-2\pi i\,\frac{a_{\nu_t}}{p}(u_{\nu_t}-u_{\nu_1})\right)\right)$$

$$-\prod_{t=3}^{j}\left(\sum_{a_{\nu_t}}^{p}\exp\left(-2\pi i\,\frac{a_{\nu_t}}{p}(u_{\nu_t}-u_{\nu_2})\right)\right)+$$

$$\cdots+\sum_{a_{\nu_j}}^{p}\exp\left(-2\pi i\,\frac{a_{\nu_j}}{p}(u_{\nu_j}-u_{\nu_{j-1}})\right).$$

Hence, in view of (2.12), (2.13) and (2.10), we obtain

$$G_{u_{\nu_1},\ldots,u_{\nu_j}}(p,p\nmid D)=\frac{\mu^{j+1}(p)}{\varphi^{j+1}(p)}\left\{\prod_{t=1}^{j}\varphi_1(u_{\nu_t},\ p)\right.$$
$$\left.+p\sum_{k=1}^{j-1}(-1)^k\prod_{t=k+1}^{j}\varphi_1(u_{\nu_t}-u_{\nu_k},\ p)\right\}. \tag{2.15}$$

The relation (2.15), in particular, allows us to obtain a nontrivial estimate of the function G under a natural restriction on $u_1,\ldots,u_m$. Namely, if $u_s\neq u_t\neq 0$ for $s\neq t$, then for each integer j in $[1,m]$, putting

$$u=u_1\ldots u_m(u_m-u_{m-1})\ldots(u_m-u_1)\ldots(u_2-u_1), \tag{2.16}$$

we have by (2.15)

$$|G_{u_{\nu_1},\ldots,u_{\nu_j}}(p;\ p\times D)|\leqslant\begin{cases}\dfrac{m}{\varphi^2(p)}\left(1+\dfrac{1}{p}\right)^3, & \text{if}\quad p\nmid u,\\[2mm] \dfrac{1}{\varphi(p)}\left(1+\dfrac{1}{p}\right)^3, & \text{if}\quad p\mid u.\end{cases} \tag{2.17}$$

From (2.17) and Lemma 2.1 we obtain

$$|G_{u_1,\ldots,u_m}(\xi_1,\ \ldots,\ \xi_m)|\leqslant\frac{m^{\vartheta(\xi_1\ldots\xi_m)}(u,\ \xi_1\ldots\xi_m)}{\varphi^2(\xi_1\ldots\xi_m)}\prod_{p\mid\xi_1\ldots\xi_m}\left(1+\frac{1}{p}\right)^3, \tag{2.18}$$

where $\vartheta(\xi_1\cdots\xi_m)$ is the number of prime divisors of $\xi_1\cdots\xi_m$.

Consequently,

$$\sum_{\substack{\xi_1,\ldots,\xi_m\\(\xi_1\ldots\xi_m,\ D)=1}}^{[1,\infty]}|G_{u_1,\ldots,u_m}(\xi_1,\ \ldots,\ \xi_m)|$$

$$=B\sum_{\xi_1,\ldots,\xi_m}^{[1,\infty]}\frac{m^{\vartheta(\xi_1\ldots\xi_m)}(u,\ \xi_1\ldots\xi_m)}{\varphi^2(\xi_1\ldots\xi_m)}\prod_{p\mid\xi_1\ldots\xi_m}\left(1+\frac{1}{p}\right)^3$$

$$=B\sum_{k=1}^{\infty}\frac{m^{\vartheta(k)}\cdot(u,\ k)\cdot\tau_m(k)}{\varphi^2(k)}\prod_{p\mid k}\left(1+\frac{1}{p}\right)^3=$$

$$= B \cdot u \cdot \sum_{k=1}^{\infty} \frac{m^{\vartheta(k)} \cdot \tau_m(k) \cdot \ln^4 k}{k^2} = B \cdot u,$$

where we have used the estimates

$$\vartheta(k) = B \frac{\ln k}{\ln \ln k}, \qquad \tau_m(k) = B \cdot k^{\xi}, \qquad \prod_{p \mid k} \left(1 + \frac{1}{p}\right) = B \ln \ln k,$$

and Lemma 1.6.

We have thus proved

LEMMA 2.3. *For any system of distinct natural numbers, the series*

$$\sigma'(u_1, \ldots, u_m; D) = \sum_{\substack{\xi_1, \ldots, \xi_m \\ (\xi_1 \ldots \xi_m, D) = 1}}^{[1, \infty]} G_{u_1, \ldots, u_m}(\xi_1, \ldots, \xi_m)$$

is absolutely convergent.

§2.3. Estimation of the partial sums of the series σ'

LEMMA 2.4. *Under the hypotheses of Lemmas* 2.2 *and* 2.3,

$$\sigma'(u_1, \ldots, u_m; D) = \prod_{p \nmid D} \frac{p - \gamma_m(p, D)}{p\left(1 - \frac{1}{p}\right)^{m+1}},$$

where the product extends over all primes p not dividing D, and $\gamma_m(p, D)$ is the number of solutions of the congruence

$$(Dz + l)(Dz + l + u_1) \ldots (Dz + l + u_m) \equiv 0 \pmod{p}.$$

PROOF. Under our hypotheses, the series $\sigma'(u_1, \ldots, u_m; D)$, according to the previous lemma, is absolutely convergent. Hence by Lemma 2.1 it can be written as an infinite product over the primes p not dividing D:

$$\sigma'(u_1, \ldots, u_m; D) = \prod_{p \nmid D} \sum_{\xi_1 \mid p} \cdots \sum_{\xi_m \mid p} G_{u_1, \ldots, u_m}(\xi_1, \ldots, \xi_m)$$

$$= \prod_{p \nmid D} \{1 + \textstyle\sum_1 G_{u_{\nu_1}}(p) + \sum_2 G_{u_{\nu_1}, u_{\nu_2}}(p) + \cdots \tag{2.19}$$

$$+ \textstyle\sum_j G_{u_{\nu_1}, \ldots, u_{\nu_j}}(p) + \cdots + G_{u_1, \ldots, u_m}(p)\},$$

where Σ_j for $1 \leqslant j \leqslant m - 1$ denotes the summation over all possible combinations $G_{u_1, \ldots, u_m}(\xi_1, \ldots, \xi_m)$ in which j of the numbers $\xi_1, \ldots, \xi_m$ are equal to p and $(m - j)$ are equal to unity.

But each such function can be calculated by (2.15), in view of which the expression under the product sign in (2.19) can be written in the form

$$1 + \sum_{j=1}^{m} \frac{\mu^{j+1}(p)}{\varphi^{j+1}(p)} \prod_{t=1}^{j} \varphi_1(u_{\nu_t}, p) - p \sum_{j=2}^{m} \frac{\mu^{j+1}(p)}{\varphi^{j+1}(p)} \sum_j W_{u_{\nu_1}, \ldots, u_{\nu_j}}(p)$$

$$= 1 + \frac{\mu(p)}{\varphi(p)} \left[\prod_{j=1}^{m} \left(1 + \frac{\mu(p)}{\varphi(p)} \varphi_1(u_{\nu_j}, p)\right) - 1\right] +$$

$$+\frac{p}{\varphi(p)}\sum_{j=2}^{m}\frac{\mu^j(p)}{\varphi^j(p)}\sum\nolimits_j W_{u_{\nu_1},\dots,u_{\nu_j}}(p).$$

Hence, introducing the notation

$$\Theta_m=1-\frac{1}{\varphi(p)}\left[\prod_{j=1}^{m}\left(1-\frac{1}{\varphi(p)}\varphi_1(u_{\nu_j},\ p)\right)-1\right],$$

$$\Theta'_m=\sum_{j=2}^{m}\frac{\mu^j(p)}{\varphi^j(p)}\Theta''_{m,\,j},\qquad \Theta''_{m,\,j}=\sum\nolimits_j W_{u_{\nu_1},\dots,u_{\nu_j}}(p),$$

we find that

$$\sum_{\xi_1\mid p}\cdots\sum_{\xi_m\mid p}G_{u_1,\dots,u_m}(\xi_1,\ \dots,\ \xi_m)=\Theta_m+\frac{p}{\varphi(p)}\Theta'_m. \tag{2.20}$$

But we have

$$\begin{aligned}\Theta_{m+1}-\left(1+\frac{1}{\varphi(p)}\right)\Theta_m&=1+\frac{1}{\varphi(p)}-\left(1+\frac{1}{\varphi(p)}\right)^2-\frac{1}{\varphi(p)}\\ &\times\prod_{j=1}^{m}\left(1-\frac{\varphi_1(u_{\nu_j},\ p)}{\varphi(p)}\right)\left(1-\frac{1}{\varphi(p)}\varphi_1(u_{m+1},\ p)-1-\frac{1}{\varphi(p)}\right)\\ =-\frac{1}{\varphi(p)}\left(1+\frac{1}{\varphi(p)}\right)&+\frac{1}{\varphi^2(p)}(1+\varphi_1(u_{m+1},\ p))\cdot\prod_{j=1}^{m}\left(1-\frac{\varphi_1(u_{\nu_j},\ p)}{p}\right).\end{aligned} \tag{2.21}$$

Moreover, since $\Theta''_{m,m-1}=0$, we obtain for $j\geqslant 2$ (see the derivation of Lemma 2.3) that

$$\Theta''_{m+1,\,j}=\Theta''_{m,\,j}+\sum\nolimits'_j W_{u_{m+1},\ u_{\nu_1},\dots,u_{\nu_j}}(p), \tag{2.22}$$

where Σ'_j denotes the summation over all distinct combinations of numbers $u_{\nu_1},\ldots,u_{\nu_j}$ taken from the system $u_1,\ldots,u_m$.

From (2.22), in view of (2.14), we obtain

$$\begin{aligned}\Theta''_{m+1,\,j}&=\Theta''_{m,\,j}+\sum\nolimits'_j\left(\prod_{t=1}^{j}\varphi_1(u_{\nu_t}-u_{m+1},\ p)-W_{u_{\nu_1},\dots,u_{\nu_j}}(p)\right)\\ &=\Theta''_{m,\,j}+\sum\nolimits'_j\prod_{t=1}^{j}\varphi_1(u_{\nu_t}-u_{m+1},\ p)-\Theta''_{m,\,j-1}.\end{aligned}$$

Consequently,

$$\begin{aligned}\Theta'_{m+1}&=\Theta'_m+\sum_{j=2}^{m+1}\frac{\mu^j(p)}{\varphi^j(p)}\sum\nolimits'_j\prod_{t=1}^{j-1}\varphi_1(u_{\nu_t}-u_{m+1},\ p)-\sum_{j=2}^{m+1}\frac{\mu^j(p)}{\varphi^j(p)}\Theta''_{m,\,j-1}\\ &=\left(1+\frac{1}{\varphi(p)}\right)\Theta'_m-\frac{1}{\varphi(p)}\left[\prod_{j=1}^{m}\left(1-\frac{1}{\varphi(p)}\varphi_1(u_{\nu_j}-u_{m+1},\ p)\right)-1\right].\end{aligned} \tag{2.23}$$

Combining (2.20), (2.21) and (2.23), we see that

$$\sum_{\xi_1 \mid p} \cdots \sum_{\xi_{m+1} \mid p} G_{u_1, \ldots, u_{m+1}}(\xi_1, \ldots, \xi_m)$$
$$-\frac{p}{\varphi(p)} \sum_{\xi_1 \mid p} \cdots \sum_{\xi_m \mid p} G_{u_1, \ldots, u_m}(\xi_1, \ldots, \xi_m) \tag{2.24}$$
$$=\frac{1}{\varphi^2(p)}(1+\varphi_1(u_{m+1}, p)) \cdot \prod_{j=1}^{m}\left(1-\frac{1}{\varphi(p)}\varphi_1(u_{\nu_j}, p)\right)$$
$$-\frac{p}{\varphi^2(p)}\prod_{j=1}^{m}\left(1-\frac{1}{\varphi(p)}\varphi_1(u_{\nu_j}-u_{m+1}, p)\right).$$

Furthermore,

$$\sum_{\xi_1 \mid p} G_{u_1}(p) = 1+\frac{\varphi_1(u_1, p)}{\varphi^2(p)}. \tag{2.25}$$

Now (2.24) and (2.25) permit us to show that

$$\sum_{\xi_1 \mid p} \cdots \sum_{\xi_m \mid p} G_{u_1, \ldots, u_m}(\xi_1, \ldots, \xi_m) = \frac{p^m}{\varphi^m(p)} \frac{p-\gamma_m(p, D)}{p-1}. \tag{2.26}$$

The assertion of Lemma 2.4 follows from (2.26), in view of (2.19).

Denoting the right- and left-hand sides of (2.26) by $\Phi_m(p)$ and $\Phi'_m(p)$ respectively, we must show, therefore, that for $p \nmid D$ we have

$$\Phi_m(p) = \Phi'_m(p) \tag{2.27}$$

for any integer $m \geqslant 1$.

Suppose $m = 1$. Then only one of the following two cases is possible:

1°. $u_1 \equiv 0 \pmod p$, and hence $\gamma_1(p, D) = 1$ and $\varphi_1(u_1, p) = p - 1$;

2°. $u_1 \not\equiv 0 \pmod p$, in which case $\gamma_1(p, D) = 2$ and $\varphi_1(u_1, p) = -1$.

Thus, (2.25) implies (2.26) for $m = 1$.

In the general case the proof is obtained by induction. So suppose (2.27) is true for a certain integer m.

Then only the following three cases are possible:

1°. $u_{m+1} \equiv 0 \pmod p$;

2°. $u_{m+1} \equiv u_j \not\equiv 0 \pmod p$ for some j, $1 \leqslant j \leqslant m$;

3°. $u_{m+1} \not\equiv u_j \pmod p$ and $u_{m+1} \not\equiv 0 \pmod p$, $1 \leqslant j \leqslant m$.

In the first case, it is not difficult to see that

$$\gamma_{m+1}(p, D) = \gamma_m(p, D), \quad \Phi_{m+1}(p) = \frac{p}{\varphi(p)}\Phi_m(p),$$

and, moreover, (2.24) yields

$$\Phi'_{m+1}(p) = \frac{p}{\varphi(p)}\Phi'_m(p).$$

Therefore,

$$\Phi'_{m+1}(p) = \frac{p}{\varphi(p)}\Phi_m(p) = \Phi_{m+1}(p).$$

Case 2° is entirely analogous. Here we have

$$1+\varphi_1(u_{m+1}, p)=0, \quad 1-\frac{1}{\varphi(p)}\varphi_1(u_j-u_{m+1}, p)=0$$

and hence also

$$\Phi'_{m+1}(p)=\Phi_{m+1}(p).$$

Finally, in case 3° we obtain

$$\gamma_{m+1}(p, D)=\gamma_m(p, D)+1,$$

and (2.24) yields

$$\Phi'_{m+1}(p)-\frac{p}{\varphi(p)}\Phi'_m(p)=-\frac{p}{\varphi(p)}\left(1+\frac{1}{p-1}\right)^{m+1}.$$

Substituting $\Phi_m(p)$ for $\Phi'_m(p)$, we have

$$\Phi'_{m+1}(p)=\frac{p}{\varphi(p)}\left(1+\frac{1}{\varphi(p)}\right)^m\left[1-\frac{\gamma_{m+1}(p, D)-1}{\varphi(p)}\right]-\frac{1}{p-1}\left(1+\frac{1}{p-1}\right)^{m+1}$$
$$=\frac{p^{m+1}}{\varphi^{m+1}(p)}\left(1-\frac{\gamma_m(p, D)}{p-}\right)=\Phi_m(p).$$

This completely proves Lemma 2.4.

We now estimate the remainder of the series

$$\sum{}'=\sum_{\substack{\xi_1>\Delta_1\\(\xi_1, D)=1}}\cdots\sum_{\substack{\xi_m>\Delta_m\\(\xi_m, D)=1}}|G_{u_1,\ldots,u_m}(\xi_1, \ldots, \xi_m)|, \tag{2.28}$$

where $\Delta_\nu = \Delta/\lambda_\nu$, λ_ν being a divisor of D.

From (2.18), under the hypotheses of Lemmas 2.2 and 2.3, putting $\Delta'_m = \Delta_1 \cdots \Delta_m$ and $\Delta''_m = \Delta'_m/d_{m+1}$, where $d_{m+1}|u$, we obtain by means of estimates applied earlier

$$\begin{aligned}\sum{}'&=B\cdot\sum_{\xi_1>\Delta_1}\cdots\sum_{\xi_m>\Delta_m}\frac{(u, \xi_1\ldots\xi_m)\cdot m^{\vartheta(\xi_1\ldots\xi_m)}}{\varphi^2(\xi_1\ldots\xi_m)}\prod_{p\mid\xi_1\ldots\xi_m}\left(1+\frac{1}{p}\right)^3\\&=B\cdot\sum_{d_{m+1}\mid u}\frac{(u, k)}{k^{2-\varepsilon}}=B\cdot\sum_{d_{m+1}\mid u}\frac{1}{d^{1-\varepsilon}_{m+1}}\sum_{s=\Delta''_m}^{\infty}\frac{1}{s^{2-\varepsilon}}\\&=B\sum_{d_{m+1}\mid u}\frac{1}{d^{1-\varepsilon}_{m+1}}\frac{1}{(\Delta''_m)^{1-\varepsilon}}=B\frac{\tau_1(u)}{(\Delta'_m)^{1-\varepsilon}}.\end{aligned} \tag{2.29}$$

Therefore,

$$\begin{aligned}&\left|\sigma_\Delta(u_1,\ldots,u_m; D)-\sum_{\lambda_1\mid D}\cdots\sum_{\lambda_m\mid D}\varphi(\lambda_1)\ldots\varphi(\lambda_m)\cdot\sigma'(u_1, \ldots, u_m; D)\right|\\&=B\cdot\frac{D^m\cdot\tau_1(u)}{\Delta^{(1-\varepsilon)m}}\sum_{\lambda_1\mid D}\cdots\sum_{\lambda_m\mid D}\varphi(\lambda_1)\ldots\varphi(\lambda_m).\end{aligned} \tag{2.30}$$

From (2.30) and Lemma 2.4 we obtain

LEMMA 2.5. *Under the hypotheses of Lemma* 2.4,

$$\sigma_\Delta(u_1, \ldots, u_m; D) = D^m \prod_{p \nmid D} \frac{p - \gamma_m(p, D)}{p\left(1 - \frac{1}{p}\right)^{m+1}} + B\frac{D^{2m} \cdot \tau_1(u)}{\Delta^{(1-\varepsilon)m}}. \tag{2.31}$$

REMARK. Since $\Delta = L^\delta$, $\delta > 0$ a constant, and it is impossible to guarantee for the function $\tau_1(u)$ an estimate better than $\tau_1(u) = B \cdot u^\epsilon$, it follows that the second term in (2.31) is not always, in fact, a remainder term. It is not possible to choose Δ so that the last expression has the expected order of growth D^m/L^{m+c}. This is related to the laws of distribution of primes in progressions with difference $D = x^\epsilon$, which is still problematical.

§2.4. The principal lemmas

From (1.1)–(1.3) we obtain

$$\begin{aligned} H(\alpha_1, \ldots, \alpha_m) = \varphi^{m+1}(D) \sum_{\substack{3 \leq p \leq x \\ p \equiv l \pmod D}} \sum_{\substack{3 \leq p_1 \leq x \\ p_1 \equiv l_1 \pmod D}} \cdots \sum_{\substack{3 \leq p_m \leq x \\ p_m \equiv l_m \pmod D}} \exp(2\pi i\alpha_1(p_1 - p)) \\ \ldots \exp(2\pi i\alpha_m(p_m - p)) = \varphi^{m+1}(D) \sum_{u_1, \ldots, u_m}^{[-x, x]} \pi(x; u_1, \ldots, u_m; D) \\ \times \prod_{\nu=1}^{m} \exp(2\pi i\alpha_\nu u_\nu). \end{aligned} \tag{2.32}$$

where

$$\pi(x; u_1, \ldots, u_m; D) = \pi(x; u_1, \ldots, u_m; D, l, l_1, \ldots, l_m)$$

denotes the number of sets of primes $p, p + u_1, \ldots, p + u_m$ in the interval $(3, x)$ belonging, respectively, to the progressions $Dn + l, Dn_1 + l_1, \ldots, Dn_m + l_m$, and the summation extends over all possible systems $u_1, \ldots, u_m$ of integers in the interval $[-x, x]$.

Also, from (1.4)–(1.6) we obtain

$$N(\alpha_1, \ldots, \alpha_m) = \sum_{u_1, \ldots, u_m}^{[-x, x]} \Psi(x; u_1, \ldots, u_m; D) \cdot \prod_{\nu=1}^{m} \exp(2\pi i\alpha_\nu \cdot u_\nu), \tag{2.33}$$

where

$$\begin{aligned} \Psi(x; u_1, \ldots, u_m; D) &= \Psi(x; u_1, \ldots, u_m; D, l, l_1, \ldots, l_m) \\ &= f(x; u_1, \ldots, u_m) \cdot \sigma_\Delta(u_1, \ldots, u_m; D), \end{aligned} \tag{2.34}$$

and

$$f(x; u_1, \ldots, u_m) = \sum_{n, n_1, \ldots, n_m}^{[3, x]} \sum_{u_1 = n_1 - n} \cdots \sum_{u_m = n_m - n} \frac{1}{\ln n \cdot \ln n_1 \ldots \ln n_m}, \tag{2.35}$$

and the expressions for $\sigma_\Delta(u_1, \ldots, u_m; D)$ are given in §2.1.

FIRST PRINCIPAL LEMMA.

$$\sum_{u_1,\ldots,u_m}^{[-x,\,x]} |\varphi^{m+1}(D)\cdot\pi(x;\,u_1,\ldots,u_m;\,D)-\Psi(x;\,u_1,\ldots,u_m;\,D)|^2=B\cdot x^{m+2}\cdot\Lambda^{-1}.$$

PROOF. The left-hand side, in view of (2.32) and (2.33), is equal to

$$\int_0^1\cdots\int_0^1 |H(\alpha_1,\ldots,\alpha_m)-N(\alpha_1,\ldots,\alpha_m)|^2\,d\alpha_1\ldots d\alpha_m.$$

This last expression, by (1.7)–(1.10), (1.41) and (1.52), is equal to $B\cdot x^{m+2}\cdot\Lambda^{-1}$, where B is a constant depending only on M, m, A and c.

SECOND PRINCIPAL LEMMA. *Let $u_1,\ldots,u_m$ be any nontrivial system of integers satisfying the following conditions:*

1°. $2\leqslant u_1,\ldots,u_m\leqslant X$;

2°. $u_\nu\equiv l_\nu-l\,(\mathrm{mod}\,D)$ $(\nu=1,\ldots,m)$;

3°. $u_s\neq u_t\neq 0$ *for* $s\neq t$;

4°. $\tau_1(u)\leqslant L^\delta$; $\delta=5m[2^{2^m}+M+(m+2)(m+c+A+1)]$.

Then the asymptotic formula

$$\Psi(x;\,u_1,\ldots,u_m;\,D)=\frac{\varphi^{m+1}(D)}{D}\prod_p\frac{p-\gamma_m(p,\,D)}{p\left(1-\frac{1}{p}\right)^{m+1}}\left(\int_2^x\frac{dz}{\ln^{m+1}z}+B\frac{x}{(\ln x)^{m+c}}\right)$$

holds, where $\gamma_m(p,D)$ denotes the number of solutions (with respect to z) of the congruence

$$(Dz+l)(Dz+l+u_1)\ldots(Dz+l+u_m)\equiv 0\,(\mathrm{mod}\,p), \tag{2.36}$$

$c>1$ is any given quantity, and B is a constant not depending on x.

REMARK. Conditions 1°–3° of the lemma are natural ones.

With regard to 4°, see the remark following Lemma 2.5.

PROOF. If $p|D$, then p divides each of the numbers $u_\nu-l_\nu+l$, $\nu=1,\ldots,m$. But $(l_\nu,D)=1$, and hence the congruence (2.36) has no solutions in this case. It follows that

$$D^m\prod_{p\nmid D}\frac{p-\gamma_m(p,\,D)}{p\left(1-\frac{1}{p}\right)^{m+1}}=\frac{\varphi^{m+1}(D)}{D}\prod_p\frac{p-\gamma_m(p,\,D)}{p\left(1-\frac{1}{p}\right)^{m+1}}, \tag{2.37}$$

where the product on the right extends over all primes without exception.

Also, condition 4° implies that

$$\frac{D^m\cdot\tau_1(u)}{\Delta^{(1-\varepsilon)m}}=B\cdot\Lambda^{-1}. \tag{2.38}$$

It remains to show that

$$f(x;\,u_1,\ldots,u_m)=\int_2^x\frac{dz}{\ln^{m+1}z}+B\cdot\frac{x}{(\ln x)^{m+c}}. \tag{2.39}$$

If $u_1, \ldots, u_m$ are given, then for any fixed integer n belonging to the segment $[3, x-u']$, where $u' = \max u_\nu$ for $\nu = 1, \ldots, m$, the numbers $n_1, \ldots, n_m$ in (2.35) are uniquely determined. Hence from (2.35) we obtain

$$f(x;\ u_1, \ldots, u_m) = \sum_{3 \leqslant n \leqslant x-u'} \left(\ln n \cdot \prod_{\nu=1}^{m} \ln (n+u_\nu)\right)^{-1}$$
$$= \Big\{ \sum_{3 \leqslant n \leqslant X} + \sum_{X < n \leqslant x-u'} \Big\} \left(\ln n \cdot \prod_{\nu=1}^{m} \ln (n+u_\nu)\right)^{-1} = \Sigma_1 + \Sigma_2. \tag{2.40}$$

But when $n > X$ we have

$$\prod_{\nu=1}^{m} \ln (n+u_\nu) = \ln^m n + B \frac{u' \cdot \ln^{m-1} n}{n}.$$

Then condition 1° of the lemma implies that

$$\Sigma_2 = \sum_{X < n \leqslant x-u'} \left(\ln n \cdot \prod_{\nu=1}^{m} \ln (n+u_\nu)\right)^{-1} = \sum_{3 \leqslant n \leqslant x-u'} (\ln n)^{-m-1}$$
$$+ B \cdot u' \cdot \sum_{n=3}^{\leqslant x} \frac{1}{n \cdot \ln^{m+1} n} - \sum_{3 \leqslant n \leqslant X} (\ln n)^{-m-1} = \int_2^x \frac{dz}{\ln^{m+1} z} + B \cdot X \tag{2.41}$$

and

$$\Sigma_1 = B \cdot X. \tag{2.42}$$

Combining (2.40)–(2.42), we obtain (2.39). And the combination of (2.37)–(2.39) with Lemma 2.5 provides the assertion of the lemma.

§2.5. Proof of the theorems

We limit ourselves to the consideration of the set of $\mathfrak{M}_{m\gamma}$-systems of numbers $u_1, \ldots, u_m$ satisfying the following four natural conditions, which we call γ-conditions:

1°. $2 \leqslant u_1, \ldots, u_m \leqslant X$;
2°. $u_\nu \equiv l_\nu - l \pmod D$ $(\nu = 1, \ldots, m)$;
3°. $u_s \neq u_t \neq 0$ if $s \neq t$;
4°. $\gamma_m(p, D) \neq p$ for any $p \leqslant m+1$.

Each such system $u_1, \ldots, u_m$ is called an *$m\gamma$-system*.

According to the First Principal Lemma we have

$$\sum_{\mathfrak{M}_{m\gamma}} |\varphi^{m+1}(D) \cdot \pi(x;\ u_1, \ldots, u_m;\ D) - \Psi(x;\ u_1, \ldots, u_m;\ D)|^2$$
$$= B \cdot x^{m+2} \cdot \Lambda^{-1}. \tag{2.42'}$$

We now split $\mathfrak{M}_{m\gamma}$ into two subsets: $\mathfrak{M}'_{m\gamma}$ and $\mathfrak{M}''_{m\gamma}$. To $\mathfrak{M}'_{m\gamma}$ we assign the $m\gamma$-systems for which

$$|\varphi^{m+1}(D) \cdot \pi(x;\ u_1, \ldots, u_m;\ D) - \Psi(x;\ u_1 \ldots, u_m;\ D)| > D^m \cdot X.$$

Let us estimate the number $I(\mathfrak{M}'_{m\gamma})$ of $m\gamma$-systems in $\mathfrak{M}'_{m\gamma}$.

It follows from (2.42′) that

$$I(\mathfrak{M}'_{m\gamma}) \cdot D^{2m} \cdot X^2 = B \cdot x^{m+2} \cdot \Lambda^{-1},$$

i.e.

$$I(\mathfrak{M}'_{m\gamma}) = B \cdot x^m \cdot D^{-m} \cdot L^{-M-m(m+c)}. \tag{2.43}$$

Also, $\mathfrak{M}''_{m\gamma}$ can be represented in the form $\mathfrak{M}''_{m\gamma} = \mathfrak{N}_{m\gamma} + \mathfrak{N}'_{m\gamma}$, where $\mathfrak{N}_{m\gamma}$ consists of those $m\gamma$-systems for which

$$\tau_1(u) \leqslant L^\delta, \quad \delta = 5m\,[2^{2m} + M + (m+2)(m+c+A+1)].$$

Let us estimate $I(\mathfrak{N}'_{m\gamma})$, the number of $m\gamma$-systems belonging to $\mathfrak{M}''_{m\gamma}$ and satisfying the condition $\tau_1(u) > L^\delta$. We have

$$I(\mathfrak{N}'_{m\gamma}) \cdot L^\delta \leqslant \sum_{u_1, \ldots, u_m}^{[2, X]} \tau_1(u). \tag{2.44}$$

But the right-hand side of (2.44) clearly does not exceed

$$\sum_{u_1}^{[2, X]} \tau_1(u_1) \sum_{u_2}^{[2, X]} \tau_1(u_2) \cdot \tau_1(u_2 - u_1) \ldots \sum_{u_m}^{[2, X]} \tau_1(u_m) \cdot \tau_1(u_m - u_{m-1}) \ldots \tau_1(u_m - u_1),$$

which in turn, by the Cauchy inequality, does not exceed

$$B \cdot \prod_{j=0}^{m-1} \prod_{\nu=1}^{m-j-1} \left(\sum_{u_{m-j}}^{[2, X]} \tau_1^{2^\nu}(u_{m-j}) \right)^{2^{-\nu}}. \tag{2.45}$$

We also make use of the well-known estimate

$$\sum_{1 \leqslant t \leqslant z} \tau_1^n(t) = B \cdot z (\ln z + 1)^{2^n - 1}.$$

Applying this to (2.45), we obtain, in view of (2.44), that

$$\sum_{u_1, \ldots, u_m}^{[2, X]} \tau_1(u) = B \cdot x^m (\ln x)^{\Theta_1}, \quad \Theta_1 = 2^{2m} + m. \tag{2.46}$$

From (2.46) and (2.44) we finally obtain

$$I(\mathfrak{N}'_{m\gamma}) = B \cdot x^m \cdot D^{-m} \cdot L^{-M-m(m+c)}. \tag{2.47}$$

Combining (2.43) and (2.47), we conclude that the $m\gamma$-systems belonging to the sets $\mathfrak{M}'_{m\gamma}$ and $\mathfrak{N}'_{m\gamma}$ occur very infrequently. In comparison with the quantity $I(\mathfrak{N}_{m\gamma})$ there are almost none of them.

But for each $m\gamma$-system in $\mathfrak{N}_{m\gamma}$, by definition, we have

$$\pi(x;\, u_1, \ldots, u_m;\, D) = \frac{\Psi(x;\, u_1, \ldots, u_m;\, D)}{\varphi^{m+1}(D)} + B \frac{D^m}{\varphi^{m+1}(D)} X,$$

where the function Ψ satisfies all of the conditions of the Second Principal Lemma.

Hence for every $m\gamma$-system, except possibly for at most

$$I(\mathfrak{M}'_{m\gamma}) + I(\mathfrak{N}_{m\gamma}) = B \cdot x^m \cdot D^{-m} \cdot L^{-M-m(m+c)}$$

of them, we have the asymptotic equality

$$\pi(x;\, u_1, \ldots, u_m;\, D) = \frac{1}{D}\prod_p \frac{p-\gamma_m(p,\, D)}{p\left(1-\frac{1}{p}\right)^{m+1}}\left(\int_2^x \frac{dz}{\ln^{m+1} z} + B\,\frac{x}{(\ln x)^{m+c}}\right), \tag{2.48}$$

and, since the system $u_1, \ldots, u_m$ is nontrivial, i.e. $\gamma_m(p, D) \neq p$ for $p \leqslant m+1$, from the obvious inequality

$$\gamma_m(p,\, D) \leqslant \min(p,\, m+1)$$

we conclude that the product over the primes in (2.48) is positive. Moreover, if we put

$$u = \prod_{\nu=1}^{m} u_\nu \prod_{j=2}^{m}\prod_{t=1}^{j-1} |u_j - u_t|,$$

then we have

$$\frac{1}{D}\prod_p \frac{p-\gamma_m(p,\, D)}{p\left(1-\frac{1}{p}\right)^{m+1}} = \frac{1}{D}\prod_{p\leqslant m+1,\ p\nmid D}\left(\frac{p}{p-1}\right)^m \frac{p-\gamma_m(p,\, D)}{p-1}$$

$$\times \prod_{p\leqslant m+1,\ p\mid D}\left(\frac{p}{p-1}\right)^m \prod_{p>m+1}\left(\frac{p}{p-1}\right)^m \frac{p-m-1}{p-1} \prod_{p>m+1,\ p\mid D}\frac{p-1}{p-m-1}$$

$$\times \prod_{\substack{p>m+1\\ p\mid u,\ p\nmid D}} \frac{p-\gamma_m(p,\, D)}{p-m-1},$$

and therefore

$$\frac{1}{D}\prod_p \frac{p-\gamma_m(p,\, D)}{p\left(1-\frac{1}{p}\right)^{m+1}} \geqslant \frac{c_m}{\varphi(D)},$$

where $c_m > 0$ depends only on m; thus (2.48) actually provides a law of distribution of the sets of primes with given nontrivial differences between them.

If the system $u_1, \ldots, u_m$ is trivial, then from the preceding estimates it follows only that

$$\pi(x;\, u_1, \ldots, u_m;\, D) = BX \cdot L^{-m-c}.$$

However, in this case $\pi(x; u_1, \ldots, u_m)$ actually does not depend on x at all. Namely, we have

LEMMA 2.6. *For any trivial system $u_1, \ldots, u_m$*

$$0 \leqslant \pi(x;\, u_1, \ldots, u_m;\, D) \leqslant 1.$$

PROOF. Under the hypotheses of the lemma there exists a prime $p \leqslant m+1$, $p \nmid D$ (when $p|D$ we have $\gamma_m(p, D) = 0$) such that

$$DZ_{\nu_1} + l + u_{\nu_1} \equiv 0 \pmod p, \ \ldots,\ DZ_{\nu_{p-1}} + l + u_{\nu_{p-1}} \equiv 0 \pmod p,$$

where $\nu_1 < \cdots < \nu_{p-1}$ belong to $[1, m]$ and $Z_{\nu_1}, \ldots, Z_{\nu_{p-1}}$ are distinct numbers in $[1, p-1]$.

Hence we have

$$u_{\nu_1} \equiv t_1 \pmod p, \ldots, \quad u_{\nu_{p-1}} \equiv t_{p-1} \pmod p,$$

where $t_1, \ldots, t_{p-1}$ are distinct numbers in $[1, p-1]$. If now g ranges over the primes of the progression $pn + j \geqslant 3$, $1 \leqslant j \leqslant p-1$, $n = 0, 1, 2, \ldots$, then among the systems $g, g + u_{\nu_1}, \ldots, g + u_{\nu_{p-1}}$ there exists at most one in which all of the numbers are primes, since among $t_1, \ldots, t_{p-1}$ occurs $t_i = p - j$, and therefore the numbers of the sequence $g + u_{\nu_i}$, except possibly for one, are composite (multiples of p and not equal to p). But the number of distinct progressions for q does not exceed $(m+1)$. This proves the lemma. The combination of Lemma 2.6 and (2.48) yields Theorems 1 and 2 and the remarks following them.

In conclusion we remark that although Theorem 3 is not a direct consequence of Theorem 2, its proof is essentially the same, and the necessary changes in the computations are easily discovered by comparing this present paper with that of N. G. Čudakov [5].

Finally, Theorem 2 generalizes to progressions with nonidentical differences; the computations are almost the same.

BIBLIOGRAPHY

1. I. M. Vinogradov, *Selected works*, Izdat. Akad. Nauk SSSR, Moscow, 1952. (Russian)

2. G. H. Hardy and J. E. Littlewood, *Some problems of 'partitio numerorum'*. III, Acta Math. **44** (1923), 1–70.

3. J. J. Sylvester, *On the partition of an even number into two primes*, Proc. London Math. Soc. **4** (1871), 4–6.

4. K. K. Mardžanišvili, *Sur un problème additif de la théorie des nombres*, Izv. Akad. Nauk SSSR Ser. Mat. **4** (1940), 193–214. (Russian; French summary)

5. N. G. Čudakov, *On the density of the set of even numbers which are not representable as a sum of two odd primes*, Izv. Akad. Nauk SSSR Ser. Mat. **2** (1938), 25–40. (Russian; English summary).

6. Hua Loo-keng, *Additive theory of prime numbers*, Trudy Mat. Inst. Steklov. **22** (1947); English transl. (of rev. Chinese ed.), Amer. Math. Soc., Providence, R. I., 1965.

7. N. G. Čudakov, *Introduction to the theory of Dirichlet's L-functions*, OGIZ, Moscow, 1947. (Russian)

8. Hans Rademacher, *Über eine Erweiterung des Goldbachschen Problems*, Math. Z. **25** (1926), 627–657.

9. A. F. Lavrik, *On the twin prime conjectures of the theory of primes by Vinogradov's method*, Dokl. Akad. Nauk SSSR **132** (1960), 1013–1015; English transl. in Soviet Math. Dokl. **1** (1960).

10. ———, *The number of k-twin primes lying on an interval of given length*, Dokl. Akad. Nauk SSSR **136** (1961), 281–283; English transl. in Soviet Math. Dokl. **2** (1961).

11. ———, *On the theory of the distribution of sets of primes with given differences between them*, Dokl. Akad. Nauk SSSR **138** (1961), 1287–1290; English transl. in Soviet Math. Dokl. **2** (1961).

12. ———, *On the distribution of k-twin primes*, Dokl. Akad. Nauk SSSR **132** (1960), 1258–1260; English transl. in Soviet Math. Dokl. **1** (1960).

13. ———, *On the Vinogradov-Goldbach theorem*, Dokl. Akad. Nauk UzSSR **1961**, no. 11, 7–10. (Russian)

14. ———, *On the representation of numbers as the sum of primes by Šnirel'man's method*, Izv. Akad. Nauk UzSSR Ser. Fiz.-Mat. Nauk **1962**, no. 3, 5–10. (Russian)

15. Viggo Brun, *Le crible d'Eratosthène et le théorème de Goldbach,* Christiania Vidensk.-Selsk. Skr. **1920**, no. 3.

16. L. Schnirelmann [L. G. Šnirel'man], *Über additive Eigenschaften von Zahlen,* Math. Ann. **107** (1932/33), 649–690.

17. I. V. Čulanovskiĭ, *Some estimates connected with Selberg's new method in elementary number theory,* Dokl. Akad. Nauk SSSR **63** (1948), 491–494. (Russian)

18. N. I. Klimov, *Combination of elementary and analytic methods in the theory of numbers,* Uspehi Mat. Nauk **13** (1958), no. 3 (81), 145–164. (Russian)

Translated by G. A. KANDALL

Amer. Math. Soc. Transl.
(2) Vol. 113, 1979

A SIMPLIFICATION OF A. SELBERG'S ELEMENTARY PROOF OF THE ASYMPTOTIC LAW OF DISTRIBUTION OF PRIME NUMBERS*

A. G. POSTNIKOV AND N. P. ROMANOV

The asymptotic law of distribution of prime numbers

$$\lim_{x\to\infty} \frac{\pi(x)}{\dfrac{x}{\ln x}} = 1$$

(where $\pi(x)$ is the number of primes in the interval $[1, x]$) was first proved in 1896 (independently by Hadamard and by de la Vallée-Poussin) using an analytic method. Finding an elementary proof of this turned out to be a difficult problem, and only in 1949 did Atle Selberg [1] give an elementary proof of the asymptotic law. Selberg's proof has technical complexities. The efforts of several mathematicians have been directed toward simplifying the proof and, by the same token, better understanding the idea behind it. Harold Shapiro [2] derived the relation

$$\sum_{p\leq x} \ln^2 p + \sum_{p,\, q\leq x} \ln p \ln q = 2x \ln x + O(x)$$

(which plays a fundamental role in Selberg's proof) in a simpler way than was done in the original paper. But the derivation of the asymptotic law from this relation remained complicated.

The asymptotic law of distribution of prime numbers has several equivalent formulations (in the sense that one can be deduced from another using elementary arguments), for example:

$$\sum_{n\leqslant x} \mu(n) = o(x), \qquad \sum_{n\leq x} \frac{\mu(n)}{n} = o(x), \qquad \sum_{p\leq x} \ln p \sim x$$

(Selberg proves the law in the last of these forms).

We noticed that, if the law is proved along the same lines as in Selberg's proof but in the form $\Sigma_{n\leqslant x}\, \mu(n) = o(x)$, then the technical side of the proof becomes much simpler. It should be noted that it is not a simple matter to derive the more usual forms of the asymptotic law from this form (see 7) and 8) in the Appendix), but the derivations have been known for a long time, and it is not there that the complexity of the elementary proof of the law is encountered.

This paper consists of the basic text, where the proof of the law is given, and several appendices. In the course of the proof we use the relation

$$\sum_{p\leq x} \ln^2 p + \sum_{p,\, q\leq x} \ln p \ln q = 2x \ln x + O(x),$$

AMS (MOS) subject classifications (1970). Primary 10H15.

*Translation of Uspehi Mat. Nauk **10** (1955), no. 4 (66), 75–87. MR **17**, 587.

which was derived by Shapiro. Some familiarity with the foundations of number theory is necessary to be able to read the article. The material in the first two chapters of [3] is completely sufficient.

We use the usual notation: p and q are prime numbers, $\mu(n)$ is the Möbius function, $\Lambda(n)$ is the Mangoldt function ($\Lambda(n) = \ln p$ for $n = p^l$, and $\Lambda(n) = 0$ for other positive integers n), and

$$\sum_{p \leqslant x} \ln p = \vartheta(x), \quad \sum_{n \leqslant x} \mu(n) = M(x).$$

Since $|\mu(n)| \leqslant 1$, we have $-x \leqslant M(x) \leqslant x$.

LEMMA 1.

$$\sum_{n \leqslant x} \frac{\mu(n)}{n} = O(1). \tag{1}$$

PROOF. See 1) in the Appendix.

LEMMA 2.

$$\ln x M(x) + \sum_{p \leqslant x} \ln p M\left(\frac{x}{p}\right) = O(x). \tag{2}$$

PROOF. In fact,

$$\ln x \sum_{n \leqslant x} \mu(n) - \sum_{n \leqslant x} \mu(n) \ln n = \sum_{n \leqslant x} \mu(n) \ln \frac{x}{n} = O\left(\sum_{n \leqslant x} \ln \frac{x}{n}\right) = O(x).$$

Furthermore,

$$\sum_{n \leqslant x} \mu(n) \ln n = \sum_{p \leqslant x} \ln p \sum_{ps \leqslant x} \mu(ps) = -\sum_{p \leqslant x} \ln p \sum_{\substack{s \leqslant \frac{x}{p} \\ (s,p)=1}} \mu(s)$$

$$= -\sum_{p \leqslant x} \ln p \left(\sum_{s \leqslant \frac{x}{p}} \mu(s) + O\left(\frac{x}{p^2}\right)\right) = -\sum_{p \leqslant x} \ln p \sum_{s \leqslant \frac{x}{p}} \mu(s) + O\left(x \sum_{2}^{\infty} \frac{\ln p}{p^2}\right).$$

But $\Sigma_{p=2}^{\infty} \ln p/p^2$ converges. Thus,

$$\ln x \sum_{n \leqslant x} \mu(n) + \sum_{p \leqslant x} \ln p \sum_{s \leqslant \frac{x}{p}} \mu(s) = O(x),$$

as required.

Changing the order of summation in the double sum, we may give Lemma 2 the form

$$\ln x \cdot M(x) + \sum_{n \leqslant x} \mu(n) \vartheta\left(\frac{x}{n}\right) = O(x). \tag{3}$$

The following lemma is a generalization of the classical discontinuity equation

$$\sum_{d | n} \mu(d) = \begin{cases} 1 & \text{for} \quad n = 1, \\ 0 & \text{for} \quad n > 1. \end{cases}$$

LEMMA 3. *The following equations hold*:

$$\sum_{d|n} \mu(d) \ln \frac{x}{d} = \begin{cases} \ln x & \text{for} \quad n=1, \\ \Lambda(n) & \text{for} \quad n>1; \end{cases} \tag{4}$$

$$\sum_{d|n} \mu(d) \ln^2 \frac{x}{d} = \begin{cases} \ln^2 x & \text{for} \quad n=1, \\ \ln p \ln \dfrac{x^2}{p} & \text{for} \quad n=p^\alpha, \ \text{where } \alpha \geqslant 1, \\ 2\ln p \ln q & \text{for} \quad n=p^\alpha q^\beta, \ \text{where } \alpha \geqslant 1 \text{ and } \beta \geqslant 1, \\ 0 & \text{for all other } n. \end{cases} \tag{5}$$

PROOF OF (4). If $n = 1$, (4) is obvious; if $n = p^\alpha$, then

$$\sum_{d|n} \mu(d) \ln^2 \frac{x}{d} = \ln x - \ln \frac{x}{p}$$

if $n = p^\alpha n'$, $n' > 1$, $(n', p) = 1$, then

$$\sum_{d|n} \mu(d) \ln \frac{x}{d} = \sum_{d|n'} \mu(d) \ln \frac{x}{d} + \sum_{d|n'} \mu(dp) \ln \frac{x}{dp} = \ln p \sum_{d|n'} \mu(d) = 0.$$

PROOF OF (5). If $n = 1$, (5) is obvious; if $n = p^\alpha$, then

$$\sum_{d|n} \mu(d) \ln^2 \frac{x}{d} = \ln^2 x - \ln^2 \frac{x}{p} = \ln p \ln \frac{x^2}{p};$$

if $n = p^\alpha q^\beta$, then

$$\sum_{d|n} \mu(d) \ln^2 \frac{x}{d} = \ln^2 x - \ln^2 \frac{x}{p} - \ln^2 \frac{x}{q} + \ln^2 \frac{x}{pq}$$

$$= \ln p \ln \frac{x^2}{p} - \ln p \ln \frac{\left(\frac{x}{q}\right)^2}{p} = 2 \ln p \ln q;$$

if $n = p^\alpha q^\beta n'$, where $(n', p) = (n', q) = 1$ and $n' > 1$, then

$$\sum_{d|n} \mu(d) \ln^2 \frac{x}{d}$$

$$= \sum_{d|n'} \mu(d) \ln^2 \frac{x}{d} + \sum_{d|n'} \mu(dp) \ln^2 \frac{x}{dp} + \sum_{d|n'} \mu(dq) \ln^2 \frac{x}{dq} + \sum_{d|n'} \mu(dpq) \ln^2 \frac{x}{dpq}$$

$$= \sum_{d|n'} \mu(d) \left(\ln^2 \frac{x}{d} - \ln^2 \frac{x}{dp} - \ln^2 \frac{x}{dq} + \ln^2 \frac{x}{dpq} \right) = 2 \ln p \ln q \sum_{d|n'} \mu(d) = 0.$$

LEMMA 4.

$$\sum_{n \leqslant x} \frac{\mu(n)}{n} \ln \frac{x}{n} = O(1). \tag{6}$$

It is well known that

$$\ln x = \sum_{v \leqslant x} \frac{1}{v} + C + O\left(\frac{1}{x}\right),$$

where C is Euler's constant (see the Appendix, 3)). Thus

$$\sum_{n\leq x}\frac{\mu(n)}{n}\ln\frac{x}{n}=\sum_{n\leq x}\frac{\mu(n)}{n}\Big(\sum_{v\leq\frac{x}{n}}\frac{1}{v}+C+O\Big(\frac{n}{x}\Big)\Big)$$

$$=\sum_{m\leq x}\frac{\sum\limits_{d\mid m}\mu(d)}{m}+O(1)+O(1)=O(1).$$

LEMMA 5.

$$\sum_{n\leq x}\Lambda(n)=O(x). \tag{7}$$

By Lemma 3 we have

$$\sum_{n\leq x}\sum_{d\mid n}\mu(d)\ln\frac{x}{d}=\ln x+\sum_{n\leq x}\Lambda(n).$$

On the other hand,

$$\sum_{n\leq x}\sum_{d\mid n}\mu(d)\ln\frac{x}{d}=\sum_{d\leq x}\mu(d)\ln\frac{x}{d}\Big[\frac{x}{d}\Big]=x\sum_{d\leq x}\frac{\mu(d)}{d}\ln\frac{x}{d}+O\Big(\sum_{d\leq x}\ln\frac{x}{d}\Big)=O(x).$$

LEMMA 6.

$$\sum_{n\leq x}\frac{\Lambda(n)}{n}=\ln x+O(1). \tag{8}$$

PROOF. See the Appendix, 5).

LEMMA 7.

$$\sum_{n\leq x}\frac{\mu(n)}{n}\ln^2\frac{x}{n}=2\ln x+O(1). \tag{9}$$

PROOF. In fact,

$$\sum_{n\leq x}\frac{\mu(n)}{n}\ln^2\frac{x}{n}=\sum_{n\leq x}\frac{\mu(n)}{n}\ln\frac{x}{n}\Big(\sum_{v\leq\frac{x}{n}}\frac{1}{v}+C+O\Big(\frac{n}{x}\Big)\Big)$$

$$=\sum_{n\leq x}\sum_{v\leq\frac{x}{n}}\frac{\mu(n)\ln\frac{x}{n}}{nv}+O(1)+O\left(\frac{\sum\limits_{n\leq x}\ln\frac{x}{n}}{x}\right)$$

$$=\sum_{n\leq x}\frac{\sum\limits_{d\mid n}\mu(d)\ln\frac{x}{d}}{n}+O(1)=\ln x+\sum_{n\leq x}\frac{\Lambda(n)}{n}+O(1)=2\ln x+O(1).$$

LEMMA 8.

$$\sum_{p\leq x}\ln^2 p+\sum_{pq<x}\ln p\ln q=2x\ln x+O(x). \tag{10}$$

PROOF. We have

$$\sum_{n\leq x}\sum_{d|n}\mu(d)\ln^2\frac{x}{d}=\sum_{d\leq x}\mu(d)\ln^2\frac{x}{d}\left[\frac{x}{d}\right]$$

$$=x\sum_{d\leq x}\frac{\mu(d)}{d}\ln^2\frac{x}{d}+O\Big(\sum_{d\leq x}\ln^2\frac{x}{d}\Big)=2x\ln x+O(x).$$

On the other hand, by (5)

$$\sum_{n\leq x}\sum_{d|n}\mu(d)\ln^2\frac{x}{d}=\ln^2 x+\sum_{p^\alpha\leq x}\ln p\ln\frac{x^2}{p}+\sum_{\substack{p^\alpha q^\beta\leq x\\ p\neq q}}\ln p\ln q.$$

But

$$\sum_{p^\alpha\leq x}\ln p\ln\frac{x^2}{p}=\sum_{p\leq x}\ln p\ln\frac{x^2}{p}+O\Big(\sum_{p^2\leq x}\ln^2 p\Big)+O\Big(\sum_{\substack{p^\alpha\leq x\\ \alpha\geq 2}}\ln p\ln x^2\Big)$$

$$=\sum_{p\leq x}\ln p\ln\frac{x^2}{p}+O\Big(\ln x\ln x\sum_{p\leq\sqrt{x}}\ln p\Big)=\sum_{p\leq x}\ln p\ln\frac{x^2}{p}+O(\sqrt{x}\ln^2 x)$$

(the sum $\ln x\,\Sigma_{p^\alpha\leqslant x,\alpha\geqslant 2}\ln p$ splits into $O(\ln x)$ terms corresponding to $\alpha=2,3,\ldots$). Moreover,

$$\sum_{\substack{p^\alpha q^\beta\leq x\\ p\neq q}}\ln p\ln q=\sum_{pq\leq x}\ln p\ln q+O\Big(\sum_{p^2\leq x}\ln^2 p\Big)+O\Big(\sum_{\substack{p^\alpha q^\beta\leq x\\ \alpha\geq 2;\ \beta\geq 1}}\ln p\ln q\Big)$$

$$=\sum_{pq\leq x}\ln p\ln q+O\Big(\ln x\sum_{p\leq\sqrt{x}}\ln p\Big)+O\Big(\sum_{\substack{p^\alpha\leq x\\ \alpha\geq 2}}\ln p\,\vartheta\Big(\frac{x}{p^\alpha}\Big)\Big)$$

$$=\sum_{pq\leq x}\ln p\ln q+O(\sqrt{x}\ln x)+O\Big(x\sum_{\substack{p^\alpha\leq x\\ \alpha\geq 2}}\frac{\ln p}{p^\alpha}\Big)=\sum_{pq\leq x}\ln p\ln q+O(x).$$

Thus

$$\sum_{p\leq x}\ln p\ln\frac{x^2}{p}+\sum_{pq\leq x}\ln p\ln q=2x\ln x+O(x).$$

Furthermore,

$$\sum_{p\leq x}\ln p\ln\frac{x^2}{p}=2\ln x\sum_{p\leq x}\ln p-\sum_{p\leq x}\ln^2 p.$$

The proof of the lemma will be complete if we show that

$$\sum_{p\leq x}\ln^2 p=\ln x\sum_{p\leq x}\ln p+O(x).$$

This is the case, because

$$\sum_{p\leq x}\ln^2 p=\ln x\sum_{p\leq x}\ln p-\sum_{p\leq x}\ln p\int_p^x\frac{du}{u}$$

$$=\ln x\sum_{p\leq x}\ln p-\int_2^x\frac{\sum_{p\leq u}\ln p}{u}\,du=\ln x\sum_{p\leq x}\ln p+O(x)$$

since $\Sigma_{p\leqslant u} \ln p = O(u)$). The lemma is proved.

From this it further follows that Lemma 8 can be written in the form

$$\ln x\,\vartheta(x) + \sum_{p\leqslant x} \ln p\,\vartheta\left(\frac{x}{p}\right) = 2x \ln x + O(x). \tag{11}$$

An analogy between inequalities (2) and (11) suggests itself. We shall use the following form of Lemma 8:

$$\vartheta(x) + \sum_{pq\leqslant x} \frac{\ln p \ln q}{\ln pq} = 2x + O\left(\frac{x}{\ln 2x}\right). \tag{12}$$

In order to obtain this, note that

$$\frac{1}{\ln x} \sum_{pq\leqslant x} \ln p \ln q - \sum_{pq\leqslant x} \frac{\ln p\cdot\ln q}{\ln pq} = \sum_{pq\leqslant x} \ln p \ln q \left(\frac{1}{\ln x} - \frac{1}{\ln pq}\right)$$

$$= -\sum_{pq\leqslant x} \ln p \ln q \int_{pq}^{x} \frac{du}{u \ln^2 u} = -\int_6^x \frac{\sum_{pq\leqslant u} \ln p \ln q}{u \ln^2 u}\, du = O\left(\int_6^x \frac{du}{\ln u}\right)$$

$$= O\left(\int_6^{\sqrt{x}} \frac{du}{\ln u}\right) + O\left(\int_{\sqrt{x}}^{x} \frac{du}{\ln u}\right) = O(\sqrt{x}) + O\left(\frac{x}{\ln \sqrt{x}}\right) = O\left(\frac{x}{\ln x}\right).$$

(We have used the relation $\Sigma_{pq\leqslant u} \ln p \ln q = O(u \ln u)$, which follows from (11).) Thus, Selberg's inequality (10) can be written in the form

$$\vartheta(x) + \sum_{pq\leqslant x} \frac{\ln p \ln q}{\ln pq} = 2x + O\left(\frac{x}{\ln x}\right),$$

and, even better, in the form (12).

LEMMA 9.

$$\ln x M(x) = \sum_{pq\leqslant x} \frac{\ln p\cdot\ln q}{\ln pq} M\left(\frac{x}{pq}\right) + O(x \ln \ln x). \tag{13}$$

In (3) we make the substitution

$$\vartheta\left(\frac{x}{n}\right) = \frac{2x}{n} - \sum_{pq\leqslant \frac{x}{n}} \frac{\ln p \ln q}{\ln pq} + O\left(\frac{x}{n \ln 2\frac{x}{n}}\right),$$

obtained from (12), and find that

$$\ln x M(x) + 2x \sum_{n\leqslant x} \frac{\mu(n)}{n} - \sum_{n\leqslant x} \mu(n) \sum_{pq\leqslant \frac{x}{n}} \frac{\ln p \ln q}{\ln pq} + O\left(x \sum_{n\leqslant x} \frac{1}{n \ln 2\frac{x}{n}}\right) = O(x).$$

Using (1) and the easily established fact that

$$\sum_{n\leqslant x} \frac{1}{n \ln 2\frac{x}{n}} = O(\ln \ln x)$$

(see the Appendix, 6)) and changing the order of summation in the double sum, we complete the proof of Lemma 9.

LEMMA 10. *The following recursion inequality holds:*

$$|M(x)| \leqslant \frac{1}{\ln x} \sum_{n \leqslant x} \left| M\left(\frac{x}{n}\right) \right| + O\left(\frac{x \ln \ln x}{\ln x}\right). \tag{14}$$

PROOF. We take absolute values in (2) and (13) and add the resulting expressions, obtaining

$$2 \ln x\, |M(x)| \leqslant \sum_{p \leqslant x} \ln p \left| M\left(\frac{x}{p}\right) \right| + \sum_{pq \leqslant x} \frac{\ln p \ln q}{\ln pq} \left| M\left(\frac{x}{pq}\right) \right| + O(x \ln \ln x).$$

Taking the Abel transform of the left side of this equation, we obtain

$$\begin{aligned} 2 \ln x\, |M(x)| &\leqslant \sum_{n \leqslant x-1} \left(\sum_{p \leqslant n} \ln p + \sum_{pq \leqslant n} \frac{\ln p \ln q}{\ln pq} \right) \left(\left| M\left(\frac{x}{n}\right) \right| - \left| M\left(\frac{x}{n+1}\right) \right| \right) \\ &\quad + \left(\sum_{p \leqslant x} \ln p + \sum_{pq \leqslant x} \frac{\ln p \ln q}{\ln pq} \right) + O(x \ln \ln x) \\ &= \sum_{n \leqslant x-1} 2n \left(\left| M\left(\frac{x}{n}\right) \right| - \left| M\left(\frac{x}{n+1}\right) \right| \right) + O\left(\sum_{n \leqslant x-1} \frac{n}{\ln 2n} \left| \left| M\left(\frac{x}{n}\right) \right| \right. \right. \\ &\qquad \left. \left. - \left| M\left(\frac{x}{n+1}\right) \right| \right| \right) + O(x) + O(x \ln \ln x). \end{aligned}$$

From the geometrically obvious inequality $||a| - |b|| \leqslant |a - b|$ we have

$$\begin{aligned} \left| \left| M\left(\frac{x}{n}\right) \right| - \left| M\left(\frac{x}{n+1}\right) \right| \right| &\leqslant \left| M\left(\frac{x}{n}\right) - M\left(\frac{x}{n+1}\right) \right| \\ &= \left| \sum_{\frac{x}{n+1} \leqslant k < \frac{x}{n}} \mu(n) \right| \leqslant \frac{x}{n} - \frac{x}{n+1} = O\left(\frac{x}{n^2}\right). \end{aligned}$$

Furthermore,

$$O\left(x \sum_{n \leqslant x} \frac{1}{n \ln 2n} \right) = O(x \ln \ln x).$$

Thus

$$2 \ln x\, |M(x)| \leqslant 2 \sum_{n \leqslant x} \left| M\left(\frac{x}{n}\right) \right| + O(x \ln \ln x),$$

as required.

If we make the estimate $|M(x/n)| \leqslant \alpha x/n$ in the right side of (14), where α is some constant (we may obviously take $\alpha = 1$), then we again obtain the inequality

$$|M(x)| \leqslant \alpha x + O\left(\frac{x \ln \ln x}{\ln x}\right),$$

i.e., there is no decrease in the constant. However, as we shall see, if (14) is combined with the reduction we have in (1), then the constant is decreased, and we thereby arrive at a proof of the law.

Applying summation by parts to (1) gives the inequality

$$\sum_{n\leq x}\frac{M(n)}{n^2}=O(1).$$

Thus, there exists a constant k such that for any x' and x''

$$\left|\sum_{x'<n\leq x''}\frac{M(n)}{n^2}\right|<k.$$

LEMMA 11. *There exist constants k_1 and $c > 1$ such that, for any $\delta \in (0, 1)$ and $x > c$ there is a y in any interval $(x, xe^{k_1/\delta})$ for which $|M(y)| < \delta y$.*

PROOF. Suppose that $M(n)$ does not change sign on (x, x'), say $M(n) \geqslant 0$. We take the minimal $M(y)/y$ on this segment:

$$k>\sum_{x\leq n\leq x'}\frac{M(n)}{n^2}>\frac{M(y)}{y}\sum_{x\leq n\leq x'}\frac{1}{n}\geqslant\frac{M(y)}{y}\left(\ln\frac{x'}{x}-c_1\right);$$

hence (using $|M(y)| \leqslant y$) we obtain

$$k_1>\frac{M(y)}{y}\ln\frac{x'}{x}.$$

If $x' > xe^{k_1/\delta}$, then $M(y) < \delta y$. If $M(n)$ changes sign in the interval $(x,xe^{k_1/\delta})$, then, since $M(n)$ is always an integer and the length of a discontinuity in $M(n)$ is one, it follows that there is a y for which $M(y) = 0$, and thus $|M(y)| < \delta y$.

LEMMA 12. *There exist constants c and k_1 such that for any $\delta \in (0, 1)$ and $x \geqslant c$ the interval $(x, xe^{k_1/\delta})$ contains a subinterval $(y_1, y_1e^{\delta/2})$ in which $|M(z)| < 3\delta z$.*

PROOF. The interval $(x, xe^{k_1/\delta})$ contains a point y at which $|M(y)| < \delta y$. Furthermore,

$$|M(a)-M(b)|=\left|\sum_{a\leq n\leq b}{}'\mu(n)\right|\leqslant b-a,$$

so that $|M(b)| \leqslant |M(a)| + b - a$. We take the interval $(ye^{-\delta/2}, ye^{\delta/2})$; for y' in this interval we have the relation (with $a = y$ and $b = y'$) $|M(y')/y'| < \delta y/y' + 1 - y/y' < \delta e^{\delta/2} + 1 - e^{-\delta/2} < 2\delta + \delta = 3\delta$. We take the part of this interval falling in $(x, xe^{k_1/\delta})$, and we let y_1 be the left endpoint of this part. All of $(y_1, y_1e^{\delta/2})$ belongs to this part: if $y_1 = ye^{-\delta/2}$, then $(ye^{-\delta/2})e^{\delta/2} = y$ belongs to $(x, xe^{k_1/\delta})$; if $y_1 = x$, then, taking $k_1 \geqslant 1$, we obtain $xe^{\delta/2} < xe^{k_1/\delta}$.

THEOREM (asymptotic law of distribution of prime numbers)

$$\sum_{n\leq x}\mu(n)=o(x). \tag{15}$$

PROOF. We let α denote any number such that, beginning with some $x \geqslant x_0(\alpha)$, we have $|M(x)| \leqslant \alpha x$. Obviously, 1 is such an α, and also

$$\overline{\lim_{x\to\infty}}\frac{|M(x)|}{x}\leqslant\alpha.$$

If we construct a sequence $\alpha_1, \alpha_2, \ldots$ such that $\alpha_n \longrightarrow 0$, then the theorem will be proved. Suppose we have proved that $|M(x)| \leqslant \alpha x$ for $x \geqslant x_0$. By (14) we have

$$|M(x)| \leqslant \frac{1}{\ln x} \sum_{n \leqslant x} \left| M\left(\frac{x}{n}\right) \right| + O\left(\frac{x}{\sqrt{\ln x}}\right)$$

$$= \frac{1}{\ln x} \sum_{n \leqslant \frac{x}{x_0}} \left| M\left(\frac{x}{n}\right) \right| + O\left(\frac{x}{\ln x} \sum_{\frac{x}{x_0} \leqslant n \leqslant x} \frac{1}{n}\right) + O\left(\frac{x}{\sqrt{\ln x}}\right)$$

$$= \frac{1}{\ln x} \sum_{n \leqslant \frac{x}{x_0}} \left| M\left(\frac{x}{n}\right) \right| + O\left(\frac{x}{\sqrt{\ln x}}\right).$$

We set $\delta = \alpha/6$ and $\rho = \epsilon^{k_1/\delta}$, and we divide up all possible values for x/n, i.e., the interval (x_0, x), into intervals using the division points $x_0, x_0\rho, x_0\rho^2, \ldots$. By Lemma 12 every such interval contains a part $(y_v, y_v e^{\delta/2})$ where the estimate for $M(x/n)$ improves. If x/n is in that part, we have the estimate

$$\left| M\left(\frac{x}{n}\right) \right| \leqslant 3 \frac{\alpha}{6} \frac{x}{n} = \alpha \frac{x}{n} - \frac{\alpha}{2} \frac{x}{n};$$

if x/n is outside, we have $|M(x/n)| < \alpha x/n$. We obtain

$$|M(x)| < \frac{\alpha x}{\ln x} \sum_{n \leqslant \frac{x}{x_0}} \frac{1}{n} - \frac{\alpha x}{2 \ln x} \sum_{v=0}^{\left[\frac{\ln \frac{x}{x_0}}{\ln \rho}\right]} \sum_{y_v \leqslant \frac{x}{n} \leqslant y_v e^{\frac{\delta}{2}}} \frac{1}{n} + O\left(\frac{x}{\sqrt{\ln x}}\right).$$

We shall use the geometrically obvious inequalities

$$-\frac{2}{\alpha} < \sum_{\alpha \leqslant n \leqslant \beta} \frac{1}{n} - \ln \frac{\beta}{\alpha} < \frac{1}{\alpha},$$

$$\sum_{\frac{x}{y_v} e^{-\frac{\delta}{2}} \leqslant n \leqslant \frac{x}{y_v}} \frac{1}{n} > \frac{\delta}{2} - \frac{2 y_v e^{\frac{\delta}{2}}}{x} > \frac{\delta}{2} - \frac{2 x_0 \rho^v}{x},$$

$$|M(x)| \leqslant \frac{\alpha x}{\ln x} \sum_{n \leqslant \frac{x}{x_0}} \frac{1}{n} - \frac{\alpha x}{2 \ln x} \sum_{v=0}^{\left[\frac{\ln \frac{x}{x_0}}{\ln \rho}\right]} \frac{\delta}{2} + \frac{\alpha x_0}{\ln x} \sum_{v=0}^{\left[\frac{\ln \frac{x}{x_0}}{\ln \rho}\right]} \rho^v + O\left(\frac{x}{\sqrt{\ln x}}\right)$$

$$\leqslant \alpha x - \frac{\alpha x \delta}{4 \ln x} \left(\left[\frac{\ln \frac{x}{x_0}}{\ln \rho} \right] + 1 \right) + \frac{\alpha x_0}{\ln x} \frac{\frac{x}{x_0} - 1}{\rho - 1} + O\left(\frac{x}{\sqrt{\ln x}}\right)$$

$$\leqslant \alpha x - \frac{\alpha x \delta}{4 \ln x} \frac{\ln x}{\ln \rho} + \frac{\alpha x \delta \ln x_0}{4 \ln x \ln \rho} + \frac{\alpha x_0}{\ln x} \frac{\frac{x}{x_0} - 1}{\rho - 1} + O\left(\frac{x}{\sqrt{\ln x}}\right)$$

$$= \alpha x \left(1 - \frac{\delta^2}{4 k_1}\right) + O\left(\frac{x}{\sqrt{\ln x}}\right) = \alpha \left(1 - \frac{\alpha^2}{4 \cdot 36 k_1}\right) x + O\left(\frac{x}{\sqrt{\ln x}}\right).$$

Thus, if the constant α_n is suitable, so is the constant

$$\alpha_{n+1} = \alpha_n \left(1 - \frac{\alpha_n^2}{200 k_1}\right).$$

We choose $\alpha_1 = 1$ and show that

$$\alpha_n \leqslant \frac{\sqrt{200\,k_1}}{\sqrt{n+3}}.$$

This is true for $n = 1$. The expression $\alpha_n(1 - \alpha_n^2/200k_1)$ increases for $\alpha_n \leqslant \sqrt{200k_1/3}$. We majorize α_n by the expression $\sqrt{200k_1}/\sqrt{n+3}$ (this can already be done for $n=1$), getting

$$\alpha_{n+1} < \frac{\sqrt{200k_1}}{\sqrt{n+3}}\left(1-\frac{1}{n+3}\right) < \frac{\sqrt{200k_1}}{\sqrt{n+4}}.$$

The theorem is proved.

APPENDIX

1) It is well known that

$$\sum_{d\mid n} \mu(d) = \begin{cases} 1 & \text{for} \quad n=1, \\ 0 & \text{for} \quad n>1; \end{cases}$$

and

$$\sum_{n\leqslant x}\sum_{d\mid n} \mu(d) = 1, \qquad \sum_{n\leqslant x} \mu(n)\left[\frac{x}{n}\right] = 1,$$

$$x\sum_{n\leqslant x}\frac{\mu(n)}{n} + O(x) = 1, \qquad \sum_{n\leqslant x}\frac{\mu(n)}{n} = O(1).$$

2) Remark on Lemma 2. We set

$$A^* = \overline{\lim_{x\to\infty}}\,\frac{M(x)}{x} \qquad \text{and} \qquad a^* = \underline{\lim_{x\to\infty}}\,\frac{M(x)}{x}.$$

We prove that Lemma 2 implies that $A^* + a^* = 0$. The inequality can be rewritten in the form

$$\frac{M(x)}{x} + \frac{1}{\ln x}\sum_{p\leqslant x}\frac{\ln p}{p}\,\frac{M\left(\frac{x}{p}\right)}{\frac{x}{p}} = O\left(\frac{1}{\ln x}\right).$$

We choose an arbitrary $\epsilon > 0$. There exists an $x_0(\epsilon)$ such that for $x > x_0$ we have $M(x)/x > a^* - \epsilon$. Furthermore,

$$O\left(\frac{1}{\ln x}\right) = \frac{M(x)}{x} + \frac{1}{\ln x}\sum_{p\leqslant\frac{x}{x_0}}\frac{\ln p}{p}\,\frac{M\left(\frac{x}{p}\right)}{\frac{x}{p}} + \frac{1}{\ln x}\sum_{\frac{x}{x_0}<p\leqslant x}\frac{\ln p}{p}\,\frac{M\left(\frac{x}{p}\right)}{\frac{x}{p}}$$

$$> \frac{M(x)}{x} + \frac{1}{\ln x}\sum_{p\leqslant\frac{x}{x_0}}\frac{\ln p}{p} - \frac{1}{\ln x}\sum_{\frac{x}{x_0}<p\leqslant x}\frac{\ln p}{p}.$$

Using Čebyšev's formula

$$\sum_{p\leqslant x}\frac{\ln p}{p} = \ln x + O(1)$$

(see 5), below), we obtain

$$O\left(\frac{1}{\ln x}\right) > \frac{M(x)}{x} + (a^* - \varepsilon) + O\left(\frac{1}{\ln x}\right).$$

Letting x approach infinity so that $M(x)/x$ approaches A^*, we have $0 \geqslant A^* + a^* - \epsilon$. But ϵ is arbitrarily small, so that $0 \geqslant A^* + a^*$. We similarly obtain $A^* + a^* \geqslant 0$. Thus, $A^* + a^* = 0$.

3) REMARK ON LEMMA 4. For a simple derivation of this relation, see [4], Chapter II, §1.1.

4) REMARK ON LEMMA 5. This derivation of the classical theorem is contained in [4], Chapter II, §3.4, formula (181) et seq.

5) By the definition of $\Lambda(n)$ we have

$$\sum_{d \mid n} \Lambda(d) = \ln n, \quad \sum_{n \leqslant x} \sum_{d \mid n} \Lambda(d) = x \ln x + O(x)$$

(a rough form of Stirling's formula), and

$$\sum_{d \leqslant x} \Lambda(d) \left[\frac{x}{d}\right] = x \ln x + O(x).$$

By Lemma 5 we have

$$x \sum_{d \leqslant x} \frac{\Lambda(d)}{d} + O(x) = x \ln x + O(x).$$

Dividing through by x, we obtain

$$\sum_{n \leqslant x} \frac{\Lambda(n)}{n} = \ln x + O(1).$$

Obviously,

$$\left|\sum_{n \leqslant x} \frac{\Lambda(n)}{n} - \sum_{p \leqslant x} \frac{\ln p}{p}\right| \leqslant \sum_{p \leqslant x} \left(\frac{1}{p^2} + \frac{1}{p^3} + \dots\right) \ln p < \sum_{p} \frac{\ln p}{p(p-1)} = O(1).$$

This proves another form of this fact:

$$\sum_{p \leqslant x} \frac{\ln p}{p} = \ln x + O(1).$$

6) We have

$$\sum_{n \leqslant x} \frac{1}{n \ln 2 \frac{x}{n}} = \frac{1}{\ln x} \sum_{n \leqslant x} \frac{1}{n} + \sum_{n \leqslant x} \frac{1}{n} \left(\frac{1}{\ln 2 \frac{x}{n}} - \frac{1}{\ln x}\right)$$

$$= \sum_{n \leqslant x} \frac{1}{n} \int_{2\frac{x}{n}}^{x} \frac{du}{u \ln^2 u} + O(1) = \int_2^x \frac{\sum_{\frac{2x}{n} \leqslant u \leqslant x} \frac{1}{n}}{u \ln^2 a}\, du + O(1) = \ln \ln x + O(1).$$

7) From $\Sigma_{n \leqslant x}\, \mu(n) = o(x)$ we derive the asymptotic law in the form $\Sigma_{n \leqslant x}\, \Lambda(n) \sim x$ using a well-known method. We consider

$$S(x) = \sum_{n \leqslant x} \mu(n) \sum_{m \leqslant \frac{x}{n}} (\ln m - \tau(m) + 2C),$$

where $\tau(m)$ is the number of divisors of m and C is Euler's constant. We have

$$S(x)=\sum_{k\leqslant x}\sum_{d\mid k}\mu(d)\left(\ln\frac{k}{d}-\tau\left(\frac{k}{d}\right)+2C\right)=\sum_{k\leqslant x}(\Lambda(k)-1)+2C,$$

since $\Sigma_{d\mid k}\,\mu(d)\ln(k/d)=\Lambda(n)$ (the derivation follows immediately from (4) and the equation $\Sigma_{d\mid k}\,\mu(d)\tau(k/d)=1$, which can be obtained, for example, using induction on the number of prime divisors of k: if $k=k'^{\alpha}_{p}$, then

$$\sum_{d\mid k}\mu(d)\tau\left(\frac{k}{d}\right)=\sum_{d\mid k'}\mu(d)\tau\left(p^{\alpha}\frac{k'}{d}\right)+\sum_{d\mid k'}\mu(dp)\tau\left(p^{\alpha-1}\frac{k'}{d}\right)$$
$$=(\tau(p^{\alpha})-\tau(p^{\alpha-1}))\sum_{d\mid k'}\mu(d)\tau\left(\frac{k'}{d}\right)=\sum_{d\mid k'}\mu(d)\tau\left(\frac{k'}{d}\right)\Big),$$

If we choose $0<\delta<1$, then

$$S(x)=S(\delta x)+\sum_{\delta x<n\leqslant x}\mu(n)\sum_{m\leqslant\frac{x}{n}}(\ln m-\tau(m)+2C)$$
$$=S(\delta x)+\sum_{n=[\delta x]+1}^{[x]}(M(n)-M(n-1))\sum_{m\leqslant\frac{x}{n}}(\ln m-\tau(m)+2C)$$
$$=S(\delta x)-M(\delta x)\sum_{n=1}^{\left[\frac{x}{1+[\delta x]}\right]}(\ln n-\tau(n)+2C)+M(x)(2C-1)$$
$$+\sum_{n=1+[\delta x]}^{[x]-1}M(n)\sum_{\frac{x}{n+1}<k\leqslant\frac{x}{n}}(\ln k-\tau(k)+2C).$$

By Dirichlet's formula (see, for example, [5], pp. 247–248) we have

$$\sum_{m\leqslant x}\tau(m)=x\ln x+(2C-1)x+O(\sqrt{x}),$$

and

$$\sum_{k\leqslant x}(\ln k-\tau(k)+2C)$$
$$=x\ln x-x+O(\ln x)-x\ln x-(2C-1)x+O(\sqrt{x})+2Cx=O(\sqrt{x}).$$

The number of nonempty sums

$$\sum_{\frac{x}{n+1}<k\leqslant\frac{x}{n}}(\ln k-\tau(k)+2C),$$

is obviously at most $x/(1+[\delta x])-x/[x]+1<1/\delta+1$. We estimate each such sum:

$$O\left(\sqrt{\frac{x}{1+[\delta x]}}\right)=O\left(\sqrt{\frac{1}{\delta}}\right).$$

We obtain

$$|S(x)|\leqslant|S(\delta x)|+|M(\delta x)|\sqrt{\frac{x}{1+\delta x}}+|M(x)|(2C-1)+o(x)\left(\frac{1}{\delta}+1\right)\sqrt{\frac{1}{\delta}}$$
$$\leqslant|S(\delta x)|+o(x)\left(\sqrt{\frac{1}{\delta}}+(2C-1)+\left(\frac{1}{\delta}+1\right)\sqrt{\frac{1}{\delta}}\right).$$

But

$$|S(\delta x)| \leqslant \sum_{n\leqslant \delta x} \Big| \sum_{m\leqslant \frac{x}{n}} \Big(\ln m - \tau(m) + 2C\Big)\Big| \leqslant k \sum_{n\leqslant\delta x} \sqrt{\frac{x}{n}}$$
$$\leqslant k' \sqrt{x}\sqrt{\delta x} = k'\sqrt{\delta}\, x.$$

Thus

$$|S(x)| \leqslant k' \sqrt{\delta}\, x + o(x)\left(\sqrt{\frac{1}{\delta}} + (2C-1) + \left(\frac{1}{\delta}+1\right)\sqrt{\frac{1}{\delta}}\right).$$

Given $\epsilon > 0$, we may then find a δ such that $k'\sqrt{\delta} < \epsilon$. Further, for $x \geqslant x_0(\delta)$ we have

$$o(x)\left(\sqrt{\frac{1}{\delta}} + (2C-1) + \left(\frac{1}{\delta}+1\right)\sqrt{\frac{1}{\delta}}\right) < \varepsilon x,$$

and consequently $|S(x)| \leqslant 2\epsilon x$. Hence

$$\overline{\lim_{x\to\infty}} \left| \frac{\sum_{n\leqslant x} (\Lambda(u)-1)}{x} \right| \leqslant 2\varepsilon,$$

but ϵ is arbitrarily small. Thus $\Sigma_{n\leqslant x} \Lambda(n) \sim x$. This can obviously be given the form $\Sigma_{p\leqslant x} \ln p \sim x$ (i.e., $\vartheta(x) \sim x$).

8) Finally, from this we obtain $\pi(x) \sim x/\ln x$. In fact,

$$\pi(x) = \sum_{p\leqslant x} 1 = \frac{1}{\ln x}\left(\sum_{p\leqslant x} \ln p + \sum_{p\leqslant x} \ln\frac{x}{p}\right)$$
$$= O\left(\frac{x}{\ln x}\right) + \frac{1}{\ln x}\sum_{p\leqslant x}\int_p^x \frac{du}{u} = O\left(\frac{x}{\ln x}\right) + \frac{1}{\ln x}\int_2^x O\left(\frac{u}{u}\right) du = O\left(\frac{x}{\ln x}\right).$$

Furthermore,

$$\left|\pi(x) - \frac{\vartheta(x)}{\ln x}\right| \leqslant \frac{1}{\ln x}\int_2^x \frac{\sum_{p\leqslant u} 1}{u}\, du = \frac{1}{\ln x}\int_2^x O\left(\frac{u}{u\ln u}\right) du = O\left(\frac{x}{\ln^2 x}\right).$$

Since $\vartheta(x) \sim x$, we have $\pi(x) \sim x/\ln x$.

BIBLIOGRAPHY

1. Atle Selberg, *An elementary proof of the prime-number theorem*, Ann. of Math. (2) **50** (1949), 305–313. MR **10**, 595.

2. Harold N. Shapiro, *On a theorem of Selberg and generalizations*, Ann. of Math. (2) **51** (1950), 485–497. MR **11**, 419.

3. I. M. Vinogradov, *Foundations of the theory of numbers*, 6th ed., GITTL, Moscow, 1952; English transl., Pergamon Press, New York, 1955. MR **15**, 601; **17**, 13.

4. A. O. Gel'fond, *The calculus of finite differences*, GITTL, Moscow, 1952; German transl., VEB Deutscher Verlag, Berlin, 1958; French transl., Dunod, Paris, 1963. MR **14**, 759; **20** #1211; **28** #376.

5. B. N. Delone, *The Petersburg school of number theory*, Izdat. Akad. Nauk SSSR, Moscow, 1947. (Russian) MR **19**, 1029.

Translated by N. KOBLITZ

Amer. Math. Soc. Transl.
(2) Vol. 113, 1979

ON THE THEORY OF INVERSE SEMIGROUPS AND GENERALIZED GROUDS*

BORIS M. SCHEIN [B. M. ŠAĬN] (a)

In this paper, which consists of three sections, certain questions of the theory of inverse semigroups and generalized grouds are considered.

In §1 various elementary systems of axioms for inverse semigroups are studied. Besides known systems of axioms [3], [4], [7], various new axiomatics are found. Interconnections between systems of axioms are considered. Then we find systems of axioms which have the form of identities for inverse semigroups considered as involuted semigroups, and solve the problem of the minimal number of such axioms. It is proved that the class of inverse semigroups cannot be singled out from the class of involuted semigroups by one identity, but can be defined by two identities. Systems of axioms for inverse semigroups considered as ordered semigroups or as ordered involuted semigroups are also found.

Then we consider semigroups which possess the weakly involutive property (if $\bar{g}_1$ and $\bar{g}_2$ are inverses(b) for g_1 and g_2 respectively, then $\bar{g}_2\bar{g}_1$ is an inverse for g_1g_2). Semigroups with the weakly involutive property turn out to be precisely those semigroups whose idempotents form a subsemigroup. The invertibility relation in semigroups with the weakly involutive property is considered, and it is proved to be difunctional.(c) Other properties of this class of semigroups are considered. In particular, it is proved that in a regular semigroup with the weakly involutive property the relation "two elements possess a common inverse" is a congruence(d) which defines an inverse quotient semigroup. As a corollary we obtain a well-known theorem on the structure of idempotent semigroups.

Then we consider a question of systems of axioms for generalized grouds. New systems of axioms are found. It is proved that the class of generalized grouds cannot

---

AMS (MOS) subject classifications (1970). Primary 20M10, 20N10.

* Translation of Theory of Semigroups and Its Applications. I, Izdat. Saratov. Univ., Saratov, 1965, pp. 286–324. MR 35 #283.

(a) This paper was submitted for publication in 1962. Since then new developments have occurred in the field covered by the paper. Since the translator happens to be the author of the paper, he has added footnotes designated by letters (in contrast to the consecutively numbered footnotes which appear in the Russian original). Items [43]–[57] of the bibliography have been added for the translation (made in 1972).

The exact translation of the title is "On the theory of generalized groups and generalized heaps", and the exact transliteration of the author's name is given in the brackets. The term "generalized group", synonymous with "inverse semigroup", has never been used in English, and by now has practically disappeared from Russian. The reasons for coining a new English term "groud" instead of "heap" are explained in the footnote (o) on pages 101–102.

(b) "generalized inverses" in the Russian original.

(c) "Quasi-single-valued" in the Russian original. The term "difunctional" is due to Jacques Riguet [25], [43], who was the first to consider this important class of binary relations in detail.

(d) "Stable equivalence relation" is used instead of "congruence" throughout the Russian text.

be singled out of the class of semigrouds by one identity, but can be defined by two identities (up to now this problem has been solved using three identities [10]). An elementary axiom which singles out generalized grouds in the class of semigrouds and which is not an identity is found.

In §2 we consider the question of mutual operations in inverse semigroups. It is known [12] that in groups there exists a derived operation (right or left division) which is expressible in terms of the basic group operations of multiplication and involution and such that the basic group operations in turn are derived operations of this derived operation (such a derived operation is called mutual). In rings there exists a ternary mutual operation [13], and in Boolean algebras there exist binary and ternary mutual operations (Sheffer's stroke; the median operation [14]). It is proved that there are no mutual operations for inverse semigroups.

The results of §2 were reported by the author at the seminar of Professor V. V. Vagner in 1957.

Several authors (for example, Charles Ehresmann [16], [17], [44]) have considered categories all of whose elements are equivalences (i.e. are invertible). Almost every author mentions that the theory of categories ought to be developed within the framework of axiomatic set theory to lessen the danger that set-theoretic paradoxes will emerge. Usually the axiomatic set theory of von Neumann-Bernays-Gödel, where the concepts of class and set differ, is chosen. However, often this theory is applied beside the point: the term "set" is replaced by the term "class", after which various constructions are considered which are inadmissable in the von Neumann-Bernays-Gödel theory (e.g. ordered pairs of classes, etc.). To lessen the number of terms used in the present paper we considered it permissible not to use the term "class" and to remain on the level of naive set theory.

If one adheres to this viewpoint, one should note that categories whose elements are all invertible have been introduced by Robert Croisot [15], so we call such categories "Croisot groupoids". A particular case of Croisot groupoids are Brandt groupoids.

The pseudogroups of transformations which are used in differential geometry are Croisot groupoids from an algebraic standpoint. However, the algebraization of the theory of pseudogroups is impeded, since their operation is not everywhere defined. The theory of inverse semigroups has been built up by V. V. Vagner in the first place in view of its applications to differential geometry as a theory intended to replace the theory of pseudogroups. Therefore, a question (considered in §3 of this paper) arises of the interconnection between the theory of inverse semigroups and theory of Croisot groupoids. A Croisot groupoid is called replenishable if its operation may be extended to an everywhere defined operation, giving an inverse semigroup as a result (the Croisot groupoid can be reconstructed from the inverse semigroup in a unique fashion). In this process one may not add new elements to the Croisot groupoid. In §3 we introduce ordered Croisot groupoids of a special type—the so-called inductive Croisot groupoids (a generalization of inductive groupoids in the sense of Ehresmann [16], [17], [44]). It is proved that a Croisot groupoid is replenishable iff(e) it can be ordered in such a way as to be

(e) By "iff" we pay tribute to the current fashion. Newspeak is not used in the Russian mathematical literature (including this paper).

turned into an inductive groupoid. In every inverse semigroup we may restrict the domain of the operation of multiplication and obtain the so-called abutting multiplication, relative to which the inverse semigroup is a Croisot groupoid. This Croisot groupoid, ordered by the canonical order relation of the inverse semigroup, is an inductive groupoid, and one can reconstruct the operation of multiplication of the inverse semigroup in a unique way using this inductive groupoid.

A Croisot groupoid is replenishable if the set of its idempotents ("identities" in another terminology) is infinite or if the groupoid contains an isolated idempotent, and is not replenishable in other cases. To prove this result we have had to introduce a special class of semigroups, which were called *left holoids* and which are a generalization of the known class of linear holoids [30]–[33]. This class of semigroups seems to be of independent interest (e.g. the set of all ordinals less than an initial ordinal forms a left holoid relative to the operation of addition, but does not form a linear holoid (if the trivial case of the set of all finite ordinals is excluded)). However, the properties of left holoids were hardly considered as irrelevant to the subject of this paper.

In the paper the symbolism of mathematical logic is used: $\wedge$, $\vee$, $\longrightarrow$ and $\leftrightarrow$ are respectively the symbols for conjunction, disjunction, implication and equivalence. We use the ordinary rules of bracketing in compound sentences. We also use the simplest concepts and notation from the theory of binary relations which may be found in V. V. Vagner's papers [10] and [21].

All concepts of the theory of semigroups which have not been defined here are defined in [2] or [6].

Some results of this paper have been published in the author's note [20].

§1. Questions and axiomatics. Inverse semigroups

A *semigroup* is an ordered pair $(G, \circ)$ where G is a nonempty set and $\circ$ is a binary associative algebraic operation on G (i.e. $\circ$ is a mapping of $G \times G$ into G such that the identity $(g_1g_2)g_3 = g_1(g_2g_3)$ is satisfied; here g_1g_2 is the result of applying the operation $\circ$ to the ordered pair (g_1, g_2) of elements of G).

In the sequel a semigroup is usually designated by the same symbol as the set of its elements (i.e. the operation will not be mentioned).

An element g of a semigroup G is called *invertible* if there exists an element $\bar{g}$ of G such that

$$g\bar{g}g = g \wedge \bar{g}g\bar{g} = \bar{g}. \tag{1.1}$$

The element $\bar{g}$ is called an *inverse* for g.

REMARK 1. *It follows from* (1.1) *that if $\bar{g}$ is an inverse for g, then g is an inverse for $\bar{g}$.*

REMARK 2. *Every idempotent (i.e. an element i for which $i^2 = i$) is its own inverse.*

REMARK 3. *It follows from* (1.1) *that if $\bar{g}$ is an inverse for g, then the elements $g\bar{g}$ and $\bar{g}g$ are idempotents.*

An element g is called *semi-invertible* if there exists an element g' such that $gg'g = g$. Clearly, if an element is invertible, then it is semi-invertible as well. The converse holds also; *every semi-invertible element is invertible* [1]. In effect, if $gg'g = g$, then the element $g'gg'$, as may be checked easily, is an inverse for g.

A semigroup is called ***weakly regular*** if the product of any two idempotents of the semigroup is semi-invertible.

A semigroup is called ***regular*** if every element of it is semi-invertible.

Taking into consideration the above connection between invertibility and semi-invertibility, we may replace the condition of semi-invertibility by that of invertibility in the two definitions given above.

Clearly, every regular semigroup is weakly regular as well. Later we shall give examples of weakly regular semigroups which are not regular.

Elements g_1 and g_2 of a semigroup G are called *commuting* if $g_1g_2 = g_2g_1$.

In the sequel we shall need a definition of the so-called singular semigroups. Their definition follows.

A semigroup is called a *singleton* if it contains precisely one element.

A semigroup is called *singular* if it is not a singleton and satisfies the identity

$$g_1g_2 = g_2 \tag{1.2}$$

(i.e. every element of the semigroup is its right zero) or the identity

$$g_1g_2 = g_1 \tag{1.3}$$

(i.e. every element of the semigroup is its left zero).

Clearly, every singular semigroup is regular, since every two elements of such a semigroup are mutually inverse.

In the sequel we shall use semigroups satisfying one of the following conditions:

A1. *Every two idempotents commute.*

A2. *For every element g and every inverse $\bar{g}$ of g the elements $g\bar{g}$ and $\bar{g}g$ commute.*

A3. *Every element possesses at most one inverse.*

A4. *Every idempotent possesses at most one inverse.*

A5. *For every two idempotents i_1, i_2 we have $i_1i_2 = i_1 \leftrightarrow i_2i_1 = i_1$.*

A6. *Every idempotent and every inverse of the idempotent commute.*

A7. *For every idempotent i and every inverse $\bar{i}$ of i the elements $i\bar{i}$ and $\bar{i}i$ commute.*

A8. *If i_1 and i_2 are idempotents, then $i_1i_2 = i_1$ and $i_2i_1 = i_2$ imply $i_1 = i_2$, and $i_1i_2 = i_2$ and $i_2i_1 = i_1$ imply $i_1 = i_2$.*

A property Φ is said to be ***stronger*** than a property Ψ if any semigroup satisfying Φ satisfies Ψ as well (i.e. if the formula $\Phi \longrightarrow \Psi$ holds in every semigroup). Properties Φ and Ψ are called *equivalent* if Φ is stronger than Ψ and Ψ is stronger than Φ. A property Φ is called *essentially stronger* than Ψ if Φ is stronger than Ψ and the properties Φ and Ψ are not equivalent.

THEOREM 1.1. *Property* **A1** *is essentially stronger than property* **A2**. *Property* **A2** *is essentially stronger than any one of the properties* **A3–A8**. *Properties* **A3–A8** *are equivalent.*

PROOF. Remark 3 implies that **A1** is stronger than **A2**.

Let us prove that **A3**–**A8** are equivalent.

A3 is obviously stronger than **A4**.

Suppose **A4** is fulfilled and $i_1 i_2 = i_1$ for idempotents i_1 and i_2. Then it can easily be checked that i_1 and $i_2 i_1$ are inverses for i_1. Therefore, $i_2 i_1 = i_1$. The second part of **A5** can be proved dually. It follows that **A4** is stronger than **A5**.(1)

Let **A5** be fulfilled, and let $i\bar{i}i = i$ for an idempotent i and an element $\bar{i}$. By **A5** and Remark 3, $i = i \cdot (\bar{i}i) = (\bar{i}i)i = \bar{i}i$ and $i = (i\bar{i})i = i(i\bar{i}) = i\bar{i}$, whence $\bar{i}i = i\bar{i}$. **A6** follows.

Let **A6** be fulfilled, and let $\bar{i}$ be an inverse for an idempotent i. Then $i\bar{i} \cdot \bar{i}i = \bar{i}ii\bar{i}$; that is, **A7** is fulfilled. Therefore, **A6** is stronger than **A7**.

Let **A7** be fulfilled, and let $i_1 i_2 = i_1$ and $i_2 i_1 = i_2$ for idempotents i_1 and i_2. Multiply the first equality by i_1 on the right and the second equality by i_2 on the right. It becomes clear that i_2 is an inverse for i_1. By **A7**, the elements $i_1 i_2 = i_1$ and $i_2 i_1 = i_2$ commute; that is, $i_1 = i_1 i_2 = i_2 i_1 = i_2$. The second part of **A8** follows dually.

Now let **A8** be fulfilled, and let the elements $\bar{g}$ and $\bar{\bar{g}}$ be inverses for an element g. Then $(g\bar{g})(g\bar{\bar{g}}) = g\bar{\bar{g}}$ and $(g\bar{\bar{g}})(g\bar{g}) = g\bar{g}$, whence, by Remark 3 and **A8**, $g\bar{\bar{g}} = g\bar{g}$. Analogously $\bar{\bar{g}}g = \bar{g}g$. Therefore $\bar{g} = \bar{g}g\bar{g} = \bar{\bar{g}}g\bar{g} = \bar{\bar{g}}g\bar{\bar{g}} = \bar{\bar{g}}$. The condition **A3** follows.

Thus we have proved that the conditions **A3**–**A8** are equivalent.

Since **A2** is obviously stronger than **A7**, it is stronger than any of the conditions **A3**–**A8**.

Let us show that **A2** is essentially stronger than any one of the conditions **A3**–**A8**. To this end it suffices to construct an example of a semigroup satisfying a condition among **A3**–**A8** and not satisfying **A2**.

If **A2** is essentially stronger than **A3**–**A8**, then, obviously, **A1** is essentially stronger than any one of the conditions **A3**–**A8**.

EXAMPLE 1. Let A be an alphabet consisting of two letters α and β. Consider the normal algorithm(f) in the alphabet A with the scheme consisting of the following two substitution formulas:(2)

$$\alpha\beta\alpha \longrightarrow \alpha, \qquad \beta\alpha\beta \longrightarrow \beta. \tag{1.4}$$

Clearly, our algorithm may be applied to any word in A a finite number of times only, and any word P will be transformed into a word $|P|$ which is called the *canonical form* of P.

---

(1) When proving that **A4** implies **A5** we have used not **A4** but a weaker condition **A4'**: *Every idempotent possesses at most one idempotent inverse*, since $i_2 i_1$ is an idempotent, as may be readily checked. It follows from Theorem 1.1 that property **A4'** is equivalent to any one of the properties **A3**–**A8**.

(f) The terms "normal algorithm" and "associative calculus" are understood as in [45].

(2) In (1.4) and (1.5) the symbols $\longrightarrow$ and $\longleftrightarrow$ signify not implication and equivalence, but are used in the sense of the theory of algorithms.

Consider the associative calculus in the alphabet A defined by the system of relations

$$\alpha\beta\alpha \leftrightarrow \alpha, \qquad \beta\alpha\beta \leftrightarrow \beta. \tag{1.5}$$

To this associative calculus corresponds a semigroup: its elements are words in A written in canonical form, and the operation is defined by the following formula:

$$|P| \cdot |R| = |PR|. \tag{1.6}$$

Let us show that this semigroup does not satisfy **A2**.

Suppose **A2** is fulfilled. Since the element $|\alpha|$ of the semigroup is an inverse for the element $|\beta|$, we have $|\alpha| \cdot |\beta| \cdot |\beta| \cdot |\alpha| = |\beta| \cdot |\alpha| \cdot |\alpha| \cdot |\beta|$, whence $|\alpha\beta\beta\alpha| = |\beta\alpha\alpha\beta|$. However, the latter equality does not hold, since if $|P| = |R|$ for two words P and R of our associative calculus, then the words P and R begin and end with the same letters (this follows from the fact that the process of applying our algorithm does not change the initial or final letters of a word, which may be seen from the scheme (1.4)).

It follows that **A2** does not hold in our semigroup.

Let us show that our semigroup satisfies condition **A5**. Indeed, let $|I_1| \cdot |I_1| = |I_1|$, $|I_2| \cdot |I_2| = |I_2|$ and $|I_1| \cdot |I_2| = |I_1|$ for some words $|I_1|$ and $|I_2|$. We will show that $|I_2| \cdot |I_1| = |I_1|$.

If $|I_2|$ is the empty word, then $|I_2| \cdot |I_1| = |I_1|$. Let $|I_2|$ be a nonempty word. It follows from the equality $\||I_1|\|I_2|\| = |I_1|$ that the word $|I_1|\|I_2|$ is not of the canonical form, since the length of $|I_1|\|I_2|$ exceeds that of $|I_1|$.

Thus one of the words $\alpha\beta\alpha$ or $\beta\alpha\beta$ occurs in $|I_1|\|I_2|$. Suppose α is the last letter of $|I_1|$; that is, $|I_1|$ is of the form $P\alpha$. Then β is the initial letter of $|I_2|$; that is, $|I_2|$ is of the form βR (otherwise the word $|I_1|\|I_2|$ would be canonical). Two cases are possible:

Case 1. $|I_1|$ ends with $\beta\alpha$, i.e. it is of the form $P_1\beta\alpha$. Then α cannot be the final letter of P_1, for otherwise the word $|I_1|$ would not be canonical. When our normal algorithm is applied, the length of a word cannot increase, as the scheme (1.4) shows. Thus the equality $S = |T|$ implies that the length of a word S cannot exceed that of the word T. Since $|I_1| = |I_1| \cdot |I_2| = |P_1\beta\alpha\beta R| = |P_1\beta R|$, it is obvious that the length of $|I_1|$ does not exceed the length of $P_1\beta R$. If the lengths of $P_1\beta R$ and $|I_1|$ are equal, then $P_1\beta R = P_1\beta\alpha$, whence $R = \alpha$ and $|I_2| = \beta\alpha$. Since $|I_1| = \||I_1|\|I_1|\| = \||P\alpha\|I_1|\|$ and the length of $|P\alpha\|I_1|$ obviously exceeds the length of $|I_1|$, the word $|P\alpha\|I_1|$ is not canonical, i.e. β is the initial letter of $|I_1|$, whence $|I_1| = \beta P_2$. Therefore, $|I_2| \cdot |I_1| = |\beta\alpha\beta P_2| = |\beta P_2| = |I_1|$.

If the length of $P_1\beta R$ exceeds the length of $|I_1|$, then the word $P_1\beta R$ is not canonical; hence P_1 ends with α. As we have already mentioned, this contradicts the fact that $|I_1|$ is canonical.

Case 2. The word $|I_2|$ begins with $\beta\alpha$, i.e. it has the form $\beta\alpha R_1$. Then $|I_1| = |I_1| \cdot |I_2| = |P\alpha\beta\alpha R_1| = |P\alpha R_1| = \||I_1|R_1|$; it is obvious that if R_1 is not empty, then the length of $|I_1|R_1$ exceeds that of $|L_1|$; that is, the word $|I_1|R_1$ is not canonical. It follows that R_1 begins with β, i.e. $|I_2|$ begins with $\beta\alpha\beta$, which contradicts the supposition that $|I_2|$ is canonical. Therefore, R_1 is empty and $|I_2| = \beta\alpha$. Hence, as we have seen in Case 1, $|I_2| \cdot |I_1| = |I_1|$.

Supposing that $|I_1|$ ends with β and $|I_2|$ begins with α, we can prove that $|I_2| \cdot |I_1| = |I_1|$, using an analogous argument.

Dually we can prove that $|I_2| \cdot |I_1| = |I_1|$ implies $|I_1| \cdot |I_2| = |I_1|$; that is, our semigroup satisfies **A5**. Therefore condition **A2** is essentially stronger than one one of the conditions **A3**–**A8**.

Now let us show that **A1** is essentially stronger than **A2**. To this end we will construct an example of a semigroup which satisfies **A2** and does not satisfy **A1**.

EXAMPLE 2. Let E_2 be a semigroup with two free idempotent generators i_1 and i_2 (i.e. a free product of two singletons $\{i_1\}$ and $\{i_2\}$ [2]).

Let $g\bar{g}g = g$ for some $g \in E_2$ and $\bar{g} \in E_2$. Then $g\bar{g}$ is an idempotent. However, it is easy to see that our semigroup has only two idempotents, namely i_1 and i_2; therefore, $g\bar{g}$ is either i_1 or i_2. It follows that g is either i_1 or i_2, and $g = \bar{g}$. Thus in E_2 condition **A2** holds, but **A1** obviously fails, since the idempotents i_1 and i_2 do not commute.

Theorem 1.1 is proved.

THEOREM 1.2. *In a weakly regular semigroup the properties* **A1**–**A8** *are equivalent.*

PROOF. Let G be a weakly regular semigroup satisfying condition **A3**, and let i_1 and i_2 be idempotents of the semigroup. If $\bar{g}$ is an inverse for $i_1 i_2$ (such an inverse exists, since G is weakly regular), then it may easily be seen that $\bar{g}i_1$ and $i_2\bar{g}$ are inverses for $i_1 i_2$ as well. By **A3**, $\bar{g} = \bar{g}i_1 = i_2\bar{g}$, whence $\bar{g}^2 = \bar{g}i_1 i_2 \bar{g} = \bar{g}$. Since $\bar{g}$ is an idempotent, by Remark 2 it is its own inverse. By Remark 1 and property **A3**, $\bar{g} = i_1 i_2$. It follows that $i_1 i_2 = i_1 i_2 i_1 = i_2 i_1 i_2$. Analogously we can prove that $i_2 i_1 = i_2 i_1 i_2 = i_1 i_2 i_1$. Therefore $i_1 i_2 = i_2 i_1$, that is, property **A1** holds. **A2** follows from **A1**.

From this fact and Theorem 1.1, Theorem 1.2 follows.

COROLLARY 1.3. *In a regular semigroup the properties* **A1**–**A8** *are equivalent.*

A regular semigroup satisfying one of the properties **A1**–**A8** is called an *inverse semigroup.*

Inverse semigroups were introudced in 1952 by V. V. Vagner [3]. To define them Vagner used **A1**.

In 1954 A. E. Liber [4] proved that in regular semigroups the properties **A1** and **A3** are equivalent; thus he gave a new axiomatics for inverse semigroups. In the same year inverse semigroups were introduced by G. B. Preston [5], who, to all appearances, was not aware of Vagner's paper. The term "inverse semigroup" is used in the monographs of Ljapin [2] and Clifford and Preston [6].

In 1955 W. D. Munn and R. A. Penrose [7] proved that in a regular semigroup the properties **A1**, **A3** and **A4** are equivalent. In 1959 this fact was rediscovered by H. I. Inasaridze [8].

Conditions **A2** and **A5**–**A8** seem to be published here for the first time.

Condition **A1** is equivalent to the condition

A1′. *The set $I(G)$ of all idempotents of a semigroup G is a commutative subsemigroup of the semigroup.*

Clearly, **A1′** implies **A1** in an obvious way. **A1** implies that $I(G)$ is a subsemigroup, since for every two idempotents i_1, i_2 we have $(i_1i_2)^2 = i_1i_2i_1i_2 = i_1i_1i_2i_2 = i_1i_2$; that is, $i_1i_2 \in I(G)$, and $I(G)$ is commutative by condition **A1**.

Condition **A8** may be replaced by the equivalent condition

A8′. *The semigroup contains no singular subsemigroups.*

In fact, if G contains a singular subsemigroup S, then for two different idempotents i_1 and i_2 of S we have $i_1i_2 = i_1$ and $i_2i_1 = i_2$, or $i_1i_2 = i_2$ and $i_2i_1 = i_1$, but $i_1 \neq i_2$, although i_1 and i_2 are idempotents. Therefore **A8** does not hold.

Conversely, if G contains no singular subsemigroups, then **A8** holds for every two idempotents i_1 and i_2 of G, since otherwise $\{i_1, i_2\}$ is a singular subsemigroup of G.

In the proof of the equivalence of **A1** with **A1′** and **A8** with **A8′** we have nowhere used regularity or weak regularity. These properties are equivalent for arbitrary semigroups.(g)

A self-map Θ of an inverse semigroup $(G, \circ)$ such that $\Theta(g)$ is the inverse for g is an involution [3], i.e.

$$(g^{-1})^{-1} = g, \quad (g_1g_2)^{-1} = g_2^{-1}g_1^{-1}, \tag{1.7}$$

where g^{-1} denotes the inverse for g, the uniqueness of g^{-1} following from **A3**. Thus the ordered triple $(G, \circ, \Theta)$ is an involuted semigroup (the definition of which may be found, e.g., in [9], §4).(h)

We are going to find a system of axioms for inverse semigroups considered as involuted semigroups.(3)

In an involuted semigroup $(G, \circ, \Theta)$ the element $\Theta(g)$ is denoted by g^{-1}.

THEOREM 1.4. *An involuted semigroup is an inverse semigroup iff it satisfies the following two identities*:

B1. $$gg^{-1}g = g.$$

B2. $$g_1g_1^{-1}g_2g_2^{-1} = g_2g_2^{-1}g_1g_1^{-1}.$$

PROOF. Let an involuted semigroup $(G, \circ, \Theta)$ satisfy **B1** and **B2**. Then $g^{-1} =$

(g) The properties A1–A8 can be written as formulas of the lower predicate calculus (LPC). A formula in LPC is called *positive* if the only sentential connectives it contains are conjunctions and disjunctions. It is well known (cf. [46], p. 280) that if a class of algebras can be axiomatized by positive axioms then the class is closed under homomorphisms. We have proved in [47] that a semigroup G is inverse iff it satisfies the following conditon A9: *For every $f \in G$ and $g \in G$ there exist inverses f^* and g^* such that $ff^*f^*f = f^*fff^*$, $gg^*g^*g = g^*ggg^*$ and at least one of the following two equalities holds*: $ff^*gg^* = gg^*ff^*$ *or* $f^*fg^*g = g^*gf^*f$. Thus the well-known theorem that a homomorphic image of an inverse semigroup is an inverse semigroup follows from the outward appearance of A9.

(h) A triple $(G, \circ, \Theta)$ is called an *involuted semigroup* if $(G, \circ)$ is a semigroup, Θ is a unary operation on G and the identities (1.7), with g^{-1} standing for $\Theta(g)$, are satisfied. The above involution Θ in an inverse semigroup is called *canonical* (in contrast to other possible involutions).

(3) By definition, an involuted semigroup $(G, \circ, \Theta)$ is an inverse semigroup iff $(G. \circ)$ is an inverse semigroup and Θ is its canonical involution.

$(gg^{-1}g)^{-1} = g^{-1}(g^{-1})^{-1}g^{-1} = g^{-1}gg^{-1}$, which, together with **B1**, signifies that g^{-1} is an inverse for g. Let i be an idempotent of our involuted semigroup. Then, substituting $g_1 = i^{-1}$ and $g_2 = i$ in **B2**, we obtain $i^{-1}iii^{-1} = ii^{-1}i^{-1}i$. Therefore, $i^{-1} = i^{-1}ii^{-1} = i^{-1}iii^{-1} = ii^{-1}i^{-1}i = i(i^{-1}i^{-1})i = i(ii)^{-1}i = ii^{-1}i = i$. Thus $ii^{-1} = i^{-1}i = i$.

Let i_1 and i_2 be two idempotents. Then $i_1i_1^{-1} = i_1$ and $i_2i_2^{-1} = i_2$. Substituting $g_1 = i_1$ and $g_2 = i_2$ in **B2**, we obtain $i_1i_2 = i_2i_1$; that is, our involuted semigroup satisfies **A1**. It follows that the involuted semigroup is an inverse semigroup.

Conversely, let $(G, \circ, \Theta)$ be an inverse semigroup. Then the conditions **B1** and **B2** are obviously satisfied (the latter follows from the fact that $g_1g_1^{-1}$ and $g_2g_2^{-1}$ are idempotents, and from **A1**).

Theorem 1.4 is proved.

Thus inverse semigroups considered as involuted semigroups admit a system of axioms consisting of two identities. It turns out that the number of identities cannot be decreased. Namely, the following theorem holds:

THEOREM 1.5. *The class of inverse semigroups considered as a class of involuted semigroups cannot be characterized by a system of axioms consisting of a single identity.*

PROOF. Suppose the opposite holds. Let

$$\varphi_1(g_1, g_2, \ldots, g_n) = \varphi_2(g'_1, g'_2, \ldots, g'_m)$$

be an identity (i.e. φ_1 and φ_2 are polynomials relative to the operations $\circ$ and Θ (cf., e.g., §2)) and assume that an involuted semigroup is an inverse semigroup iff it satisfies our identity. Suppose neither φ_1 nor φ_2 is a single letter (i.e. is an individual variable). Consider an involuted semigroup $(G, \circ, \Theta)$, where $g_1g_2 = 0$ for every $g_1, g_2 \in G$, where 0 is a zero of the involuted semigroup G and $g^{-1} = g$. In this semigroup $\varphi_1(g_1, \ldots, g_n) = 0$ and $\varphi_2(g'_1, \ldots, g'_m) = 0$, i.e. it satisfies our identity. However, if our involuted semigroup is not a singleton, it does not satisfy **B1**, whence it is not an inverse semigroup. Therefore at least one of the polynomials φ_1 or φ_2 is a letter, i.e. our identity has the form $\varphi(g_1, \ldots, g_n) = g^{-1}$ or $\varphi(g_1, \ldots, g_n) = g$. In the former case the identity may be transformed into the form $(\varphi(g_1, \ldots, g_n))^{-1} = g^{-1}$, which is equivalent to $\varphi(g_1, \ldots, g_n) = g$. Therefore, we may suppose that our identity has the form

$$\varphi(g_1, \ldots, g_n) = g^{-1}. \tag{1.8}$$

By Lemma 2.2, whose proof is independent of Thereom 1.5, it follows that the polynomial φ may be reduced to the form $g^{-1}gg^{-1}g \cdots g^{-1}gg^{-1}$, i.e. the identity (1.8) may be written in the form

$$g^{-1}gg^{-1} \cdots gg^{-1} = g^{-1}. \tag{1.9}$$

Let A be a set containing more than one element. On its cartesian square $A \times A$ define a multiplication and involution as follows: $(a_1, a_2)(a_3, a_4) = (a_1, a_4)$ and $(a_1, a_2)^{-1} = (a_2, a_1)$. It is a matter of straightforward computation to check that $A \times A$ turns out to be an involuted semigroup. All elements of the involuted semigroup are idempotent, but the involuted semigroup is not commutative and, by **A1**, is not an inverse semigroup. However, it satisfies (1.9), as may be easily seen. This contradiction completes the proof of Theorem 1.5.

THEOREM 1.6. *An involuted semigroup* $(G, \circ, \Theta)$ *satisfying the conditon* **B1** *of Theorem* 1.4 *is an inverse semigroup iff it satisfies either one of the following conditions*:

B2′. $gg^{-1}g^{-1}g = g^{-1}ggg^{-1}$;

B2″. *Every idempotent is a fixed point of the involution.*

PROOF. Clearly, if g_1^{-1} is substituted for g_2 in **B2**, then **B2′** is obtained. If an idempotent i is substituted for g in **B2′**, then, since i^{-1} is an idempotent as well, applying **B1** we obtain $i = i^{-1}$; that is, **B2″** is fulfilled.

Now Theorem 1.6 is an obvious consequence of the following lemma:

LEMMA 1.7. *A regular semigroup is inverse iff it possesses an involution for which all idempotents are fixed points.*

PROOF. It suffices to prove that if $(G, \circ, \Theta)$ is an involuted regular semigroup and all idempotents are fixed by Θ, then $(G, \circ)$ is an inverse semigroup. Let $i_1 i_2 = i_1$ for idempotents i_1 and i_2. Applying the involution to both sides of the equality and taking into consideration the fact that idempotents are fixed points, we obtain $i_2 i_1 = i_1$. The latter equality in turn implies $i_1 i_2 = i_1$. Therefore, the semigroup $(G, \circ)$ is inverse, since it satisfies **A5**.

It is stated in [6], p. 62, that Lemma 1.7 was proved by W. D. Munn. However, a proof of the lemma has never been published.

In the sequel we will consider regular ordered semigroups, i.e. the regular semigroups endowed with a stable order relation.(i) For an order relation we use an arithmetic notation: $g_1 \prec g_2$ means that the ordered pair (g_1, g_2) belongs to the order relation.

Every inverse semigroup can be considered as a semigroup ordered by its canonical order [3]. A regular ordered semigroup is called the inverse semigroup if the semigroup is inverse and its order relation coincides with the canonical order relation of the inverse semigroup.(j)

An order relation is called *steady* [22]–[24] if it satisfies the condition(4)

$$z \prec xv,\ uv,\ uy \longrightarrow z \prec xy. \tag{1.10}$$

THEOREM 1.8. *A regular ordered semigroup is inverse iff its order relation is steady.*

PROOF. The canonical order relation of an inverse semigroup is steady [23]. Conversely, let G be a regular semigroup with a steady order relation. Then the semigroup is isomorphically embeddable in an inverse semigroup [23], whence G satisfies **A1**, since this condition is an obvious necessary condition for embeddability in an inverse

(i) In this paper an order relation signifies a partial order relation, i.e. an order need not be linear. An order relation $\prec$ is called *stable* if $g_1 \prec g_2$ and $g_3 \prec g_4$ imply $g_1 g_3 \prec g_2 g_4$ for all g_1, g_2, g_3, g_4.

(j) There exist many equivalent definitions of the canonical order relation on an inverse semigroup, e.g. $g_1 \prec g_2$ iff $g_1 = g_1 g_1^{-1} g_2$.

(4) x, y, u, v may be empty symbols provided (1.9) makes sense. This condition was not mentioned in [20].

semigroup. Therefore, G is an inverse semigroup. It remains to mention that the only stable and steady order relation on an inverse semigroup is the canonical order relation of the inverse semigroup [24].

An involuted semigroup is called *ordered* if it is ordered and its order relation is involutively invariant, i.e. $g_1 \prec g_2$ implies $g_1^{-1} \prec g_2^{-1}$.

THEOREM 1.9. *An ordered involuted semigroup is inverse iff it satisfies the conditions* **B1** *and*

C1. *If i is an idempotent, then $gi \prec g$ for every $g \in G$.*

PROOF. Let $z \prec xv, uv, uy$. Then

$$z = zz^{-1}z \prec (xv)(uv)^{-1}(uy) = xvv^{-1}u^{-1}uy \prec xu^{-1}uy \prec xy,$$

since vv^{-1} and $u^{-1}u$ are idempotents by **B1**. It follows that the order relation is steady. By Theorem 1.8, G is an inverse semigroup.

A semigroup G is called *weakly involutive* if the fact that $\bar{g}_1$ and $\bar{g}_2$ are inverses for g_1 and g_2 implies that $\bar{g}_2\bar{g}_1$ is an inverse for g_1g_2.

THEOREM 1.10. *A semigroup is weakly involutive iff the product of any two idempotents of the semigroup is an idempotent.* (k)

PROOF. Let the product of any two idempotents of a semigroup G be an idempotent, and let $\bar{g}_1$ and $\bar{g}_2$ be inverses for g_1 and g_2 respectively. Then, by Remark 3, $\bar{g}_1g_1g_2\bar{g}_2$ is an idempotent, i.e. $\bar{g}_1g_1g_2\bar{g}_2\bar{g}_1g_1g_2\bar{g}_2 = \bar{g}_1g_1g_2\bar{g}_2$. Multiplying this equality by g_2 on the right and by g_1 on the left and using (1.1), we obtain $g_1g_2\bar{g}_2\bar{g}_1g_1g_2 = g_1g_2$. In the same way we obtain $\bar{g}_2\bar{g}_1g_1g_2\bar{g}_2\bar{g}_1 = \bar{g}_2\bar{g}_1$. Therefore, $\bar{g}_2\bar{g}_1$ is an inverse for g_1g_2, i.e. G is weakly involutive.

Conversely, suppose G is weakly involutive and i_1 and i_2 are idempotents. Since every idempotent is its own inverse, i_2i_1 is an inverse for i_1i_2; that is, $i_1i_2 \cdot i_2i_1 \cdot i_1i_2 = i_1i_2$ or $i_1i_2i_1i_2 = i_1i_2$, which means that i_1i_2 is an idempotent, q.e.d.

It should be noted that *a weakly involutive semigroup is weakly regular.*

Let Θ be the *invertibility relation* of a semigroup G, i.e. Θ is a binary relation between the elements of the set G and $(g_1, g_2) \in \Theta$ iff g_2 is an inverse for g_1.

REMARK 4. *The invertibility relation is symmetric, i.e.* $\overset{-1}{\Theta} = \Theta$.

This fact follows from Remark 1.

A binary relation φ between the elements of two sets A and B is called *difunctional* if $\varphi \circ \overset{-1}{\varphi} \circ \varphi = \varphi$.

It follows from the symmetry of Θ that Θ is difunctional iff $\overset{3}{\Theta} = \Theta$, where $\overset{3}{\Theta} = \Theta \circ \Theta \circ \Theta$.

PROPOSITION 1.11. *In a weakly involutive semigroup the invertibility relation is difunctional.*

PROOF. Let $(g_1, g_2), (g_2, g_3), (g_3, g_4) \in \Theta$. It follows from weak involutivity

(k) This theorem in the case of regular semigroups was proved in the paper [48], whose authors did not know of our results. In [49] weakly involutive regular semigroups were christened "orthodox".

that g_1g_4 is an inverse for g_3g_2. By the same property, $g_4g_1g_4$ is an inverse for $g_3g_2g_3 = g_3$. Therefore, $g_3g_4g_1g_4g_3 = g_3$. Multiplying this equality by g_4 on both sides and using the fact that $(g_3, g_4) \in \Theta$, we obtain $g_4g_1g_4 = g_4$. Analogously, $g_1g_4g_1 = g_1$. Therefore, $(g_1, g_4) \in \Theta$, i.e. $\overset{3}{\Theta} = \Theta$ and Θ is difunctional.(1)

REMARK 5. *Weak involutivity does not follow from the difunctionality of the invertibility relation in general.*

In fact, in the semigroup E_2 considered in Example 2 the only invertible elements are i_1 and i_2. Difunctionality of Θ can be easily deduced from this fact; however, i_1i_2 is not an idempotent and, by Theorem 1.10, weak involutivity fails.

THEOREM 1.12. *A semigroup is weakly involutive iff it is weakly regular and its invertibility relation is difunctional.*

PROOF. By Proposition 1.11, it suffices to prove that difunctionality of Θ and weak regularity imply weak involutivity. Let i_1 and i_2 be idempotents, and let g be an inverse for i_1i_2. Then it can be easily verified that $(g, i_1i_2), (i_1i_2, gi_1), (gi_1, i_2gi_1i_2) \in \Theta$, whence $(g, i_2gi_1i_2) \in \Theta$. It follows that $gi_2g = g$ and i_2g is an idempotent. Then we can easily check that $(i_1i_2, i_2g), (i_2g, i_2g), (i_2g, i_1i_2gi_1) \in \Theta$. It follows that $(i_1i_2, i_1i_2gi_1) \in \Theta$. Therefore, $i_1i_2 = i_1i_2 \cdot i_1i_2gi_1 \cdot i_1i_2 = i_1i_2 \cdot i_1i_2$; that is, i_1i_2 is an idempotent. By Theorem 1.10, weak involutivity follows.

PROPOSITION 1.13. *In a weakly involutive semigroup the relation* $\overset{2}{\Theta}$, *which is called the* equi-invertibility relation, *is a stable relation of partial equivalence.* (5)

PROOF. Transitivity of $\overset{2}{\Theta}$ follows from difunctionality of Θ. Therefore, $\overset{2}{\Theta}$ is a partial equivalence relation. Let $(g_1, g_2), (g_3, g_4) \in \overset{2}{\Theta}$. Then for some g_0 and g we have $(g_1, g_0), (g_2, g_0) \in \Theta$ and $(g_3, g), (g_4, g) \in \Theta$; it follows from weak involutivity that $(g_1g_3, gg_0), (g_2g_4, gg_0) \in \Theta$ and $(g_1g_3, g_2g_4) \in \overset{2}{\Theta}$, i.e. $\overset{2}{\Theta}$ is stable.

$(g_1, g_2) \in \overset{2}{\Theta}$ means that g_1 and g_2 possesses a common inverse.

REMARK 6. *The relation of equi-invertibility of a wealky involutive semigroup need not be either right or left regular.* (m)

In fact, let G be a semigroup without idempotents. Adjoin an identity element e to G. Let G^1 be the resulting semigroup. In this semigroup $\Theta = \{(e, e)\}$. Therefore, $\overset{2}{\Theta} = \{(e, e)\}$. If $\overset{2}{\Theta}$ were right regular, we would have $(eg, eg) \in \overset{2}{\Theta}$ for every $g \in G$, i.e. $(g, g) \in \overset{2}{\Theta}$, which is not true.

The very definition of weak involutivity implies that in a weakly involutive semigroup the set of all invertibile elements forms a regular subsemigroup. It can be easily proved that the equi-invertibility relation is regular iff the subsemigroup consisting of

(1) We have proved that $\overset{2}{\Theta} \subset \Theta$. However, $\varphi \subset \varphi \circ \overset{-1}{\varphi} \circ \varphi$ for every binary relation φ, whence $\Theta \subset \overset{3}{\Theta}$.

(5) A relation of partial equivalence is a symmetric and transitive binary relation. It differs from an equivalence relation, since the union of the equivalence classes modulo a partial equivalence need not cover the whole underlying set.

(m) A binary relation φ on a semigroup G is called *right regular* if $(g_1, g_2) \in \varphi$ implies $(g_1g, g_2g) \in \varphi$ for all $g, g_1, g_2 \in G$. *Left regularity* is defined dually. φ is called *regular* if it is both right and left regular. φ is called *stable* if $(g_1, g_2), (g_3, g_4) \in \varphi$ imply $(g_1g_3, g_2g_4) \in \varphi$. For transition binary relations, regularity implies stability; for reflexive binary relations, stability implies regularity.

all invertible elements is an ideal of the semigroup.

In a regular semigroup every element is invertible; therefore, the equi-invertibility relation of a regular semigroup is a congruence. It is easily seen that the quotient semigroup modulo this congruence is inverse.(n)

By Theorem 1.10, a semigroup is weakly involutive iff the set of idempotents of the semigroup forms a subsemigroup. In particular, the set of idempotents of a weakly involutive regular semigroup forms a semigroup. Let I be an idempotent semigroup. Then it is weakly involutive and regular in an obvious fashion. By Remark 2, the invertibility relation Θ of the semigroup is reflexive. From this fact and Theorem 1.12, $\overset{2}{\Theta} \subset \overset{3}{\Theta} = \Theta$, i.e. Θ is transitive. Therefore, the invertibility relation of an idempotent semigroup is a congruence (since, by Proposition 1.13, $\overset{2}{\Theta}$ is stable and $\Theta = \overset{2}{\Theta}$). The quotient semigroup of I modulo Θ is, as has already been mentioned, inverse and, obviously, idempotent. The following known result follows: *every idempotent semigroup is a semilattice of anticommutative semigroups* [26], [27].

Generalized grouds

A *semigroud* is an ordered pair $(K, \circ)$ such that K is a set and $\circ$ is a ternary algebraic operation given on K and satisfying the condition

$$[[k_1k_2k_3]\,k_4k_5] = [k_1\,[k_4k_3k_2]\,k_5] = [k_1k_2\,[k_3k_4k_5]], \tag{1.11}$$

where $[k_1k_2k_3]$ denotes the result of applying the operation $\circ$ to the ordered triple (k_1, k_2, k_3) of elements from K, i.e. $[k_1k_2k_3] = \circ\,((k_1, k_2, k_3))$. The result of consecutively applying the operation $\circ$ to an ordered system $(k_1, \ldots, k_{2n+1})$ consisting of $2n + 1$ elements of K is denoted by $[k_1 \cdots k_{2n+1}]$.

A semigroud will be denoted frequently in the same fashion as the set of its elements, i.e. without the operation being mentioned.(o)

(n) In order that the reader may better comprehend the real meaning of the results 1.11–1.13 we cite here certain known results [25], [43], [21] on symmetric difunctional binary relations. Θ is such a relation on a set G iff $\Theta = \bigcup_{i\in I}(H_i \times H_{\pi(i)})$, where $(H_i)_{i\in I}$ is an indexed family of pairwise disjoint nonempty subsets of G and π is an involution on I (i.e. π is a permutation on I such that $\pi(\pi(i)) = i$ for every $i \in I$). Thus $\overset{2}{\Theta} = \bigcup_{i\in I}(H_i \times H_i)$. If Θ is the invertibility relation on a weakly involutive semigroup G, then every element of $H_{\pi(i)}$ is an inverse of every element of H_i. If G is regular, then a trivial argument shows that $\overset{2}{\Theta}$ is the smallest element of the set of all congruences ϵ such that G/ϵ is inverse. In fact, if $(g_1, g_2) \in \overset{2}{\Theta}$, then g_1 and g_2 have a common inverse in G. If G/ϵ is an inverse semigroup, the images of g_1 and g_2 in G/ϵ must necessarily coincide, i.e. $(g_1, g_2) \in \epsilon$ and $\overset{2}{\Theta} \subset \epsilon$. For every semigroup G there exists a smallest element $\hat{\epsilon}$ in the set of all congruences defining the quotient semigroups embeddable in inverse semigroups [23], [24], [50]. If G is regular, then $\hat{\epsilon}$ is the smallest congruence defining the inverse quotient semigroup. Our result that in a weakly involutive regular semigroup $\overset{2}{\Theta}$ is the smallest congruence defining the inverse quotient semigroup has recently been obtained in [49].

(o) Semigrouds satisfying the identity $[k_1k_2k_2] = k_1$ were introduced by Heinz Prüfer [51] who used the term "Schar" (German). In his book [38] A. K. Suškevič translated this term by the Russian word груда (meaning "heap", "pile"). The translation was not accurate but it had the advantage that it was close phonetically to группа (the Russian word for "group"); in fact the theories of груда and groups are intimately connected. Later the term груда was translated into other languages; the translations had meanings quite remote from the original "Schar". We wish to coin a new term "groud" instead of "heap". The advantages of the new term are that it is not overloaded semantically, it is phonetically similar to "group", and it is more euphonious than

An element k of a semigroud K is called *idempotent* if

$$[kkk] = k. \tag{1.12}$$

A semigroud is called *idempotent* if all its elements are idempotent.

Elements k_1 and k_2 of a semigroud K are called *right bicommuting* if for all $k \in K$

$$[kk_1k_1k_2k_2] = [kk_2k_2k_1k_1], \tag{1.13}$$

and *left bicommuting* if for all $k \in K$

$$[k_2k_2k_1k_1k] = [k_1k_1k_2k_2k]. \tag{1.14}$$

Right and left bicommuting elements are called *bicommuting.*

A semigroud is called *right* (*left*) *bicommutative* if every two elements of it are right (left) bicommuting, and *bicommutative* if it is both right and left bicommutative.

LEMMA 1.14. *If for elements k_1 and k_2 of a semigroud $(K, \circ)$*

$$[k_1k_2k_2] = [k_1k_2k_2k_1k_1], \tag{1.15}$$

then k_1 and k_2 are right bicommuting. If

$$[k_2k_2k_1] = [k_1k_1k_2k_2k_1], \tag{1.16}$$

then k_1 and k_2 are left bicommuting.

PROOF. Let (1.15) hold for k_1 and k_2. Then for every $k \in K$

$$[kk_1k_1k_2k_2] = [kk_1[k_1k_2k_2]] = [kk_1[k_1k_2k_2k_1k_1]]$$

$$= [kk_1k_1k_2k_2k_1k_1] = [k[k_1k_2k_2k_1k_1]k_1] = [k[k_1k_2k_2]k_1] = [kk_2k_2k_1k_1],$$

i.e. k_1 and k_2 are right bicommuting. The second part of Lemma 1.14 follows dually.

THEOREM 1.15. *In an idempotent semigroud two elements k_1 and k_2 are right bicommuting iff* (1.15) *holds and are left bicommuting iff* (1.16) *holds.*

PROOF. By Lemma 1.14, if k_1 and k_2 satisfy (1.15), then they are right bicommuting.

Conversely, let the elements k_1 and k_2 of an idempotent semigroud K be right bicommuting, and substitute $k = k_1$ in (1.13). Using the fact that k is idempotent, we obtain (1.15).

The second part of the theorem follows dually.

COROLLARY 1.16. *An idempotent semigroud is right bicommutative iff it satisfies* (1.15), *and left bicommutative iff it satisfies* (1.16).

This result is due to H. N. Inasaridze [53].

A bicommutative idempotent semigroud is called a *generalized groud.* Generalized grouds were introduced for the first time by V. V. Vagner. As a system of axioms for

---

"heap". "Semigrouds" are to grouds more or less as semigroups are to groups. We discard the term "semiheap" and "generalized heap" and use "semigroud" and "generalized groud" in this paper.

generalized grouds Vagner considered the identities (1.11)–(1.14). As we have seen, the identities (1.11), (1.12), (1.15) and (1.16) can be taken as a system of axioms.

THEOREM 1.17. *An idempotent semigroud is a generalized groud iff it satisfies the identitiy*

$$[k_1k_1k_2k_2kk_3k_3k_4k_4] = [k_2k_2k_1k_1kk_4k_4k_3k_3], \tag{1.17}$$

or the identity

$$[k_1k_1k_2k_3k_3] = [k_2k_2k_1k_1k_2k_3k_3k_2k_2], \tag{1.18}$$

or the identity

$$[k_1k_1k_2k_3k_3k_2k_2] = [k_2k_2k_1k_1k_2k_3k_3]. \tag{1.19}$$

PROOF. Using idempotence and bicommutativity, it may be proved easily that any generalized groud satisfies (1.17)–(1.19).

Supposing $k_1 = k_2 = k$ in (1.17) and using idempotence, one obtains an identity of the form (1.13), i.e. right bicommutativity. It may be proved analogously that (1.17) implies left bicommutativity.

If $k_1 = k_2$ in (1.18) or (1.19), then by idempotence an identity of the form (1.15) is obtained. It may be proved analogously that each of the identities (1.18) and (1.19) implies (1.16). By Corollary 1.16, we obtain the statement of our theorem.

Thus a semigroud is a generalized groud if it satisfies the two identities (1.12) and any one of the identities (1.17)–(1.19).

THEOREM 1.18. *For the class of semigrouds which are generalized grouds there are no systems of axioms consisting of a single identity.*

PROOF. Suppose the opposite holds. Let

$$[k_1k_2 \cdots k_n] = [k_1'k_2' \cdots k_m'] \tag{1.20}$$

be an identity and suppose a semigroud is a generalized groud iff it satisfies (1.20).

Consider a semigroud $(K, \circ)$ in which $[k_1k_2k_3] = 0$ for all $k_1, k_2, k_3 \in K$ (0 is a zero of K; for a definition see, for example, [9], §1). If the semigroud contains more than one element it is not a generalized groud, since it is not idempotent. If both sides of (1.20) do not reduce to one letter, then this identity obviously holds in our semigroud, since both sides of the identity are always equal to zero. Therefore, (1.20) must be of the form

$$[k_1k_2 \cdots k_n] = k. \tag{1.21}$$

Let $(K, \circ)$ be a generalized groud with zero containing more than one element. Suppose a letter different from k occurs in the left-hand side of (1.21). Substituting 0 for this letter, we obtain $0 = k$. Since this is not true, the left-hand side of (1.21) cannot contain occurrences of letters different from k, i.e. (1.21) has the form

$$[kk \cdots k] = k. \tag{1.22}$$

It is obvious that (1.12) implies (1.22). Therefore, every idempotent semigroud is a generalized groud. Consider a semigroud $(K, \circ)$ containing more than one element, in which for every $k_1, k_2, k_3 \in K$ the following identity holds:

$$[k_1 k_2 k_3] = k_1. \tag{1.23}$$

Clearly, $(K, \circ)$ is an idempotent semigroud; however, it is not a generalized groud, since it does not satisfy (1.14). Therefore, (1.12) is not a system of axioms for those semigrouds which are generalized grouds, whence the assumption stated at the beginning of the proof of our theorem is false. Theorem 1.18 follows.

Consider a binary relation ω between elements of a semigroud $(K, \circ)$ defined as follows:

$$(k_1, k_2) \in \omega \leftrightarrow [k_1 k_2 k_1] = k_1. \tag{1.24}$$

Clearly, $(K, \circ)$ is idempotent iff ω is reflexive.

THEOREM 1.19. *A semigroud $(K, \circ)$ is a generalized groud iff the binary relation ω between its elements is reflexive and antisymmetric, i.e. iff $(K, \circ)$ is idempotent and satisfies the condition*

$$[k_1 k_2 k_1] = k_1 \wedge [k_2 k_1 k_2] = k_2 \longrightarrow k_1 = k_2. \tag{1.25}$$

PROOF. The relation ω in generalized grouds was introduced by Vagner [10]. He proved that in generalized grouds ω is an order relation, i.e. ω is reflexive, transitive and antisymmetric.

Let the relation ω between the elements of $(K, \circ)$ be reflexive and antisymmetric, i.e. $(K, \circ)$ is idempotent and satisfies (1.25). Then

$$\begin{aligned}
[[k_1k_2k_2][k_1k_2k_2k_1k_1][k_1k_2k_2]] &= [k_1k_2k_2k_1k_1k_2k_2k_1[k_1k_2k_2]] \\
&= [k_1k_2k_2k_2k_2k_1k_1k_2k_2k_2k_2k_1[k_1k_2k_2]] \\
&= [[k_1k_2k_2][k_1k_2k_2][k_1k_2k_2][k_1k_2k_2]] = [k_1k_2k_2],
\end{aligned}$$

$$\begin{aligned}
[[k_1k_2k_2k_1k_1]\ [k_1k_2k_2]\ [k_1k_2k_2k_1k_1]] &= [k_1k_2k_2k_1k_1k_2k_2k_1[k_1k_2k_2k_1k_1]] \\
&= [k_1k_2k_2k_1k_1k_1k_1k_2k_2k_1[k_1k_2k_2k_1k_1]] \\
&= [[k_1k_2k_2k_1k_1][k_1k_2k_2k_1k_1][k_1k_2k_2k_1k_1]] = [k_1k_2k_2k_1k_1].
\end{aligned}$$

By (1.25), $[k_1k_2k_2] = [k_1k_2k_2k_1k_1]$, i.e. the semigroud satisfies (1.13). By Lemma 1.14, $(K, \circ)$ is right bicommutative. Dually, $(K, \circ)$ is left bicommutative, i.e. it is a generalized groud.

Theorem 1.19 is proved.

As we have just seen, an idempotent semigroud need not be a generalized groud, i.e. reflexivity of ω does not imply antisymmetry. Antisymmetry of ω does not imply reflexivity, as an example of a free semigroud in which ω is always antisymmetric (since $[k_1k_2k_1]$ is never equal to k_1) and is not reflexive shows. However, the following corollary holds.

COROLLARY 1.20. *If the binary relation ω between elements of a semigroud $(K, \circ)$ is reflexive and antisymmetric, then it is transitive.*

PROOF. If ω is reflexive and antisymmetric, then by the above theorem $(K, \circ)$ is a generalized groud. As we have already mentioned, in a generalized groud ω is an order relation, whence it is transitive.

THEOREM 1.21. *A semigroud $(K, \circ)$ is a generalized groud iff the relation ω on it satisfies the condition*

$$\omega \cap \overset{-1}{\omega} = \Delta_K \text{(6)} \tag{1.26}$$

or, which is the same, when it satisfies the condition

$$[k_1 k_2 k_1] = k_1 \wedge [k_2 k_1 k_2] = k_2 \leftrightarrow k_1 = k_2. \tag{1.27}$$

The proof follows from Theorem 1.19 and the fact that (1.26) means that ω is reflexive and antisymmetric.(P)

§2. Derived operations in inverse semigroups

Everywhere in this section inverse semigroups are considered as involuted semigroups, i.e. as sets with two operations: binary (multiplication) and unary (involution).

Starting with these two operations we may construct various derived operations which are expressed as polynomials in two basic operations of an inverse semigroup (see [11] and [46] for a definition of the polynomial and the derived operation in abstract algebras).

Using the property (1.7) of involution we may write down every polynomial in inverse semigroups in the form $g_1^{\epsilon_1} \cdots g_n^{\epsilon_n}$, where the individual variables $g_1, \ldots, g_n$ need not be different and the symbol ϵ_i is either 1 or -1.

A derived operation of an abstract algebra is called *mutual* if the basic operations of the algebra can be expressed as derived operations of the derived operation.

For example, in groups considered as involuted semigroups there exist mutual binary operations, namely, the operations $g_1 g_2^{-1}$ and $g_1{}^{-1} g_2$ [12]. In rings considered as algebras with three operations $(g_1 + g_2, g_1 g_2, -g)$ there exist no mutual binary operations, but there exist mutual ternary operations [13]. In Boolean algebras there exist mutual operations, for example, Sheffer's stroke or the median operation (cf. [14], pp. 157 and 158).

The main result of §2 is

THEOREM 2.1. *There are no mutual operations in inverse semigroups.*

To prove the theorem we deduce several lemmas.

---

(6) Δ_K, as usual, denotes the identical binary relation on K.

(P) The theory of generalized grouds seems to be an exotic branch of modern algebra. We give the *raison d'être* for this theory.

A one-to-one partial mapping of a set A into a set B is called a *map* in A with coordinates in B. If k is a map and $k(a) = b$, then b is the coordinate of the point a in the map k. If k_1, k_2, k_3 are three maps, then their ternary product $[k_1 k_2 k_3] = k_3 \circ \overset{-1}{k_2} \circ k_1$ is a map as well. Any set of maps closed under this ternary product is a generalized groud; and, conversely, every generalized groud is isomorphic to a generalized groud of maps (to an "atlas") [10]. All representations of generalized grouds as "atlases" are described in [54]. V. V. Vagner [55] stated that generalized grouds can be applied in differential geometry.

LEMMA 2.2. *Let $\varphi(g_1, \ldots, g_n)$ be a polynomial, and suppose the identity*

$$\varphi(g_1, \ldots, g_n) = g^{-1} \tag{2.1}$$

holds in every inverse semigroup. Then $\varphi(g_1, \ldots, g_n)$ has the form

$$g^{-1}gg^{-1} \cdots gg^{-1}.$$

PROOF. Let a variable different from g occur in $\varphi(g_1, \ldots, g_n)$. Consider a nonsingleton inverse semigroup with zero and substitute in (2.1) a nonzero element for g and zero for all other variables. Then the left-hand side of the identity is zero and the right-hand side is different from zero. This contradiction shows that $\varphi(g_1, \ldots, g_n)$ contains no occurrences of variables different from g. Therefore (2.1) has the form

$$g^{\epsilon_1}g^{\epsilon_2} \cdots g^{\epsilon_m} = g^{-1}, \tag{2.2}$$

where ϵ_i is either 1 or -1.

Consider the inverse semigroup G_2 consisting of five elements: zero 0, an element a, an element a^{-1} which is an inverse for a, and idempotents i_1 and i_2, so that $aa^{-1} = i_1$, $a^{-1}a = i_2$ and $i_1i_2 = 0$.[q] Then $a^2 = (a^{-1})^2 = 0$. (2.2) must be valid in this inverse semigroup. Substitute a for g in (2.2). Then the right-hand side of (2.2) is not zero. If in the left-hand side of (2.2) $\epsilon_i = \epsilon_{i+1}$ for some i, then $a^{\epsilon_i}a^{\epsilon_{i+1}} = 0$ and the left-hand side of (2.2) is zero. Therefore, $\epsilon_i \neq \epsilon_{i+1}$ for all i.

Let $\epsilon_1 = 1$. Then the left-hand side of (2.2) has the form $gg^{-1}g \cdots g^{-1}g$ or $gg^{-1}g \cdots gg^{-1}$, i.e. it is equal either to g or to gg^{-1}; but neither $g = g^{-1}$ nor $gg^{-1} = g^{-1}$ holds in our inverse semigroup. Therefore, $\epsilon_1 = -1$. Analogously we can prove that $\epsilon_m = -1$, i.e. the polynomial $g^{\epsilon_1} \cdots g^{\epsilon_m}$ has the form mentioned in Lemma 2.2. Lemma 2.2 is proved.

We say that the polynomial $g_1^{\epsilon_1} \cdots g_m^{\epsilon_m}$ has a *positive beginning* (*positive end*) if $\epsilon_1 = 1$ ($\epsilon_m = 1$). The *length* of the polynomial is m.

We can construct the following derived polynomials from our polynomial:

1. *$g_1^{\epsilon_1} \cdots g_m^{\epsilon_m}$ is a polynomial derived from the polynomial $g_1^{\epsilon_1} \cdots g_m^{\epsilon_m}$.*

2. *If $\varphi(g_1', \ldots, g_n')$ is a polynomial derived from $g_1^{\epsilon_1} \cdots g_m^{\epsilon_m}$, then every polynomial obtained as a result of substituting any variables $g_1'', \ldots, g_n''$ for the variables $g_1', \ldots, g_n'$ is a derived polynomial of $g_1^{\epsilon_1} \cdots g_m^{\epsilon_m}$ as well* (*it is assumed that if $g_i' = g_j'$ then the variables g_i' and g_j' are replaced by the same variable*).

3) *If φ and ψ are polynomials derived from $g_1^{\epsilon_1} \cdots g_m^{\epsilon_m}$, then the polynomial obtained as a result of substituting ψ for any variable occurring in φ is a polynomial derived from $g_1^{\epsilon_1} \cdots g_m^{\epsilon_m}$.*

4. *All polynomials derived from $g_1^{\epsilon_1} \cdots g_m^{\epsilon_m}$ can be obtained in the above way.*

LEMMA 2.3. *Every polynomial derived from a polynomial with a positive beginning (end) has a positive beginning (end).*

PROOF. If $\varphi(g_1', \ldots, g_n')$ is a polynomial with a positive beginning, then the operations 2 and 3 on polynomials cannot produce a polynomial which does not have

[q] In other words, G_2 is a Brandt semigroup of order 5 which is neither a group nor a group with zero.

a positive beginning. Consequently our lemma follows from 4. The proof for polynomials with a positive end is analogous.

LEMMA 2.4. *Every polynomial derived from a polynomial of odd length has an odd length.*

PROOF. If the length of a polynomial $\varphi(g'_1, \ldots, g'_n)$ is odd, then the operations 2 and 3 cannot make it even, since 2 does not change the length of a polynomial and 3 replaces a letter in a polynomial by an odd number of letters, i.e. a polynomial of odd length is transformed into a polynomial of odd length as a result of applying the operation 3.

PROOF OF THEOREM 2.1. Suppose the opposite holds: let there exist a mutual operation for inverse semigroups. This operation is determined by a polynomial. Let the polynomial be

$$\varphi(g_1, \ldots, g_n) = g_1^{\epsilon_1} g_2^{\epsilon_2} \cdots g_n^{\epsilon_n}. \tag{2.3}$$

Then the involution is a derived operation from the polynomial φ, i.e. there exists a polynomial $\psi(g'_1, \ldots, g'_m)$ derived from φ and such that in all inverse semigroups the identity $\psi = g^{-1}$ holds. By Lemma 2.2, ψ has the form $g^{-1}gg^{-1} \cdots gg^{-1}$. If the polynomial φ had a positive beginning or a positive end, by Lemma 2.3 the polynomial ψ derived from φ would have a positive beginning or a positive end as well. Since the beginning and end of ψ are not positive, the beginning and the end of φ are not positive either, i.e. $\epsilon_1 = \epsilon_n = -1$.

The binary operation of multiplication is, by our supposition, a derived operation from φ; therefore, there exists a polynomial $\chi(g'_1, \ldots, g'_p)$ derived from φ and such that the identity

$$\chi(g'_1, \ldots, g'_p) = g_1 g_2 \tag{2.4}$$

holds in all inverse semigroups.

We shall prove that χ has an even length.

Let A be an alphabet consisting of two letters α and β. Define $\alpha^{-1} = \beta$, $\beta^{-1} = \alpha$ and $(\gamma_1 \cdots \gamma_k)^{-1} = \gamma_k^{-1} \cdots \gamma_1^{-1}$, where the γ_i are letters in the alphabet A.

Consider an associative calculus with a set of defining relations of the form

$$\begin{gathered} PP^{-1}P \leftrightarrow P, \\ PP^{-1}RR^{-1} \leftrightarrow RR^{-1}PP^{-1}, \end{gathered} \tag{2.5}$$

where all words in the alphabet A are taken as P and R. The semigroup corresponding to this associative calculus is an inverse semigroup. This inverse semigroup is called a *free inverse semigroup with one generator*. The identity (2.4) is fulfilled in the inverse semigroup, so that $\chi(\alpha, \ldots, \alpha) = \alpha$, i.e. $\alpha^{\epsilon_1} \cdots \alpha^{\epsilon_q} = \alpha\alpha$. But in our semigroup the word $\alpha^{\epsilon_1} \cdots \alpha^{\epsilon_q}$ can be transformed into $\alpha\alpha$ only if it has an even length, since the first defining relation of (2.5) cannot transform words of an even length into words of odd length, and the second of the defining relations (2.5) does not change the length of a word. Therefore, the length of χ is even.

It follows that φ has an even length, since the polynomial χ derived from φ has an even length (cf. Lemma 2.4).

Since in the polynomial (2.3) $\epsilon_1 = \epsilon_n = -1$ and n is even, there obviously exists an i such that $\epsilon_i = \epsilon_{i+1}$; otherwise the length of φ could easily be proved to be odd.

Consider an inverse semigroup constructed in the proof of Lemma 2.2. Since $a^2 = (a^{-1})^2 = 0$, it follows that $\varphi(a, \ldots, a) = 0$. Applying the above rules for constructing polynomials derived from φ, we obtain that for every polynomial ψ the equality $\psi(a, \ldots, a) = a$ holds. By our assumption, there exists a derived polynomial ψ such that $\psi(g_1, \ldots, g_n) = g^{-1}$. It follows that $0 = a^{-1}$. This contradiction implies the falsity of our assumption, i.e. there exist no derived operations on inverse semigroups.

§3. Replenishable Croisot and Brandt groupoids

An inverse semigroup is called a *Croisot semigroup* if the product of any two different idempotents of the semigroup is zero. Clearly, if a Croisot semigroup does not contain a zero, then it contains only one idempotent, i.e. it is a group.

A *binary operative* is an ordered pair $(G, \circ)$ where G is a set and $\circ$ is a binary operation on G (which need not be defined everywhere). Adding an element 0 to G (provided $0 \notin G$) and extending the operation $\circ$ to the resulting set G^0 in such a way that $g_1g_2 = 0$ for those $g_1, g_2 \in G^0$ for which the operation $\circ$ is not defined, we obtain a binary operative $(G^0, \circ^0)$ with an everywhere defined operation and with a zero 0.

A binary operative is called a *Croisot groupoid* if after adjoining a zero we obtain a Croisot semigroup.

Croisot groupoids were introduced for the first time by Roger Croisot [15], who called them "partial groups". Later Croisot groupoids were introduced by Charles Ehresmann [16], who called them "groupoids".

Note that Croisot groupoids can be defined without adjoining a zero element to a binary operative [15], [16].

It is known [21] that for every binary operative $(G, \circ)$ we can introduce a *connectivity relation* π which is a partial equivalence relation between elements of G. Equivalence classes modulo π are called *connected components* of the operative $(G, \circ)$. If g_1 and g_2 belong to different connected components, the product g_1g_2 is not defined. It can be proved that π is the smallest partial equivalence relation having this property.(r)

If $(G, \circ)$ is a Croiset groupoid, it can be easily proved that the connectivity relattion π is an equivalence relation and

$$\pi = \overset{-1}{\iota_d} \circ \iota_d \circ \overset{-1}{\iota_s} \circ \iota_s, \tag{3.1}$$

where ι_s and ι_d are mappings of G into the set I of idempotents of the groupoid $(G, \circ)$ (an idempotent of $(G, \circ)$ is an element which is an idempotent in the Croisot semigroup corresponding to $(G, \circ)$) which are called the *left* and *right projection mappings* respectively and are defined by the formulas

(r) π is the smallest symmetric and transitive relation on G such that if the product g_1g_2 exists, then $(g_1, g_2) \in \pi$.

$$\iota_s(g) = gg^{-1},\ \iota_d(g) = g^{-1}g. \tag{3.2}$$ (7)

An element $g^{-1}g$ is called a *right projection* of g, and gg^{-1} is called a *left projection* of g.

A Croisot groupoid $(G, \circ)$ is called a *Brandt groupoid* if it is connected, i.e. if its connectivity relation is universal: $\pi = G \times G$ (in other words, G is a connected component). Brandt groupoids were introduced for the first time by Heinrich Brandt [18], who defined them in a way different from ours.

Inverse semigroups obtained from Brandt groupoids after they are provided with an adjoined zero were considered by G. B. Preston [19].

It can be proved without difficulty that *every connected component of a Croisot groupoid is a Brandt groupoid relative to the induced operation.* It can be proved that this decomposition of a Croisot groupoid into Brandt groupoids is unique.

Let $(G, \circ)$ be an inverse semigroup. Let us define a partial binary operation $\circ_{\rm a}$ on G which is defined for elements g_1 and g_2 iff $\iota_d(g_1) = \iota_s(g_2)$ and which coincides with $\circ$ in the case when $\circ_{\rm a}$ is defined (in other words, we restrict the domain of the operation $\circ$). The operation $\circ_{\rm a}$ is called the *abutting multiplication.* (8) Thus to every inverse semigroup $(G, \circ)$ there corresponds a binary operative $(G, \circ_{\rm a})$. We will show this operative is a Croisot groupoid. Adjoin a zero to $(G, \circ_{\rm a})$. The operative $(G^0, \circ_{\rm a}^0)$ is obtained. We have to prove that $(G^0, \circ_{\rm a}^0)$ is a Croisot semigroup. It is not difficult to check associativity of the multiplication in $(G^0, \circ_{\rm a}^0)$, i.e. $(G^0, \circ_{\rm a}^0)$ is an inverse semigroup, since the element g^{-1} is an inverse for the element g and the idempotents of the semigroup $(G^0, \circ_{\rm a}^0)$ commute under multiplication; the latter follows from the fact that if i is an idempotent different from zero, then $i^2 = i$, so that $i \in G$ and $i^2 = i$ in the inverse semigroup $(G, \circ)$. If i_1 and i_2 are two different idempotents of $(G^0, \circ_{\rm a}^0)$, then $\iota_d(i_1) = i_1 \neq i_2 = \iota_s(i_2)$, whence $i_1 i_2$ is not defined in $(G, \circ_{\rm a})$. Therefore, $i_1 i_2 = 0$ in $(G^0, \circ_{\rm a}^0)$, q.e.d. Thus to every inverse semigroup there corresponds a Croisot groupoid $(G, \circ_{\rm a})$ which is called the Croisot groupoid *associated* with the inverse semigroup $(G, \circ)$.

In this section we are concerned with the converse problem: for a given Croisot groupoid find, if possible, an inverse semigroup with which the groupoid is associated.

A Croisot groupoid is called *replenishable* if it is associated with an inverse semigroup.

It is known [3], [10] that on every inverse semigroup $(G, \circ)$ we can define a stable and involutively invariant order relation ω, which is called the *canonical order relation* of the inverse semigroup $(G, \circ)$ and is defined by the formula

$$\overset{-1}{\omega}\langle g\rangle = \{g\}I = I\{g\} = I\{g\}I, \tag{3.3}$$ (9)

(7) Left and right projections are sometimes called left and right identities respectively. It can be proved [15]–[17] that the product g_1g_2 is defined iff $\iota_d(g_1) = \iota_s(g_2)$ and, this being the case, $\iota_s(g_1g_2) = \iota_s(g_1)$, $g_2^{-1}g_1^{-1}$ is defined and $g_2^{-1}g_1^{-1} = (g_1g_2)^{-1}$. We shall use these facts in the sequel without special reference.

(8) The term was proposed by V. V. Vagner.

(9) $\{g\}I$ denotes the product of the subsets $\{g\}$ and I.

where $\overset{-1}{\omega}\langle g\rangle$ is the set of all minorants of g, i.e. $g_1 \in \overset{-1}{\omega}\langle g\rangle \leftrightarrow g_1 \prec g$, where $g_1 \prec g$ means that $(g_1, g) \in \omega$.

It is not difficult to verify that the canonical order relation ω of an inverse semigroup $(G, \circ)$ satisfies the following properties:

I1. *If $g_1 \prec g_2$ and $g_3 \prec g_4$ and the abutting products g_1g_3 and g_2g_4 are defined, then $g_1g_3 \prec g_2g_4$.*

I2. *If $g_1 \prec g_2$, then $g_1^{-1} \prec g_2^{-1}$.*

I3. *If i_1 and i_2 are idempotents, then their greatest minorant $i_1 \curlywedge i_2$ (i.e. the greatest element of the set $\overset{-1}{\omega}\langle i_1\rangle \cap \overset{-1}{\omega}\langle i_2\rangle$) exists.*

I4. *If i is an idempotent and $i \prec \iota_s(g)$, then there exists a unique element $i * g$ such that $i * g \prec g$ and $\iota_s(i * g) = i$.*

EXPLANATION.(10) **I1** means that ω is stable relative to the operation of abutting multiplication. This follows from stability of ω. **I2** means that ω is involutively invariant. In **I3**, i_1i_2 should be taken as $i_1 \curlywedge i_2$. In **I4**, *ig* should be taken as $i * g$.

It should be noted that in the properties **I1–I4** we use the concepts of abutting multiplication and involution; the latter may be defined in terms of abutting multiplication. Therefore, we can consider Croisot groupoids whose sets of elements are endowed with order relations satisfying the axioms **I1–I4.**

An ordered triple of the form $(G, \circ, \omega)$, where $(G, \circ)$ is a Croisot groupoid, (G, ω) is an ordered set and the axioms **I1–I4** are fulfilled, is called an *inductive Croiset groupoid*; the order relation ω (satisfying **I1–I4**) is called an *inductor.* The element $i * g$ is said to be *induced* on an idempotent i by an element g.

The concept of inductive groupoid was introduced by Charles Ehresmann [16], [17], [44]. Instead of axioms **I1–I4**, he introduced other axioms equivalent to ours and required that the ordered set (G, ω) be conditionally complete (for a definition of conditional completeness, see [28], p. 39). For our aims conditional completeness is an unjustifiable restriction.

PROPOSITION 3.1. *Let $(G, \circ, \omega)$ be an inductive Croisot groupoid. Then*:

3.1.1. $g_1 \prec g_2 \longrightarrow \iota_s(g_1) * g_2 = g_1$.

3.1.2. $\overset{-1}{\omega} \circ \omega \cap \overset{-1}{\iota_s} \circ \iota_s \subset \Delta_G$.

3.1.3. $\overset{-1}{\omega}(I) \subset I$.

PROOF. Let $g_1 \prec g_2$. By axiom **I2**, $g_1^{-1} \prec g_2^{-1}$; by **I1**, $g_1g_1^{-1} \prec g_2g_2^{-1}$, i.e. $g_1g_1^{-1} \prec \iota_s(g_2)$. By axiom **I4** there exists a unique minorant $\iota_s(g_1) * g_2$ of g_2 whose left projection is $\iota_s(g_1)$. However, g_1 satisfies these properties. Therefore, 3.1.1 holds.

At the same time we have proved that $g_1 \prec g_2 \longrightarrow \iota_s(g_1) \prec \iota_s(g_2)$.

Let $(g_1, g_2) \in \overset{-1}{\omega} \circ \omega \cap \overset{-1}{\iota_s} \circ \iota_s$. This means that for some $g \in G$ we have $g_1 \prec g \wedge g_2 \prec g \wedge \iota_s(g_1) = \iota_s(g_2)$. By 3.1.1, $g_1 = g_2$.

Let $g \prec i$. Then, by **I2** and **I1**, $g^{-1} \prec i^{-1}$ and $gg^{-1} \prec ii^{-1} = i$. Moreover, $\iota_s(g) = gg^{-1} = \iota_s(gg^{-1})$. By 3.1.2, $g = gg^{-1}$, i.e. $g \in I$.

(10) Clearly, the axioms I1–I4 are connected with the operation of restrictive multiplication [21]. These axioms could be expressed in terms of the operation of restrictive multiplication.

COROLLARY. *If i_1 and i_2 are idempotents, so is $i_1 \curlywedge i_2$.*

PROOF. The element $i_1 \curlywedge i_2$ is smaller than the idempotent i_1, and by 3.1.3 it is an idempotent.

Let $g_1, g_2 \in G$ and $i = \iota_d(g_1) \curlywedge \iota_s(g_2)$. Then, by **I4** and (3.2),

$$\iota_d((i * g_1^{-1})^{-1}) = \iota_s(i * g_1^{-1}) = i = \iota_s(i * g_2);$$

therefore the product of $(i * g_1^{-1})^{-1}$ and $(i * g_2)$ is defined. Introduce the definition

$$g_1 + g_2 = (i * g_1^{-1})^{-1}(i * g_2). \tag{3.4}$$

Then $+$ is an everywhere defined binary operation on the set G.

Let the product of the elements g_1 and g_2 be defined. Then $\iota_d(g_1) = \iota_s(g_2)$, whence $i = \iota_g(g_1) = \iota_s(g_2)$, $(i * g_1^{-1})^{-1} = (g_1^{-1})^{-1} = g_1$, $i * g_2 = g_2$ and $g_1 + g_2 = g_1 g_2$. Therefore, *the operation $+$ coincides with the operation $\circ$ of our groupoid whenever $\circ$ is defined.*

LEMMA 3.2. *If i is an idempotent and $i \prec \iota_s(g)$, then $i + g = i * g$. If i_1 and i_2 are idempotents, then $i_1 + i_2 = i_1 \curlywedge i_2$.*

PROOF. Let $i \prec \iota_s(g)$. Then $\iota_d(i) \curlywedge \iota_s(g) = i \curlywedge \iota_s(g) = i$. Therefore, $(i * i^{-1})^{-1} = (i^{-1})^{-1} = i$ and $i + g = (i * i^{-1})^{-1}(i * g) = i(i * g) = i * g$. Also, $i = \iota_d(i_1) \curlywedge \iota_s(i_2) = l_1 \curlywedge i_2$, whence $(i * i_1^{-1})^{-1} = i$, $i * i_2 = i$ and $i_1 + i_2 = ii = i$.

COROLLARY. *Idempotents of an inductive Croisot groupoid $(G, \circ, \omega)$ commute under the operation $+$.*

PROOF. $i_1 + i_2 = i_1 \curlywedge i_2 = i_2 \curlywedge i_1 = i_2 + i_1$.

PROPOSITION 3.3. *Let $+$ be the operation defined on the set G of elements of an inductive Croisot groupoid $(G, \circ, \omega)$ by* (3.4). *Then $+$ is associative.*

PROOF. Let $g_1, g_2, g_3 \in G$. Introduce the following notation:

$$i = \iota_d(g_1) \curlywedge \iota_s(g_2), \tag{3.5}$$

$$i_2 = \iota_d(i * g_2) \curlywedge \iota_s(g_3), \tag{3.6}$$

$$i_1 = \iota_d(i_2 * g_2^{-1}). \tag{3.7}$$

Since $i \prec i_s(g_2)$, it follows by **I4** that $i * g_2$ exists; therefore i_2 exists. Now $i_2 \prec \iota_d(i * g_2) \prec \iota_d(g_2) = \iota_s(g_2^{-1})$ (the second part of the inequality follows from **I4**, **I2** and **I1**); therefore $i_2 * g_2^{-1}$ exists, and hence i_1 exists. By (3.2) and the Corollary to Proposition 3.1, the elements i, i_1 and i_2 are idempotents.

Now $i_2 \prec \iota_d(i * g_2) = \iota_s((i * g_2)^{-1})$, and hence, by **I4**, $i_2 * (i * g_2)^{-1}$ exists, and it follows from **I4** and **I2** that $i_2 * (i * g_2)^{-1} \prec (i * g_2)^{-1} \prec g_2^{-1}$. Since $i_2 * g_2^{-1} \prec g_2^{-1}$ and $\iota_s(i_2 * g_2^{-1}) = i_2 = \iota_s(i_2 * (i * g_2)^{-1})$, we obtain, by 3.1.2, that $i_2 * g_2^{-1} = i_2 * (i * g_2)^{-1}$, whence

$$i_1 = \iota_d(i_2 * g_2^{-1}) = \iota_d(i_2 * (i * g_2)^{-1}) \prec \iota_d((i * g_2)^{-1}) = \iota_s(i * g_2) = i.$$

The inequality in this formula follows from **I4**, **I2** and **I1**. Since $i \prec \iota_d(g_1) = \iota_s(g_1^{-1})$, we obtain that $i_1 \prec \iota_s(g_1^{-1})$ exists, whence, by **I4**, $i_1 * g_1^{-1}$ exists and

$$\iota_d((i_1 * g_1^{-1})^{-1}) = \iota_s(i_1 * g_1^{-1}) = i_1. \tag{3.8}$$

Now $i_1 = \iota_d(i_2 * g_2^{-1}) \prec \iota_d(g_2^{-1}) = \iota_s(g_2)$, where the inequality follows from **I4**, **I2** and **I1**. Therefore, $i_1 * g_2$ exists and

$$\iota_s(i_1 * g_2) = i_1. \tag{3.9}$$

By (3.9) and (3.7), $\iota_s(i_1 * g_2) = i_1 = \iota_s((i_2 * g_2^{-1})^{-1})$ and $i_1 * g_2 \prec g_2$, $(i_2 * g_2^{-1})^{-1} \prec (g_2^{-1})^{-1} \prec g_2$ (the latter follows from **I2**, **I4**); hence by 3.1.2

$$i_1 * g_2 = (i_2 * g_2^{-1})^{-1}. \tag{3.10}$$

Since $i_2 \prec \iota_s(g_3)$, we see by **I4** that $i_2 * g_3$ exists and

$$\iota_s(i_2 * g_3) = i_2. \tag{3.11}$$

By (3.4) and (3.6),

$$i_2 \prec \iota_d(i * g_2) = \iota_d(g_1 + g_2) = \iota_s((g_1 + g_2)^{-1}).$$

It follows from this fact and **I4** that $i_2 * (g_1 + g_2)^{-1}$ exists. By (3.5), $i * g_2$ exists. By the formula $i_1 \prec i$, which has been proved earlier, and **I4**, $i_1 * (i * g_2)$ exists. Since $i_1 * (i * g_2) \prec g_2$ (which can be obtained by applying **I4** twice), $i_1 * g_2 \prec g_2$ and $\iota_s(i_1 * (i * g_2)) = i_1 = \iota_s(i_1 * g_2)$, we obtain by 3.12 that $i_1 * (i * g_2) = i_1 * g_2$, whence $i_1 * g_2 \prec i * g_2$. In the same way we can prove that $i_1 * (i * g_1^{-1}) = i_1 * g_1^{-1}$, which implies $i_1 * g_1^{-1} \prec i * g_1^{-1}$. By **I1**,

$$(i_1 * g_2)^{-1}(i_1 * g_1^{-1}) \prec (i * g_2)^{-1}(i * g_1^{-1}) = (g_1 + g_2)^{-1}.$$

The equality follows from the definition (3.4) and the fact that $^{-1}$ is an involution. Applying **I4** and (3.10), we obtain

$$\iota_s((i_1 * g_2)^{-1}(i_1 * g_1^{-1})) = \iota_s((i_1 * g_2)^{-1}) = \iota_s(i_2 * g_2^{-1}) = i_2.$$

It follows from 3.1.2 that

$$i_2 * (g_1 + g_2)^{-1} = (i_1 * g_2)^{-1}(i_1 * g_1^{-1}). \tag{3.12}$$

Let $i_3 = \iota_d(g_2) \curlywedge \iota_s(g_3)$. Since, as we have repeatedly seen above, ι_d is isotone, it follows that $\iota_d(i * g_2) \prec \iota_d(g_2)$ and $i_2 \prec i_3$. Since ι_d is isotone, we have

$$i_1 = \iota_d(i_2 * g_2^{-1}) \prec \iota_d(i_3 * g_2^{-1}) = \iota_s((i_3 * g_2^{-1})^{-1}),$$

which implies the existence of

$$i_1 * ((i_3 * g_2^{-1})^{-1}(i_3 * g_s)) = i_1 * (g_2 + g_3).$$

Since $i_2 \prec i_3$, we obtain, because ι_s and ι_d are isotone, that

$$i_2 * g_2^{-1} \prec i_3 * g_2^{-1} \tag{3.13}$$

and $i_2 * g_3 \prec i_3 * g_3$, whence, by **I2** and **I1**,

$$(i_2 * g_2^{-1})^{-1}(i_2 * g_3) \prec (i_3 * g_2^{-1})^{-1}(i_3 * g_3) = g_2 + g_3.$$

From this fact and the relations

$$i_1 * (g_2 + g_3) \prec g_2 + g_3$$

and

$$\iota_s((i_2 * g_2^{-1})^{-1}(i_2 * g_3)) = \iota_s((i_2 * g_2^{-1})^{-1}) = \iota_s(i_1 * g_2) = i_1 = \iota_s(i_1 * (g_2 + g_3)),$$

which follow from **I4** and (3.10), we obtain by 3.1.2 that

$$i_1 * (g_2 + g_3) = (i_2 * g_2^{-1})^{-1}(i_2 * g_3). \tag{3.14}$$

Now

$$\iota_d(g_1 + g_3) = \iota_d((i * g_1^{-1})^{-1}(i * g_2)) = \iota_d(i * g_2).$$

Therefore, by (3.4), (3.6), (3.12) and (3.14),

$$\begin{aligned}(g_1 + g_2) + g_3 &= (i_2 * (g_1 + g_2)^{-1})^{-1}(i_2 * g_3)\\ &= ((i_1 * g_2)^{-1}(i_1 * g_1^{-1}))^{-1}(i_2 * g_3)\\ &= ((i_1 * g_1^{-1})^{-1}(i_1 * g_2))(i_2 * g_3) = (i_1 * g_1^{-1})^{-1}((i_1 * g_2)(i_2 * g_3)).\end{aligned}$$

By (3.7), (3.10), (3.12), **I2** and the fact that ι_s is isotone, we have

$$i_1 = \iota_s((i_2 * g_2^{-1})^{-1}) \prec \iota_s((i_3 * g_2^{-1})^{-1}),$$

whence, by the inequality $i_1 \prec i$ which has been obtained above,

$$\begin{aligned}i_1 \prec i \wedge \iota_s((i_3 * g_2^{-1})^{-1}) \prec \iota_d(g_1) \wedge \iota_s((i_3 * g_2^{-1})^{-1}(i_3 * g_3))\\ = \iota_d(g_1) \wedge \iota_s(g_2 + g_3).\end{aligned}$$

Thus by (3.4), **I1**, **I2**, the formula for $(g_1 + g_2) + g_3$ which has been deduced above, (3.10) and (3.14) we have

$$(g_1 + g_2) + g_3 = (i_1 * g_1^{-1})^{-1}(i_1 * (g_2 + g_3)) \prec g_1 + (g_2 + g_3). \tag{3.15}$$

By (3.4), the properties of involution and the formula $\iota_s(g) = \iota_d(g^{-1})$, we obtain

$$(g_1 + g_2)^{-1} = g_2^{-1} + g_1^{-1}. \tag{3.16}$$

Since $(g_1 + g_2) + g_3 \prec g_1 + (g_2 + g_3)$ is true for any elements g_1, g_2, g_3, we may replace them by $g_3^{-1}, g_2^{-1}, g_1^{-1}$ respectively and then (by I2) apply involution to both sides of the inequality. We obtain

$$((g_3^{-1} + g_2^{-1}) + g_1^{-1})^{-1} \prec (g_3^{-1} + (g_2^{-1} + g_1^{-1}))^{-1}$$

or, applying (3.16) twice on each side of the inequality,

$$g_1 + (g_2 + g_3) \prec (g_1 + g_2) + g_3, \tag{3.17}$$

which, together with (3.15), implies

$$g_1 + (g_2 + g_3) = (g_1 + g_2) + g_3. \tag{3.18}$$

Therefore, the operation + is associative, q.e.d.

Ehresmann [16], [17] introduced the operation + in inductive Croisot groupoids (which were understood, as we have noted, in a narrower sense than ours) and stated Proposition 3.3 without proof.

THEOREM 3.4. *Every inductive Croisot groupoid can be replenished up to an inverse semigroup whose canonical order relation coincides with the inductor of the inductive Croisot groupoid. The operation in any inverse semigroup, obtained as indicated above, necessarily coincides with the operation + of* (4.4).

PROOF. Let $(G, \circ, \omega)$ be an inductive Croisot groupoid. Then $(G, +)$ is a semigroup with commuting idempotents (Proposition 3.3 and the Corollary to Lemma 3.2). Since for every element g of the semigroup there exists an inverse g^{-1}, $(G, +)$ is an inverse semigroup. We noted above that the operation + is a replenishment of $\circ$. It is easily seen that $\circ$ coincides with the operation of abutting multiplication for the inverse semigroup $(G, +)$. Therefore, the Croisot groupoid $(G, \circ)$ is replenishable. It follows from 3.1.1 and **I4** that every element g_1 from $\overset{-1}{\omega}\langle g\rangle$ is of the form $i * g$, where i is an idempotent such that $i \prec \iota_s(g)$. By Lemma 3.2, $i * g = i + g$ and

$$i_1 + g = i_1 + (\iota_s(g) + g) = i_1 + \iota_s(g) + g = (i_1 \curlywedge \iota_s(g)) * g \prec g.$$

Therefore it follows from (3.3) that the inductor ω is the canonical order relation of the inverse semigroup.

Let $(G, \bar{\circ}, \omega)$ be an inverse semigroup for which $\circ$ is the abutting multiplication operation and ω is the canonical order relation. The operation in this inverse semigroup will be designated by a dot. Then for every $g_1, g_2 \in G$ we have

$$\begin{aligned} g_1 \cdot g_2 &= g_1 \cdot (g_1^{-1} \cdot g_1 \cdot g_2 \cdot g_2^{-1}) \cdot g_2 = g_1 \cdot (\iota_d(g_1) \cdot \iota_s(g_2)) \cdot g_2 \\ &= g_1 \cdot (i \cdot i) \cdot g_2 = (g_1 \cdot i) \cdot (i \cdot g_2) = (g_1 \cdot i)(i \cdot g_2) = (i \cdot g_1^{-1})^{-1}(i \cdot g_2). \end{aligned}$$

We have used the fact that $\iota_d(g_1) \cdot \iota_s(g_2) = \iota_d(g_1) \curlywedge \iota_s(g_2) = i$. From the fact that the canonical order relations in the inverse semigroup $(G, \bar{\circ}, \omega)$ and $(G, +, \omega)$ coincide, and from **I4**, it follows that $i \circ g_j^{-1} = i * g_j^{-1}$, $j = 1, 2$, whence

$$g_1 \cdot g_2 = (i * g_1^{-1})^{-1}(i * g_2) = g_1 + g_2.$$

Therefore $+ = \bar{\circ}$, and our inverse semigroups coincide.

COROLLARY. *An inductive Croisot groupoid can be replenished in a unique way up to an inverse semigroup with the canonical order relation coinciding with the inductor of the inductive groupoid.*

This corollary is a restatement of Theorem 3.4.

An idempotent i of a Croisot groupoid $(G, \circ)$ is called *isolated* whenever i is the only idempotent of the connected component $\pi\langle i\rangle$ (when this is the case, the component is a group).

Choose a single idempotent in every connected component of a Croisot groupoid $(G, \circ)$ (this can be done, since every connected component contains idempotents; we use the axiom of choice without limitation). The set of chosen idempotents is denoted by A. If $g \in G$, then $|g|$ denotes the idempotent chosen from the connected component containing g. Then $|g_1| = |g_2|$ means that g_1 and g_2 belong to the same connected component.

Let ω_A be an order relation on A and suppose the ordered set (A, ω_A) has the following properties:

1. *It is a minorant semilattice* [29], *i.e. for every* $i_1, i_2 \in A$ *there exists the greatest minorant* $i_1 \curlywedge i_2$.

2. *For every* $i \in A$ *there exists a unique predecessor* i^* *(that is,* $i^* \neq i$, $i^* \prec i$ *and if* $i_1 \prec i$, *then* $i_1 = i \vee i_1 \prec i^*$*).*

3. *If there is a smallest element, it is an isolated idempotent.*

Define a binary relation ω on the set G by the formula

$$(g_1, g_2) \in \omega \leftrightarrow g_1 = g_2 \vee (g_1 \in A \wedge (g_1, |g_2|^*) \in \omega_A). \tag{3.19}$$

It can easily be checked that ω is reflexive, transitive and antisymmetric, i.e. it is an order relation. It follows from (3.19) that ω induces ω_A on A. Instead of $(g_1, g_2) \in \omega$ we shall write $g_1 \prec g_2$.

We are going to show that $(G, \circ, \omega)$ *is an inductive Croisot groupoid.* To this end we must check that ω is an inductor, i.e. the axioms **I1**–**I4** are satisfied.

Verification of **I1**. Let $g_1 \prec g_2$ and $g_3 \prec g_4$, and suppose the products g_1g_3 and g_2g_4 exist. Then $|g_1| = |g_3|$ and $|g_2| = |g_4|$. If $g_1 = g_2$, then $|g_3| = |g_4|$, which, together with $g_3 \prec g_4$, implies $g_3 = g_4$ and $g_1g_3 \prec g_2g_4$. Let $g_1 \neq g_2$. Then $g_3 \neq g_4$, whence $g_3 \in A$. From this fact and $g_1 \prec g_2$ we see that $g_1 = g_3$. Since $|g_2| = |g_2g_4|$, we obtain that $|g_2|^* = |g_2g_4|^*$ and $g_1g_3 = g_1$. Thus $|g_1g_3| = |g_1| \prec |g_2|^* = |g_2g_4|^*$, whence $g_1g_3 \prec g_2g_4$.

Verification of **I2**. This follows from the fact that $i = i^{-1}$ and $|g| = |g^{-1}|$.

Verification of **I3**. Let i_1 and i_2 be idempotents. We shall define $i_1 \curlywedge i_2$ in different cases.

Case 1. Let $|i_1| \prec |i_2|^*$. If $i_1 \in A$, suppose $i_1 \curlywedge i_2 = i_1$. If $i_1 \notin A$, suppose $i_1 \curlywedge i_2 = |i_1|^*$.

Case 2, Let $|i_2| \prec |i_1|^*$. If $i_2 \in A$, suppose $i_1 \curlywedge i_2 = i_2$. If $i_2 \notin A$, suppose $i_1 \curlywedge i_2 = |i_2|^*$.

Case 3. Let $|i_1| = |i_2|$. Suppose $i_1 \curlywedge i_2 = |i_1|^*$, if $i_1 \neq i_2$.

Case 4. Suppose the Cases 1–3 do not hold, i.e. the elements $|i_1|$ and $|i_2|$ are incomparable relative to ω_A. $i_1 \curlywedge i_2 = |i_1| \curlywedge |i_2|$.

It can be checked easily that $i_1 \curlywedge i_2$ is the greatest minorant of the elemnts i_1 and i_2.

Verification of **I4**. Let $g \in G$, and let i be an idempotent, $i \prec \iota_s(g)$. Let $i * g = g$ in case $\iota_s(g) = i$ and $i * g = i$ otherwise. Obviously $i * g \prec g$, $\iota_s(i * g) = i$ and $i * g$ is the only element satisfying these properties, as follows from (3.19).

The definition (3.19) could have been given without referring to $|g_2|^*$, since $g_1 \prec |g_2|^*$ means that $g_1 \prec |g_2|$ and $g_1 \neq |g_2|$. The properties **I1**, **I2** and **I4** of ω have been proved without any use of the properties 1–3 of ω_A, which are necessary only for the proof of **I3**.

To proceed further we need to consider a special class of semigroups. A semigroup $(G, \circ)$ is called a *left holoid* if it satisfies the following axioms:

LH1. *$(G, \circ)$ is left cancellative, i.e.* $gg_1 = gg_2 \longrightarrow g_1 = g_2$.

LH2. *For any two elements of $(G, \circ)$, one is a left divisor of the other, i.e. for every g_1, g_2 there exists g such that $g_1 g = g_2$ or $g_2 g = g_1$.*

LH3. *If two elements of $(G, \circ)$ are left divisors of each other, then the elements coincide, i.e.* $g_1 g_0 = g_2 \wedge g_2 g = g_1 \longrightarrow g_1 = g_2$.

If $g_1 g = g_2$, then, by **LH1**, g is uniquely defined by the elements g_1 and g_2. Let $g = g_2 \cdot g_1$. Then, as can easily be seen, the following formula holds:

$$(g_2 \cdot g_3)(g_1 \cdot g_2) = g_1 \cdot g_3. \tag{3.20}$$

Every left holoid contains an identity. In fact, let $ge_g = g$ (existence of e_g follows from **LH2**). Multiplying this equality by g_1 on the right and then cancelling g, we obtain $e_g g_1 = g_1$; that is, e_g is a left identity of the semigroup. In particular, $e_g e_{g_1} = e_{g_1}$. Analogously $e_{g_1} e_g = e_g$. Therefore, by **LH3**, $e_g = e_{g_1}$ and $g_1 e_g = g_1$; that is, e_g is a right identity of the semigroup as well.

If a left holoid is commutative, we obtain a semigroup which has been considered by Fritz Klein-Barmen [30], [31], who called it a linear holoid. In this connection cf. also [32], [33] and [2].

Right holoids may be defined dually. A left holoid need not be a right holoid, and it need not be right cancellative. An example will be considered below.

Let $(G, \circ)$ be a left holoid. Define a binary relation ω on G as follows: $(g_1, g_2) \in \omega$ iff $g_1 g = g_2$ for some $g \in G$. Then **LH2** means that $\omega \cup \overset{-1}{\omega} = G \times G$, and **LH3** means that $\omega \cap \overset{-1}{\omega} = \Delta_G$, i.e. ω is antisymmetric. Reflexivity of ω follows from **LH2** (existence of an identity). Transitivity of ω follows from (3.20). Therefore ω is a linear order relation on G; it will be called the *natural order relation* of the left holoid $(G, \circ)$. Instead of $(g_1, g_2) \in \omega$ we shall write $g_1 \prec g_2$.

Let A be a set. Define a partial binary operation on the set $A \times A$ as follows: the product $(a_1, a_2)(a_3, a_4)$ is defined iff $a_2 = a_3$, and it is then equal to (a_1, a_4). Then $A \times A$ is a Brandt groupoid (cf. [34], Chapter I, §6, Exercise 23); it is called the *simple Brandt groupoid over the set A*.

Let $(A, \circ)$ be a left holoid. Define a binary relation $\bar{\omega}$ on $A \times A$ as follows:

$$((a_1, a_2), (a_3, a_4)) \in \bar{\omega} \leftrightarrow a_3 \prec a_1 \wedge a_4 \prec a_2 \wedge a_1 \cdot a_3 = a_2 \cdot a_4. \tag{3.21}$$

It can be seen from the definition that $\bar{\omega}$ is reflexive.

Let $((a_1, a_2), (a_3, a_4)) \in \bar{\omega}$ and $((a_3, a_4), (a_5, a_6)) \in \bar{\omega}$. By (3.20) and (3.21),

$$a_3 \prec a_1 \wedge a_4 \prec a_2 \wedge a_5 \prec a_3 \wedge a_6 \prec a_4 \wedge a_1 \cdot a_5$$
$$= (a_3 \cdot a_5)(a_1 \cdot a_3) = (a_4 \cdot a_6)(a_2 \cdot a_4) = a_2 \cdot a_6.$$

It follows that $((a_1, a_2), (a_5, a_6)) \in \bar{\omega}$, i.e. $\bar{\omega}$ is transitive. Antisymmetry of $\bar{\omega}$ follows from antisymmetry of ω. Therefore, $\bar{\omega}$ is an order relation. Instead of $((a_1, a_2), (a_3, a_4)) \in \bar{\omega}$ we shall write $(a_1, a_2) \prec (a_3, a_4)$. This will not lead to confusion, since in this case *ordered pairs* of elements of A appear on both sides of the symbol $\prec$, while if $\prec$ is used instead of ω , elements of A appear on both sides of $\prec$.

We now show that $\bar{\omega}$ is an inductor on the Brandt groupoid $A \times A$.

Verification of **I1**. Let

$$(a_1, a_2) \prec (a_3, a_4) \wedge (a_2, a_5) \prec (a_4, a_6),$$

i.e.

$$a_3 \prec a_1 \wedge a_4 \prec a_2 \wedge a_1 \cdot a_3 = a_2 \cdot a_4 \wedge a_4 \prec a_2 \wedge a_6 \prec a_5 \wedge a_2 \cdot a_4 = a_5 \cdot a_6.$$

It follows that

$$a_3 \prec a_1 \wedge a_6 \prec a_5 \wedge a_1 \cdot a_3 = a_5 \cdot a_6,$$

i.e. $(a_1, a_5) \prec (a_3, a_6)$, q.e.d.

Verification of **I2**. This follows from $(a_1, a_2)^{-1} = (a_2, a_1)$ and (3.21).

Verification of **I3**. Let (a_1, a_1) and (a_2, a_2) be two idempotents of the groupoid $A \times A$. By (3.21),

$$(a_1, a_1) \prec (a_2, a_2) \leftrightarrow a_2 \prec a_1. \tag{3.22}$$

Therefore the order relation $\bar{\omega}$ induces an order relation isomorphic to $\overset{-1}{\omega}$ on the set of idempotents of $A \times A$, i.e. it induces a linear order. Therefore, a greatest minorant exists for any two idempotents of $A \times A$.

Verification of **I4**. Let $(a, a) \prec \iota_s((a_1, a_2))$. From the formula $\iota_s((a_1, a_2)) = (a_1, a_1)$ and (3.22) we obtain $a_1 \prec a$. Let $(a, a) * (a_1, a_2) = (a, a_2(a \cdot a_1))$. Then $\iota_s((a, a) * (a_1, a_2)) = (a, a)$ and $(a, a) * (a_1, a_2) \prec (a_1, a_2)$. If $(a, a_3) \prec (a_1, a_2)$, then $a_2 \prec a_3$ and $a \cdot a_1 = a_3 \cdot a_2$, whence $a_3 = a_2(a \cdot a_1)$. Thus $(a, a_3) = (a, a) * (a_1, a_2)$.

If an additive semigroup of nonnegative integers is taken in place of $(A, \circ)$ (such a semigroup is a left holoid) and the Brandt groupoid $A \times A$ is replenished up to an inverse semigroup in such a way that the inductor $\bar{\omega}$ is the canonical order relation of the inverse semigroup (this is possible, by Theorem 3.4), then the inverse semigroup is an elementary inverse semigroup of the type $\left[\begin{smallmatrix}1 & 1 & 1\\ 0 & 0 & 0\end{smallmatrix}\right]$ (see [35] and [36]). This inverse semigroup is considered in [6] and called *bicyclic*. It was considered by E. S. Ljapin in [37].(s,t)

LEMMA 3.5. *Let $(G, \circ)$ be a Croisot groupoid, G' a connected component of it and $(G', \circ')$ an induced Brandt groupoid on the component. If $(G', \circ', \omega')$ is an inductive Brandt groupoid whose set of idempotents possesses a greatest element relative*

(s) A left holoid A is isomorphic to the left unit subsemigroup [52] of the inverse semigroup which is the replenishment of the Brandt groupoid $A \times A$ with the inductor $\bar{\omega}$.

(t) (Added August, 1976). Suppose A is an ordinal number, which we identify with the set of all smaller ordinals. Then, by definition, A is closed under the addition of ordinals iff A is a prime component ([40], p. 282). In other words, $A = \omega^\alpha$ for some ordinal α ([40], Theorem 1 on p. 323]. So A, considered as an additive semigroup of ordinals, is a left holoid. The Brandt groupoid $A \times A$ can be replenished up to an inverse semigroup in such a way that the inductor $\bar{\omega}$ is the canonical order relation of the inverse semigroup. Thus we obtain precisely the α-bicyclic semigroups of Hogan [56] (see also [57]). A particular case of such semigroups when A is an in initial ordinal is considered by us in the proof of Lemma 3.6 of this paper.

to ω', then there exists an inductor ω on the Croisot groupoid, and the inductor induces ω' on G'.

PROOF. Let $(G', \circ', \omega')$ be an inductive Brandt groupoid satisfying the above properties, and let e be its greatest idempotent. Define ω as follows:

$$(g_1, g_2) \in \omega \leftrightarrow (g_1, g_2) \in \omega' \vee (g_1 \in I' \wedge g_2 \notin G') \vee g_1 = g_2,$$

where I' is the set of idempotents of $(G', \circ')$. Using the definition of ω and the fact that ω' is an inductor, we can verify easily that ω is an inductor. In particular, if i_1 and i_2 are idempotents which do not belong to G', then $i_1 \curlywedge i_2 = e$.

LEMMA 3.6. *On every Brandt groupoid which contains an infinite set of idempotents there exists an inductor such that the set of idempotents of the groupoid possesses a greatest ele ment.*

PROOF. Let $(G, \circ)$ be a Brandt groupoid containing an infinite set of idempotents. It is known that every Brandt groupoid can be decomposed into a direct product of the form $G_0 \times (A \times A)$, where G_0 is a group and $A \times A$ is a simple Brandt groupoid over a set A (cf. [38], where a proof for the finite case is given: the proof can be extended easily to the general case). The multiplication in $G_0 \times (A \times A)$ is given as follows: $(g_1, (a_1, a_2))(g_2, (a_3, a_4))$ is defined iff $a_2 = a_3$, and in this case is equal to $(g_1 g_2, (a_1, a_4))$. Decomposing our groupoid $(G, \circ)$ as above, we see that $A \times A$ must contain an infinite set of idempotents, i.e. the set A must be infinite. Let ω_α be the initial ordinal number corresponding to the cardinality of A (cf., for example, [39], p. 71, or [40], p. 389]). It is known ([39], p. 72) that the sum of any two ordinals which are smaller than ω_α is an ordinal smaller than ω_α, i.e. the set of all ordinals beginning with 0 and smaller than ω_α is closed under addition. A has the same cardinality as the set of all ordinals smaller than ω_α. We identity A with this set of ordinals. Then A is a left holoid relative to addition of ordinals ([40], Chapter 14). This holoid is not right cancellative if A is infinite. In fact, $0 + \omega = 1 + \omega$, where ω is the ordinal number of the set of nonnegative integers ordered in the usual way.

Therefore, as we have seen above, there exists an inductor $\bar{\omega}$ on the Brandt groupoid $A \times A$, and, as is seen from the definition of $\bar{\omega}$, $(0, 0)$ is the largest idempotent of $A \times A$.

Introduce a binary relation ω on $G_0 \times (A \times A)$ as follows:

$$((g_1, (a_1, a_2)), (g_2, (a_3, a_4))) \in \omega \leftrightarrow g_1 = g_2 \wedge ((a_1, a_2), (a_3, a_4)) \in \bar{\omega}.$$

Clearly ω is an inductor, and $(e, (0, 0))$, where e is the identity of the group G_0, is the largest idempotent of $G_0 \times (A \times A)$. Lemma 3.6 is proved.

THEOREM 3.7. *A Croisot groupoid cannot be replenished up to an inverse semigroup iff the set of its idempotents is finite and no idempotent is isolated.*

PROOF. Suppose an inverse semigroup $(G, \circ)$ contains only a finite number of idempotents, let i_0 be the product of all its idempotents, and let $(G, \circ_a)$ be a Croisot groupoid associated with $(G, \circ)$. By A.E. Liber's result [4], i_0 is an isolated idempotent of $(G, \circ_a)$. Therefore, a Croisot groupoid having a finite number of idempotents and no isolated idempotents is not replenishable.

Let $(G, \circ)$ be a Croisot groupoid containing an isolated idempotent i. Then the structure of a Brandt groupoid $(G_i, \circ_i)$ is induced on the connected component G_i containing i, which Brandt groupoid is a group, since G_i contains a single idempotent. The canonical order relation Δ_{G_i} of the group $(G_i, \circ_i)$ is its inductor, and i is the greatest idempotent of $(G_i, \circ_i)$. By Lemma 3.5, there exists an inductor on the Croisot groupoid $(G, \circ)$. By Theorem 3.4, $(G, \circ)$ is replenishable.

Let $(G, \circ)$ be a Croisot groupoid and G' a connected component containing an infinite set of idempotents. The structure of a Brandt groupoid $(G', \circ')$ is induced, and this groupoid contains an infinite set of idempotents. By Lemmata 3.6, 3.5 and Theorem 3.4, $(G, \circ)$ is replenishable.

Let $(G, \circ)$ be a Croisot groupoid containing an infinite set of idempotents, but in which each connected component contains only finitely many idempotents. Then the set G/π of all connected components of the groupoid is infinite. Let A be a set of idempotents selected one from each connected component of $(G, \circ)$, and let ω_A be an order relation on A such that $(A, \overset{-1}{\omega}_A)$ is a well-ordered set. Then the ordered set (A, ω_A) satisfies the above conditions 1–3, and therefore, as we have seen, there exists an inductor on $(G, \circ)$. By Theorem 3.4, $(G, \circ)$ is replenishable.

Thus a Croisot groupoid is replenishable if its set of idempotents is infinite or if it contains an isolated idempotent.

Theorem 3.7 is proved. We have seen in the proof that if a Croisot groupoid is replenishable, it can be replenished up to an inverse semigroup with a fairly simple canonical order relation.

COROLLARY. *Every Brandt groupoid containing an infinite set of idempotents can be replenished up to an inverse semigroup containing an identity whose canonical order relation induces an order of type ω_α^* on the set of all idempotents, where ω_α is an initial ordinal and ω_α^* is the ordinal type inverse to that of ω_α. A Brandt groupoid which contains only a finite number of idempotents and which is not a group cannot be replenished up to an inverse semigroup.*

It should be noted that every Croisot groupoid can be replenished up to a semigroup (by defining its operation where it has not already been defined, but not adjoining any new elements to it). In fact, let $G_0 \times (A \times A)$ be a Brandt groupoid (G_0 a group and $A \times A$ the simple Brandt groupoid over a set A). As we noted above, every Brandt groupoid is isomorphic to such a groupoid. Define

$$(g_1, (a_1, a_2))(g_2, (a_3, a_4)) = (g_1 g_2, (a_1, a_4)).$$

Then we obtain a semigroup. It is easily seen that this operation coincides with the operation of $G_0 \times (A \times A)$ whenever the latter is defined.

Let $(G, \circ)$ be a Croisot groupoid. Replenish each of its connected components up to a semigroup in the way just described. Let $g_1, g_2 \in G$. If g_1 and g_2 belong to different components, define $g_1 g_2 = g_1$. As a result we obtain a semigroup whose operation is an extension of the operation of $(G, \circ)$.

If a Croisot groupoid is a group or consists of just two connected components, one of which is a singleton and the other contains only a finite number of idempotents, then the Croisot groupoid can be replenished up to an inverse semigroup in a unique way (with the single exception of the Croisot groupoid consisting of just two elements, both of which are idempotents; this groupoid can be replenished in two ways: one of the elements is defined to be a zero and the other one an identity; but the inverse semigroups obtained after the groupoid is replenished are isomorphic).

Every Croisot groupoid which does not belong to the two above types can be replenished in two ways leading to nonisomorphic inverse semigroups. The rather long proof of this statement is omitted. It is based on the following lemmata:

1. *Every set containing more than two elements can be ordered in two nonisomorphic ways so that the order relations satisfy the conditions* 1–3 *given on p.* 116.

2. *On every infinite set two left holoid structures can be given in such a way that their order relations are not isomorphic.*

It follows from Theorem 3.4 that an inverse semigroup is completely determined by its abutting multiplication and canonical order relation. In Theorem 3.7 necessary and sufficient conditions were found in order that a partial binary operation be the operation of abutting multiplication of an inverse semigroup. It would be interesting to find necessary and sufficient conditions in order that a given order relation ω be the canonical order relation of an inverse semigroup. At present this problem is far from being solved. A necessary condition is known: $\overset{-1}{\omega} \circ \omega \subset \omega \circ \overset{-1}{\omega}$ [10].(u) It is not known whether the condition is sufficient.

BIBLIOGRAPHY

1. Gabriel Thierrin, *Sur les éléments inversifs et les éléments unitaires d'un demi-groupe inversif*, C. R. Acad. Sci. Paris **234** (1952), 33–34.

2. E. S. Ljapin, *Semigroups*, Fizmatgiz, Moscow, 1960; English transl., Amer. Math. Soc., Providence, R. I., 1963, 1968, 1972.

3. V. V. Vagner, *Generalized groups*, Dokl. Akad. Nauk SSSR **84** (1952), 1119–1122. (Russian)

4. A. E. Liber, *On the theory of generalized groups*, Dokl. Akad. Nauk SSSR **97** (1954), 25–28. (Russian)

5. G. B. Preston, *Inverse semi-groups*, J. London Math. Soc. **29** (1954), 396–403.

6. A. H. Clifford and G. B. Preston, *The algebraic theory of semigroups*. Vol. I, Amer. Math. Soc., Providence, R. I., 1961.

7. W. D. Munn and R. Penrose, *A note on inverse semigroups*, Proc. Cambridge Philos. Soc. **51** (1955), 396–399.

8. H. N. Inasaridze, *On certain questions in the theory of semigroups*, Tbiliss. Gos. Univ. Trudy **76** Ser. Fiz.-Mat. Nauk **1** (1959), 247–260. (Russian)

9. B. M. Šaĭn, *Atomic semigrouds and involutory semigroups*, Izv. Vysš. Učebn. Zaved. Matematika **1965**, no. 3 (46), 172–184. (Russian)

10. V. V. Vagner, *The theory of generalized grouds and generalized groups*, Mat. Sb. **32** (74) (1953), 545–632. (Russian)

11. A. I. Mal'cev, *On the general theory of algebraic systems*, Mat. Sb. **35** (77) (1954), 3–20; English transl. in Amer. Math. Soc. Transl. (2) **27** (1963).

12. Graham Higman and B. H. Neumann, *Groups as groupoids with one law*, Publ. Math. Debrecen **2** (1952), 215–221.

13. Ju. I. Sorkin, *Rings as sets with one operation subject to a unique identity*, Uspehi Mat. Nauk **12** (1957), no. 4 (76), 357–362. (Russian)

(u) I.e. if $x \prec z$ and $y \prec z$, then there exists u such that $u \prec x$ and $u \prec y$.

14. Garrett Birkhoff, *Lattice theory*, 2nd rev. ed., Amer. Math. Soc., Providence, R. I., 1948.

15. Robert Croisot, *Une interprétation des relations d'équivalence dans un ensemble*, C. R. Acad. Sci. Paris **226** (1948), 616–617.

16. Charles Ehresmann, *Gattungen von lokalen Strukturen*, Jber. Deutsch. Math. Verein. **60** (1957), Abt. 1, 49–77.

17. ———, *Catégories inductives et pseudo-groupes*, Ann. Inst. Fourier (Grenoble) **10** (1960), 307–332.

18. Heinrich Brandt, *Über eine Verallgemeinerung des Gruppenbegriffs*, Math. Ann. **96** (1926) 360–366.

19. G. B. Preston, *Congruences on Brandt semigroups*, Math. Ann. **139** (1959), 91–94; cor-correction, **143** (1961), 456.

20. B. M. Šaĭn, *On the theory of generalized groups*, Dokl. Akad. Nauk SSSR **153** (1963), 296–299; English transl. in Soviet Math. Dokl. **4** (1963).

21. V. V. Vagner, *The theory of relations and the algebra of partial mappings*, Theory of Semigroups and Its Applications, Vyp. 1, Izdat. Saratov. Univ., Saratov, 1965, pp. 3–178. (Russian)

22. B. M. Šaĭn, *Embedding a semigroup in a generalized group*, Uspehi Mat. Nauk **16** (1961), no. 2 (98), 218–219. (Russian)

23. ———, *Embedding of semigroups in generalized groups*, Mat. Sb. **55** (**97**) (1961), 379–400. (Russian)

24. ———, *The abstract theory of semigroups of one-to-one transformations*, Candidate's Dissertation, Saratov State Univ., Saratov, 1962. (Russian)

25. Jacques Riguet, *Relations binaires, fermetures, correspondances de Galois*, Bull. Soc. Math. France **76** (1948), 114–155.

26. A. H. Clifford, *Bands of semigroups*, Proc. Amer. Math. Soc. **5** (1954), 499–504.

27. David McLean, *Idempotent semigroups*, Amer. Math. Monthly **61** (1954), 110–113.

28. M.-L. Dubreil-Jacotin, L. Lesieur and R. Croisot, *Leçons sur la théorie des treillis des structures algébriques et des treillis géométriques*, Gauthier-Villars, Paris, 1953.

29. V. V. Vagner, *Semigroups associated with generalized grouds*, Mat. Sb. **52** (**94**) (1960), 597–628; English transl. in Amer. Math. Soc. Transl. (2) **36** (1964).

30. Fritz Klein-Barmen, *Über gewisse Halbverbände und kommutative Semigruppen*. II, Math. Z. **48** (1942/43), 715–734.

31. ———, *Ein Beitrag zur Theorie der linearen Holoide*, Math. Z. **51** (1948), 355–366.

32. A. H. Clifford, *Naturally totally ordered commutative semigroups*, Amer. J. Math. **76** (1954), 631–646.

33. ———, *Totally ordered commutative semigroups*, Bull. Amer. Math. Soc. **64** (1958), 305–316.

34. N. Bourbaki, *Algèbre*, Chap. 1, Actualités Sci. Indust. no. 934, Hermann, Paris, 1942; rev. English transl., Hermann, Paris, Addison-Wesley, Reading, Mass., 1974.

35. L. M. Gluskin, *Elementary generalized groups*, Mat. Sb. **41** (**83**) (1957), 23–36. (Russian) (Russian)

36. ———, *On inverse semigroups*, Zap. Meh.-Mat. Fak. i Har'kov. Mat. Obšč. (4) **28** (1961), 103–110. (Russian)

37. E. S. Ljapin, *A canonical form of the elements of a certain associative system given by defining relations*, Leningrad. Gos. Ped. Inst. Učen. Zap. **89** (1953), 45–54. (Russian)

38. A. K. Suškevič, *The theory of generalized groups*, Gos. Naučno-Tehn. Izdat. Ukraine, Kharkov, 1937. (Russian)

39. Felix Hausdorff, *Mengenlehre*, 3rd ed., De Gruyter, Berlin, 1937; English transl., Chelsea, New York, 1957, 1962.

40. Wacław Sierpiński, *Cardinal and ordinal numbers*, PWN, Warsaw, 1958.

41. L. Dubikajtis, *Groupoïde inductif élémentaire et pseudogroup élémentaire*, Bull. Acad. Polon. Sci. Sér. Sci. Math. Astronom. Phys. **11** (1963), 469–471.

42. A. A. Vinogradov, *Remarks on the theory of partially ordered groups and semigroups*, Algebra i Logika 1 (1962), no. 2, 22–29. (Russian)

43. Jacques Riguet, *Fondements de la théorie de relations binaires*, Doctoral Dissertation, Univ. of Paris, Paris, 1951; abstract published in Ann. Univ. Paris **22** (1953), 132–134.

44. Charles Ehresmann, *Catégories et structures*, Dunod, Paris, 1965.

45. A. A. Markov, *The theory of algorithms*, Trudy Mat. Inst. Steklov. **42** (1954); English transl., Israel Program for Scientific Translations, Jerusalem, 1961.

46. George Grätzer, *Universal algebra*, Van Nostrand, Princeton, N. J., 1968.

47. B. M. Schein [Šaĭn], *Positive axioms for generalized groups*, Rev. Roumaine Math. Pures Appl. **10** (1965), 589–590.

48. N. R. Reilly and H. E. Scheiblich, *Congruences on regular semigroups*, Pacific J. Math. **23** (1967), 349–360.

49. T. E. Hall, *On regular semigroups whose idempotents form a subsemigroup*, Bull. Austral. Math. Soc. **1** (1969), 195–208; addendum, **3** (1970), 287–288.

50. B. M. Šaĭn, *A system of axioms for semigroups embeddable in generalized groups*, Dokl. Akad. Nauk SSSR **134** (1960), 1030–1033; English transl. in Soviet Math. Dokl. **1** (1960).

51. Heinz Prüfer, *Theorie der Abelschen Gruppen*. I, Math. Z. **20** (1924), 165–187.

52. A. H. Clifford, *A class of d-simple semigroups*, Amer. J. Math. **75** (1953), 547–556.

53. H. N. Inasaridze, *On the theory of generalized grouds*, Tbiliss. Gos. Univ. Trudy Ser. Mat. Nauk **102** (1964), 207–210. (Russian)

54. B. M. Šaĭn, *Representation of generalized grouds*, Izv. Vysš. Učebn. Zaved. Matematika **1961**, no. 6 (25), 142–154. (Russian)

55. V. V. Vagner, *The foundations of differential geometry and modern algebra*, Proc. Fourth All-Union Math. Congr. (Leningrad, 1961), vol. I, Izdat. Akad. Nauk SSSR, Leningrad, 1963, pp. 17–29. (Russian)

56. J. W. Hogan, *Bisimple semigroups with idempotents well-ordered*, Semigroup Forum **6** (1973), 298–316.

57. G. L. White, *The dual ordinal of a bisimple inverse semigroup*, Semigroup Forum **6** (1973), 295–297.

Note added in Proof (of the Russian original). After this paper had been submitted to the publishers we learned of the paper [42], where left holoids are introduced as well. Also the note [41] has been published; in it Croisot groupoids (called "elementary inductive groupoids") are considered.

Translated by the author

Amer. Math. Soc. Transl.
(2) Vol. 113, 1979

LECTURES ON SEMIGROUPS OF TRANSFORMATIONS*

B. M. SCHEIN [ŠAĬN]

TABLE OF CONTENTS

Introduction

As is well known, the theory of semigroups is an abstract study of transformations, and semigroups of transformations comprise the central part of this theory. Due to limitations imposed by the size of this work, the author was not able to encompass the subject with any considerable degree of generality, and he makes no claim to have achieved this.

This lectures provide an introduction to the range of ideas and methods, and to the problems of one facet of the theory of semigroups of transformations. They are connected with the classical study of logic (especially algebraic logic) and the theory of relations, which evolved basically during the sixties and led to the solution of a number of important and long-standing problems of the theory.

The author tried not so much to expound a possibly larger number of separate results as to acquaint the reader with new viewpoints, ways of setting the problems, and the existing methods for their solution. Since it turned out that hitherto the most fruitful method permitting one to approach a series of problems on semigroups of transformations was the method of determinative pairs, particular attention is given to this approach. To aid the reader in actively mastering this method we find by means of determinative pairs a detailed solution to a few problems. In fact the solution of one of the problems uses the rudiments of the method of vector-complexes

AMS (MOS) subject classifications (1970). Primary 20M20, 08A05, 20M30; Secondary 20A50.

*Translation of a volume of lecture notes, Izdat. Saratov. Univ., Saratov, 1970; MR 53 #681.

in addition to that of determinative pairs. These solutions can serve both as an illustration of the results and a reflection of the spirit of the entire direction discussed in the lectures.

Due to the relative youth of the theory of semigroups, this subject is still insufficiently known outside the circle of specialists. While every mathematician is familiar with the foundations of the kindred theories of groups and rings, the foundations of semigroups do not always enter into curricula of mathematics departments of universities. Regrettably, time and again one still has to encounter wrong impressions of semigroups. Therefore, besides the basic purpose, which is to serve to some extent as an introduction to the subject for a young algebraist who wants to study the theory of semigroups of transformations and try his or her strength in it, the lectures can also aid a reader who specializes in another branch of mathematics and who is interested in how things stand in nonclassical domains of algebra to become acquainted with some of the ideas of the theory of semigroups.

The text represents material developed and expanded by the author from lectures delivered to professors and students at the Vinnica Pedagogical Institute in April, 1966. The author used the script based on the tape-record compiled by the instructors of the institute: V. S. Garvac′kiĭ, V. T. Kulik, S. Leĭferman, and V. S. Trohimenko, under the guidance of Associate Professor P. M. Oloničev. The author wishes to extend his sincere thanks to all of these persons.

**Note added in translation.* In this translation some additions have been inserted; they are marked by asterisks. A few items ([30]–[34]) have been added to the bibliography.

§§10 and 11 were written by the author in English for this translation; they are absent from the Russian original and are based on results obtained after the Russian original was published.

The author thanks Professor D. E. Zitarelli, who translated these "Lectures" as well as some of his other papers, and also the American Mathematical Society for including these "Lectures" in their Selected Translations series.

§1. Algebraic systems

An *algebraic system* is an ordered triple $\mathfrak{A} = (A, (o_i)_{i\in I}, (\rho_j)_{j\in J})$, where A is a nonempty set, each component of the family $(o_i)_{i\in I}$ is an m_i-ary operation on A, and each component of the family $(\rho_j)_{j\in J}$ is an n_j-ary relation on A. The ordered pair $((m_i)_{i\in I}, (n_j)_{j\in J})$ is called the *signature* of $\mathfrak{A}$. Two systems with the same signature are said to be *similar.* In what follows, when considering classes of algebraic systems we will regard any two systems in the same class as being similar.

If $J = \varnothing$ (i.e., only operations are given on A), $\mathfrak{A}$ is called an *algebra*; if $I = \varnothing$ (i.e., only relations are given on A) it is called a *polyrelative* (or simply a relative). If $I = J = \varnothing$, we obtain arbitrary nonempty sets.

A class of algebraic systems is said to be abstract if it contains all systems which are isomorphic to any system in it. Note that the class of semigroups of transformations is not abstract, since it is not closed under isomorphisms.

Every class of algebraic systems K can be enlarged to an abstract class $I(K)$ consisting of the isomorphic images of all systems in K.

A property of an algebraic system $\mathfrak{A}$ is called ***abstract*** if it holds for every system isomorphic to $\mathfrak{A}$ (i.e., if the class of all algebraic systems having the given property is abstract). Since isomorphic algebraic systems vary only as to the nature of their elements, abstract properties are those which do not depend on the concrete nature of the elements of a given algebraic system. Modern algebra is primarily concerned with abstract properties of algebraic systems, i.e., as a rule isomorphic systems are not distinguished and investigations are carried out "up to isomorphism". This principle (inherent not only in algebra but in all of mathematics) provides for vast generality and strength of the results obtained, in that all abstract properties of any concrete algebraic system are properties of an infinite number of systems isomorphic to it.

In order to construct algebraic theories it is necessary to make use of a definite language as a vehicle for presenting the reasoning and notation of the results obtained. In principle it is possible to confine oneself to the usual spoken language, but the inconvenience this causes is well known. The language for expressing algebraic results must permit one to write statements about the properties of these systems, and the form of the notation must be suitable for tracing easily the logical dependence of different statements. In addition, the language must be as economical as possible, containing nothing "superfluous". For example, it is desirable that only the abstract properties of the systems given might be written in such a language, since this is precisely what interests us. All of these demands are satisfied by the existing varieties of the predicate calculus or the theory of types languages. The simplest of these languages is the first order language, the so-called language of the lower predicate calculus with equality (LPC). The expressive possibilities of this language are limited, for only elementary properties of an algebraic system can be written in it, i.e., only properties of the elements of the system with respect to its operations and relations. Nevertheless, in a large number (maybe a majority) of cases, this language turns out to be quite sufficient.

Let $((m_i)_{i\in I}, (n_j)_{j\in J})$ be a given signature. An alphabet of the LPC language of a given signature consists of an infinite collection of letters $x_1, x_2, \ldots$, called individual variables (sometimes we will use the letters x, y, z); for each $i \in I$, there is given a letter O_i which is called a functional constant; for each $j \in J$, a letter P_j is given which is called a predicate constant. The letters $\neg$, $\wedge$, $\vee$, $\rightarrow$, $\leftrightarrow$, $\bigwedge$ and $\bigvee$, called negation, conjunction, disjunction, implication, (logical) equivalence, universal quantifier, and existential quantifier, respectively, are also in the alphabet. Also the letter $=$, called equality, is in the alphabet, as are the technical symbols: parentheses and commas. Usually we are limited by these letters. However, our alphabet will sometimes contain an infinite number of letters $\varphi_1, \varphi_2, \ldots$, called binary predicate variables (sometimes the letters φ, ψ will be used).

A finite sequence of letters in the alphabet is called a *word*. Among the words we single out some special ones, called *terms* (or *polynomials*), *atomic formulas* and *formulas*.

The aim of this lecture is to fix our notation, emphasize some highlights, and recall some results; it is not to present the various chapters of mathematical logic. We assume that the reader is familiar with the fundamentals of mathematical logic, in particular, with the definitions of terms and formulas, with the usual conventions regarding parentheses in formulas (by convention, many of the parentheses are generally omitted), and with the free or bound occurrences of variables in formulas. A formula not containing any free occurrences of individual variables is called *closed* (such formulas are sometimes called axioms). A set of closed formulas, called *LPC axioms*, is selected. We assume that the reader is acquainted with the different systems of axioms for LPC. If α and $(\alpha) \longrightarrow (\beta)$ are closed formulas, we say that formula β is obtained from these formulas by *modus ponens.* If α is a formula with a free variable x, then we say that the formula $(\bigwedge x)[\alpha]$ is obtained from α by *generalization.* By a (formal) *deduction* of a closed formula α in a set of closed formulas Φ we mean a finite sequence of formulas $\alpha_1, \alpha_2, \ldots, \alpha$ in which each one is either an LPC axiom, or belongs to Φ, or can be obtained from the two preceding formulas by modus ponens, or can be obtained from a preceding formula by generalization (in the case of generalization the variable x which is to be bound by the universal quantifier is not supposed to occur in Φ). In this case α is called (syntactically) deducible from Φ, denoted by $\Phi \vdash \alpha$. If $\Psi \vdash \alpha$ implies that $\alpha \in \Psi$ (i.e., all formulas which can be deduced from Ψ are contained in Ψ), then Ψ is called an (elementary) *theory*. In particular, the set $T(\Phi)$ of all closed formulas which can be deduced from Φ is a theory. In this case Φ is called a *system of axioms* for the theory $T(\Phi)$.

The rules for constructing terms, formulas, and rules of deduction are parts of the syntax of the LPC language. Now we consider questions of semantics in our language, i.e., questions of imparting meaning to expressions in our language (recall that until now formulas have been considered as being meaningless chains of meaningless letters).

An algebraic system $\mathfrak{A} = (A, (o_i)_{i \in I}, (\rho_j)_{j \in J})$ of the given signature is called a *model* of the language. Individual variables are interpreted as variables over the set A, the functional constants O_i as symbols for the operations o_i, and the predicate constants P_j as symbols for the relations ρ_j. We will interpret logical connectives, quantifiers, parentheses, and the equality symbol in the usual way. When predicate variables are in the alphabet of the language, then by a model of the language we will mean a pair $(\mathfrak{A}, \Phi)$ consisting of an algebraic system $\mathfrak{A}$ and a nonempty set Φ of binary relations on A. In such a case the predicate variables are interpreted as variables over the set Φ. Formulas of the form $P_j(t_1, \ldots, t_{n_j})$ and $\varphi_i(t_1, t_2)$ mean that $(t_1, \ldots, t_{nj}) \in \rho_j$ and $(t_1, t_2) \in \varphi_i$, respectively (here the t_k are terms).

Given a model of the language, every closed formula not containing free predicate variables becomes a proposition which may be either true or false. For example, if the language contains but one functional constant O_1, $m_1 = 2$, and there are no predicate constants, then the models of the language will be binary operatives, i.e., algebras of the form $\mathfrak{A} = (A, o_1)$, where o_1 is a binary operation on A. The closed formula

$$\alpha: (\bigwedge x_1, x_2, x_3)[O_1(O_1(x_1, x_2), x_3) = O_1(x_1, O_1(x_2, x_3))]$$

turns by the interpretation into the proposition that the operation o_1 is associative. This proposition is true precisely when $\mathfrak{A}$ is a semigroup. Connected with this is the set $T(\{\alpha\})$, called the (elementary) *theory of semigroups,* and the formula α constitutes a system of axioms for the theory of semigroups.

The LPC axioms are always converted into true propositions under an interpretation. If all of the formulas in a set of closed formulas Φ are valid in a model of $\mathfrak{A}$, then we will say that $\mathfrak{A}$ is a *model of* Φ. The class of all models of Φ will be denoted by $\mathfrak{M}(\Phi)$. For example, $\mathfrak{M}(\{\alpha\})$ is the class of all semigroups. Φ is called a *system of axioms* for the class $\mathfrak{M}(\Phi)$.

A class K of algebraic systems is called *axiomatizable* if there exists a system of axioms for it, i.e., $K = \mathfrak{M}(\Phi)$ for some Φ. Of course an axiomatizable class can have different systems of axioms, so the problem of finding possibly simple systems of axioms is of interest.

Let K be a class of algebraic systems. If we are interested only in the abstract properties of systems in K, then naturally we consider the abstract closure $I(K)$ of K instead of K. An *abstract characterization* of K is a system of axioms for the class $I(K)$. If $I(K)$ is axiomatizable, we say that K admits an elementary abstract characterization. This case deserves particular attention.

Indeed, if we want to characterize systems in K "up to isomorphism", i.e., find conditions which characterize precisely those algebraic systems which are in $I(K)$, then such conditions often can be written in the language of the theory of types. It is always desirable to determine whether there are simpler nontrivial conditions. Indeed, the class $I(K)$ always admits the trivial characterization: "$\mathfrak{A} \in I(K)$ if and only if $\mathfrak{A}$ is isomorphic to a system in K"; since K is given (its systems are being described in a certain way), the above phrase does indeed characterize $I(K)$. It is clear that since the LPC language, being a first order language, is the simplest of the predicate calculus languages, an elementary abstract characterization, if it exists, has a definite value.

Why is finding abstract characterizations necessary?

We proffer some reasons here (others will be mentioned later).

1) An abstract characterization permits one to visualize all systems in the class $I(K)$ in a simple way. The conditions in an abstract characterization are properties of a system in K, and they form a complete system of properties in the sense that every algebraic system having these properties is isomorphic to a system in K.

2) Any property which can be formally deduced from an abstract characterization of K remains valid for each system in K. The importance of discovering an abstract characterization which is elementary follows from the completeness theorem for LPC: every closed LPC formula which holds for every algebraic system in some axiomatizable class can be derived formally from the axioms of this class. Thus if K admits an elementary abstract characterization, then all elementary properties of systems in K can be derived from this abstract characterization, i.e., it is not necessary to consider concrete systems in K for their proof.

3) The fact that a class of algebraic systems is axiomatizable imposes certain limitations on the class. Therefore, knowing that a class K can be characterized by a system of axioms of a particular form, and using results from model theory, one can draw certain conclusions about properties of K. As is well known, any formula is equivalent to a formula in which all quantifiers are placed at the beginning; that is, to a formula in prenex form. If a formula in prenex form contains no existential quantifiers, it is called ***universal.*** A class of algebraic systems is called ***universal*** if it possesses a system of axioms consisting of universal formulas. Among all axiomatizable classes, universal classes are characterized by the property that they are hereditary (or closed under subsystems), i.e., that they contain, together with any system, all of its subsystems. When writing a universal axiom we usually omit the quantifiers in its prefix; as an example, the quantifiers are usually omitted when describing associativity: $(x_1x_2)x_3 = x_1(x_2x_3)$ (where x_1x_2 denotes the term $O_1(x_1x_2)$). The class of semigroups and the class of ordered sets serve as examples of universal classes. The class of groups (considered as an algebra with one binary operation) is not universal.

Universal classes and various specializations of this concept have been studied extensively. Therefore it is usually of interest to determine systems of universal axioms for certain classes of algebraic systems.

The reader should consult any text on mathematical logic to become acquainted with LPC language (cf., for example, [4] or [16]). For the basic concepts of model theory, see [10] or [12] (cf. also the corresponding chapters in [5] and [8]).[1]

§2. Relation algebras

Let K be a fixed class of algebraic systems (for example, the class of all nonempty sets). We will consider binary relations on an arbitrary system $\mathfrak{A} \in K$. We are only interested here in those binary relations which satisfy a given sets of conditions M (either finite or infinite). Thus we might consider only single-valued binary relations (i.e., partial transformations), reflexive relations, endomorphisms of $\mathfrak{A}$, or any pair of relations must commute under multiplication, etc. What is important for us is that each condition in M be written in terms of LPC formulas with free binary predicate variables. For example, the condition that a relation is single-valued can be expressed by the formula

$$(\bigwedge x, y, z)[\varphi(x, y) \wedge \varphi(x, z) \longrightarrow y = z]. \tag{1}$$

If K is a class of semigroups and O_1 is a functional constant designating an operation on a semigroup, then the condition "is an endomorphism of" can be written by means of three formulas: formula (1) and the following two conditions:

---

(1) The author mentioned the monographs which were available in Russian translation. The reader may substitute for them any other texts on mathematical logic or model theory.

$$(\bigwedge x)(\bigvee y)[\varphi(x, y)], \tag{2}$$

$$(\bigwedge x, y, z, w)[\varphi(x, y) \wedge \varphi(z, w) \longrightarrow \varphi(O_1(x, z), O_1(y, w))] \tag{3}$$

(condition (2) means that the transformation is everywhere defined, i.e., is defined on the entire semigroup).

Let O_r and P_s be LPC formulas defining operations and relations, respectively, on binary relations. The ordered pair $\sigma = ((O_r)_{r\in R}, (P_s)_{s\in S})$ is called a *genre.* Let Φ be a nonempty set of binary relations on a system in the class K, where relations in Φ satisfy all conditions in M, and are stable (i.e., closed) with respect to the operations defined by the formulas O_r in some fixed genre σ. Each formula O_r defines an operation o_r on Φ, and each formula P_s defines a relation ρ_s on Φ. The algebraic system $F = (\Phi, (o_r)_{r\in R}, (\rho_s)_{s\in S})$ so obtained is called a *concrete algebra of (binary) M-relations of the genre σ over K.* The class of all concrete relation algebras is denoted by $\mathfrak{R}(K, M, \sigma)$, or simply $\mathfrak{R}(\sigma)$. An *abstract relation algebra* is an algebraic system isomorphic to a concrete relation algebra. The class of all abstract relation algebras of a given genre σ over a given class K and with a given set of formulas M is denoted by $R(K, M, \sigma)$ on simply $R(\sigma)$. Clearly $R(\sigma) = I(\mathfrak{R}(\sigma))$. One can introduce n-ary relation algebras in exactly the same way.

Let us consider some examples.

1. Let σ consist of the formula

$$(\bigvee z)[\varphi(x, z) \wedge \psi(z, y)], \tag{4}$$

defining the operation of the usual multiplication of binary relations: $\psi \circ \varphi = \{(x, y) \mid O(x, y)\}$, where $O(x, y)$ denotes formula (4) (i.e., $(x, y) \in \psi \circ \varphi$ if and only if the elements x and y satisfy (4)). Let K be a class of sets.

a) If M consists only of formula (1), which defines single-valued binary relations, then $\mathfrak{R}(\sigma)$ is the class of all semigroups of single-valued partial transformations (in that which follows we shall call "partial transformations" simply "transformations").

b) If M consists of formula (1) and

$$(\bigwedge x, y, z)[\varphi(x, y) \wedge \varphi(z, y) \longrightarrow x = z], \tag{5}$$

which define one-to-one relations, then $\mathfrak{R}(\sigma)$ is the class of all semigroups of invertible (i.e., one-to-one) transformations.

c) If M consists of (1), (2), (5), and $(\bigwedge x)(\bigvee y)[\varphi(y, x)]$, which define permutations on a set, then $\mathfrak{R}(\sigma)$ is the class of all semigroups of permutations.

2. Let $\sigma = (O_1, O_2)$, where O_1 defines the binary operation of multiplication of binary relations ($\circ$) and O_2 defines the unary operation of inversion of binary relations ($^{-1}$). Consider the class $\mathfrak{R}(\sigma)$ of concrete algebras $(\Phi, \circ, {}^{-1})$ of the genre $\sigma = (O_1, O_2)$.

a) If M consists of the formulas defining one-to-one binary relations, then $\Re(\sigma)$ represents the class of all inverse semigroups of invertible transformations.

b) If M consists of the formula defining permutations, then $\Re(\sigma)$ is the class of permutation groups.

c) If K is a class of algebraic systems and M consists of both the formulas defining permutations and the condition "is an automorphism of", then $\Re(\sigma)$ is the class of groups of (not necessarily all) automorphisms of systems in K.

3. Let K be the class of all abelian groups, let M consist of the formulas defining endomorphisms, and let $\sigma = (O_1, O_2, O_3)$, where O_1 defines multiplication of binary relations ($\circ$), and O_2 and O_3 define the operations $+$ and $-$, of addition and subtraction of endomorphisms, respectively:

$$(\bigwedge a)\,[(\varphi \pm \psi)(a) = \varphi(a) \pm \psi(a)]\,,$$

which is shorthand for an obvious formula in LPC. E. g., $\varphi(a)$ denotes the b such that $(a, b) \in \varphi$, i.e. $\varphi(a, b)$. By M, such a b is unique and always exists. Consider the class $\Re(\sigma)$ of concrete algebras $(\Phi, \circ, +, -)$ of a given genre. Then $\Re(\sigma)$ is a class of rings of endomorphisms (not necessarily all endomorphisms) of abelian groups.

4. Let $\sigma = (P)$, where P defines the binary relation of inclusion ($\subset$) of binary relations. Consider the class $\Re(\sigma)$ of concrete algebras $(\Phi, \subset)$ of the genre $\sigma = (P)$. If M is empty, then $\Re(\sigma)$ is the class of all ordered sets of binary relations.

5. Let $\sigma = (O, P)$, where O defines the binary operation of multiplication ($\circ$) of binary relations and P is as in 4. Consider the class $\Re(\sigma)$ of concrete algebras $(\Phi, \circ, \subset)$ of the given genre (O, P). If M is empty, then $\Re(\sigma)$ is the class of all inclusion-ordered semigroups of binary relations. If we take different sets for M, we obtain different classes of inclusion-ordered semigroups of concrete transformations.

Let us consider the classes of abstract relation algebras which are associated with the foregoing examples.

1.a) $R(\sigma)$ is the class of semigroups which are isomorphic to semigroups of single-valued transformations, i.e., the class of all semigroups.

b) $R(\sigma)$ is the class of all semigroups which are isomorphic to semigroups of one-to-one transformations. This abstract class coincides with the class of all semigroups which can be imbedded in inverse semigroups. A system of axioms for this class is given in [17].

c) $R(\sigma)$ is the class of all semigroups which can be imbedded in groups. A system of axioms for this class can be found in [11].

2.a) $R(\sigma)$ is the class of all inverse semigroups. A system of axioms for inverse semigroups considered as algebras with two operations is given in [20].

b) $R(\sigma)$ is the class of all groups, which follows from Cayley's theorem on representations of groups by permutations.

c) $R(\sigma)$ is the class of groups which are isomorphic to the group of (not necessarily all) automorphisms of a system in K.

3. $R(\sigma)$ is the class of all associative rings.

4. $R(\sigma)$ is the class of all ordered sets.

5. $R(\sigma)$ is the class of all fundamentally ordered semigroups. A system of axioms for this class is given in [6] for the case $M = \varnothing$. The case $M \neq \varnothing$ is treated in these lectures.

We now mention some examples of unary relation algebras. If K is a class of algebraic systems, unary relations are considered only on algebras in K. Special types of unary relations are singled out by means of certain conditions M. For example, if K is a class of semigroups, we can take as unary relations its left or right ideals; while if K is a class of groups, we can consider its subgroups or only its normal subgroups. The conditions in M must be expressed in LPC language.

After the properties M which are of interest to us have been chosen, we define certain operations and relations on subsets by means of LPC formulas. For example, we can thus define the Boolean operations of intersection, union, and complemention and the relations of inclusion, nonempty intersection, etc.

Hence let Φ be a set of subsets of a system $\mathfrak{A} \in K$ satisfying conditions in M.

1. If we introduce just the binary relation of inclusion of subsets ($\subset$), and M is the empty set, then we obtain a concrete unary relation algebra which is ordered by inclusion, i.e., the set of subsets, ordered by inclusion.

In this case the class of abstract algebras is the class of all ordered sets. The system of axioms for this class is well known.

2. If the operations of union and intersection are introduced on Φ, and M is empty, then we obtain a concrete unary relation algebra, called the *ring of sets*. The corresponding class of abstract algebras is the class of all distributive lattices [1].

3. If we introduce all three Boolean operations on Φ—union, intersection, and complementation—and M is empty, we obtain a concrete unary relation algebra called the *field of sets*. The corresponding class of abstract algebras is the class of all Boolean algebras [14].

4. Let K be a class of groups, and let M consist of the condition "is a normal subgroup of the group". If we introduce two operations on Φ, intersection and multiplication (by the product of two normal subgroups A and B we mean the normal subgroup AB), then a concrete unary relation algebra is obtained. It is a lattice of normal subgroups of a group in K. Every lattice of normal subgroups of a group in K forms a class of concrete algebras. In such a case, the only abstract relation algebras are those lattices which can be represented isomorphically as lattices of normal subgroups of a group in the given class K. The problem of determining an abstract characterization for these lattices has not yet been resolved, even for the class of abelian groups.

5. A *Tarski relation algebra* is an algebraic system $(\Phi, \circ, ^{-1}, \cap, \cup, ', \Delta, U, \varnothing)$, where Φ is a set of binary relations, on which are defined the binary operation of multiplication ($\circ$), the unary operation of inversion ($^{-1}$), the Boolean binary relations of intersection ($\cap$) and union ($\cup$), the binary relation of complementation ($'$), and also the nullary operations: the identity relation (the diagonal, Δ), the universal relation (U) and the empty relation $\varnothing$. A. Tarski [15] has proved that there exists a

system of axioms which describes the class of abstract Tarski relation algebras. The corresponding system of elementary axioms was found by R. C. Lyndon [9].

6. A *Jónsson relation algebra* is an abstract system $(\Phi, \circ, ^{-1}, \cap, \Delta)$, where Φ is a set of binary relations, on which are defined the binary operation of multiplication $(\circ)$, the unary operation of inversion $(^{-1})$, the Boolean operation of intersection $(\cap)$, and the nullary operation of the identity relation (the diagonal, Δ). An abstract characterization of this algebra was found in 1959 by B. Jónsson [7]. In addition, Jónsson [7] also gave an abstract characterization of such algebras when Φ is a set of equivalence relations. The abstract algebras in this case turn out to be lattices of a special type.

One can consider n-ary relation algebras on n different algebraic systems. V. V. Vagner [2] investigated semigrouds([2]) of binary relations (algebras of nonhomogeneous binary relations ([3]) with a ternary operation of the form $\varphi_3 \circ \overset{-1}{\varphi_2} \circ \varphi_1$).

FUNDAMENTAL THEOREM ON RELATION ALGEBRAS. *Let K be an axiomatizable class of algebraic systems, M a set of LPC formulas which indicate properties of binary relations on systems in K, and σ an arbitrary genre. Then the class R(K, M, σ) of all abstract binary M-relation algebras of σ over K is universal. Moreover, if K is recursively axiomatizable, and M and σ are recursive, then R(K, M, σ) is recursively universally axiomatizable.*

The recursive axiomatization of the class indicates that there exist a system of axioms for this class and an algorithm for deciding whether any given LPC formula is an axiom for a given system. If a class is not recursively axiomatizable, no system of axioms for this class can be of any practical use. (Every such system is necessarily infinite, since a finite system of axioms is trivially recursive, and so we are unable to write out all of the axioms in an explicit manner, since there does not exist any constructive procedure which permits one to survey all of the axioms and write down any of them).

For an extensive account of relation algebras and a proof of the fundamental theorem, see [23], [25], [26] and [29].

§3. An application of relation algebras to semigroups of transformations

The role which the concept of a function, or transformation, plays in modern mathematics is commonly known. However, the study of the algebraic properties of functions was begun comparatively recently, along with the development of Galois theory. At first, only systems of one-to-one transformations forming permutation

(2) *Editor's note.* The usual terminology is "semiheap", but see the author's views on this matter as expressed on p. 101.

(3) A binary relation $\rho \subset A \times B$ on the two sets A and B is called *homogeneous* if $A = B$.

groups were considered. This concept gradually evolved into the concept of an abstract group. It would be superfluous to mention how important this concept is in modern mathematics. For the most part, the reason for this lies in the fact that group theory is a natural instrument for making algebraic studies of groups of one-to-one transformations of a set.

However, transformations other than one-to-one mappings of a set onto itself also play an important role, especially those which are not necessarily one-to-one, everywhere defined, or even single-valued. Those transformations which are closed under composition form a system more general than a group, namely a semigroup. Conversely, every semigroup is isomorphic to a semigroup of transformations. Thus the concept of a semigroup corresponds precisely to that of a set of transformations which is closed under composition. Neither the concept of a group nor the related notion of a ring is adequate for studying transformations of a general type. This explains the importance of the theory of semigroups and the necessity for developing such a theory, particularly that part which deals with semigroups of transformations.

The most general type of transformation of a set A is a multi-valued transformation, i.e., a binary relation on A. If $\varphi \subset A \times A$ is a binary relation and $(a_1, a_2) \in \varphi$, then we say that a_2 is an image of the element a_1 under φ. From this point of view, those functions on A satisfying (1) are called single-valued (partial) transformations. Our interest may lie not in arbitrary multi-valued functions, but in certain transformations of algebraic systems which belong to a fixed axiomatizable class K. Suppose those special properties of transformations we are interested in can be written in the form of a set M of LPC formulas. By "transformation" we mean a binary relation on an algebraic system in K satisfying the conditions in M.

One loses a great amount of information when employing only algebraic methods in studying semigroups of transformations, since semigroups are considered up to isomorphism, i.e., they are abstracted from all properties which are not preserved under isomorphism. However, two isomorphic semigroups of transformations can be quite different from each other. For example, one can consist of everywhere defined transformations and the other partial transformations, or one can be transitive and the other intransitive, etc. It is desirable to take all these singular cases into account, and yet at the same time retain an algebraic point of view. To this end we will introduce some special objects which yield additional information about semigroups of transformations.

Usually such objects are relations and operations on transformations. For example, one additional object might be the relation of inclusion of binary relations. This object carries supplementary information on the structure of Φ, since two isomorphic semigroups of transformations may be ordered in a nonisomorphic manner. From the abstract point of view we obtain semigroups ordered in a special way.

For examples of other additional objects on semigroups of transformations, we mention the operations of inversion and restrictive multiplication of binary relations,([4])

([4])If $\rho, \sigma \subset A \times B$ are binary relations on the sets A and B then $\rho \rhd \sigma = (\mathrm{pr}_1\rho \times B) \cap \sigma$ and $\rho \lhd \sigma = \rho \cap (A \times \mathrm{pr}_2\sigma)$ are restrictive products of ρ and σ. Here $\mathrm{pr}_1\rho$ is the domain of ρ and $\mathrm{pr}_2\, \sigma$ the range of σ.

the Boolean operations, and various relations between binary relations, e.g., the unary relation "is reflexive", and others. Composition of binary relations can also be regarded as such an object.

From the abstract point of view, semigroups of transformations with the selected additional objects are algebraic systems on which one of the operations is associative, and the remaining operations and relations correspond to those additional objects which we agree to consider on semigroups of transformations, and which carry supplementary information. We do not choose these operations and relations (which are the additional objects) arbitrarily, but so that they somehow impinge on the intrinsic essence of the transformations themselves. In otherwords, they must be "natural". For example, the operation of composition

$$(a_1, a_3) \in \varphi_2 \circ \varphi_1 \longleftrightarrow (\bigvee a_2)[(a_1, a_2) \in \varphi_1 \wedge (a_2, a_3) \in \varphi_2],$$

the operation of inversion

$$(a_1, a_2) \in \overset{-1}{\varphi} \longleftrightarrow (a_2, a_1) \in \varphi,$$

the Boolean operation of union

$$(a_1, a_2) \in \varphi_1 \cup \varphi_2 \longleftrightarrow ((a_1, a_2) \in \varphi_1 \vee (a_1, a_2) \in \varphi_2),$$

and the relation of inclusion

$$\varphi_1 \subset \varphi_2 \longleftrightarrow (\bigwedge a_1, a_2)[(a_1, a_2) \in \varphi_1 \longrightarrow (a_1, a_2) \in \varphi_2],$$

are intrinsically associated with the transformations themselves. Obviously the right-hand sides of the equivalences defining these operations and relations are LPC formulas. The operations and relations on binary relations mentioned above can be thought of as being "natural", as can the relation on binary relations in the set Φ:

$$(\varphi_1, \varphi_2) \in \zeta \longleftrightarrow (\bigvee a)[(a, a) \in \varphi_1 \wedge (a, a) \in \varphi_2].$$

(That is, the relation "have a common fixed point".)([5])

Thus when considering semigroups of transformations, we restricted ourselves by the LPC language. This leads us to consider a special type of relation algebras: semigroups with additional relations defined on them. Now, among such semigroups it is those we are interested in that are abstract analogs of semigroups of transformations with the chosen additional objects. It is obvious that to this end the additional relations on the semigroups must satisfy certain special necessary and sufficient conditions (i.e., axioms) which reflect specific properties of the semigroups of transformations. These axioms reflect additional information on the structure of semigroups of transformations, and therefore the search for such axioms is of paramount interest.

Therefore applying the notion of relation algebras to the study of semigroups of transformations leads to a new general point of view and to new form of setting a

([5]) This and similar relations on semigroups of transformations were characterized in [34].

vast class of natural problems. It follows from the fundamental theorem on relation algebras that these problems are posed properly, in the sense that there is always a solution. That is, there exist necessary and sufficient conditions reflecting the specific properties of the chosen objects on the semigroups of transformations, and they can be written in LPC language. In other words, all additional information on the structure of semigroups of transformations, obtained at the expense of imposing on the semigroups additional objects which are defined by a first-order language, can be written in this same language.

In spite of the conceptual importance of the fundamental theorem on relation algebras, it is a purely existential theorem: it states the existence of a system of universal axioms, and indicates nothing whatsoever about how to find these axioms. Regarding semigroups of transformations as relation algebras gives a general perspective and method of setting the problems. So the task of searching for methods of solving these problems becomes of foremost importance.

A typical problem concerning semigroups of transformations looks as follows: find an abstract characterization of a class of concrete semigroups of transformations having the special properties. This means one must find conditions for isomorphic representability of an abstract semigroup by these concrete transformations. Naturally, in order to solve the problem it is desirable to know how to construct all representations of semigroups, and be able to construct representations of semigroups with preassigned properties. Moreover, it is important to know the structure not only of isomorphic but also of homomorphic representations of semigroups, since it turns out that by "pasting together" sufficiently many homomorphic representations we can obtain isomorphic representations.

Of those methods which permit one to solve problems of representing semigroups of transformations with superimposed conditions, the method of determinative pairs may be mentioned in the first place. By means of this method the problem of determining a special representation of a semigroup of transformations is reduced to the following purely semigroup-theoretic one: for an abstract semigroup one must find out under what conditions there exists a collection of determinative pairs (ϵ, W) possessing certain properties.

We shall consider some examples of concrete problems of this type. Of the operations which are considered on semigroups of transformations, most often we choose the binary operation of multiplication and the unary operation of inversion. Among the relations we single out inclusion of transformations (ζ), inclusion of their first projections $(\underset{1}{\chi})$, inclusion of their second projections $(\underset{2}{\chi})$, equality of the first projections $(\underset{1}{\pi})$, and equality of the second projections $(\underset{2}{\pi})$.(6)

For semigroups of binary relations and semigroups of transformations, the problems of finding abstract characterizations of one or more of the above-mentioned relations have all been solved. Some of these solutions have appeared in [21]–[26] and [29]. For a discussion of semigroups of transformations as relation algebras and

(6) E.g., if φ and ψ are transformations, then $(\varphi, \psi) \in \underset{2}{\chi}$ means that $\mathrm{pr}_2\varphi \subset \mathrm{pr}_2\psi$.

the problems so engendered, see [23], [25], [26] and [29], where further references are also given.

§4. Method of determinative pairs

Let S be a semigroup and A be a set. We will denote by F_A the set of all partial transformations of A. Obviously F_A is a semigroup under multiplication of binary relations. Therefore one can consider homomorphic representations of S into F_A, which we shall call simply *representations.*(7) Our problem is to describe all representations of the abstract semigroup S. In order to solve it we must introduce some definitions and consider properties of representations of semigroups.

Two representations $P_1 : S \longrightarrow F_A$ and $P_2 : S \longrightarrow F_B$ are said to be *similar* if there exists a one-to-one mapping φ of A onto B such that $(a_1, a_2) \in P_1(s) \longleftrightarrow (\varphi(a_1), \varphi(a_2)) \in P_2(s)$ for all $a_1, a_2 \in A$ and $s \in S$. Similar representations are sometimes called equivalent or nondistinguishable. We will consider representations up to similarity.

Let there be given a representation $P : S \longrightarrow F_A$ and a set $\mathfrak{A} \subset A$. Let $P_{\mathfrak{A}}(s)$ denote $P(S) \cap \mathfrak{A} \times \mathfrak{A}$. It is clear that $P_{\mathfrak{A}}(s)$ is a partial transformation of A, but in general $P_{\mathfrak{A}}$ is not a homomorphism. However, if $P_{\mathfrak{A}}$ is a homomorphism it is called a *restriction* of P to $\mathfrak{A}$, and $\mathfrak{A}$ is called *homomorphic.*

We will denote the image of a semigroup S under a homomorphism P by $\check{P}(S)$. The binary relation $\tau_P = \bigcup_{s \in S} P(s)$ (or $\tau_P = \bigcup \check{P}(S)$) is called the *transitivity relation* of the representation P.

Next we consider a family of representations $\{P_i\}_{i \in I}$ of S by means of partial transformations of A, and form the binary relation $P(s) = \bigcup_{i \in I} P_i(s)$ on A. When P is a representation we call it the *join* of the representations, and call the family itself *joinable.* Let there be given a family of representations $\{P_i\}_{i \in I}$, where $P_i : S \longrightarrow F_{A_i}$, and let $A = \bigcup_{i \in I} A_i$ and $P(s) = \bigcup_{i \in I} P_i(s)$. It is easy to prove that if the A_i are mutually disjoint, then $P : S \longrightarrow F_A$ is a representation. We call P the *sum* of the given family of representations.

*We can form the sum of any family of representations, replacing some of them, if necessary, by similar ones in order to make the sets S_i mutually disjoint.

To every representation P there corresponds an equivalence relation ϵ_P of S, called the *kernel* of the given representation, defined by

$$(s_1, s_2) \in \epsilon_p \longleftrightarrow P(s_1) = P(s_2).$$

One can show that ϵ_P is a congruence and that $\epsilon_P = \overset{-1}{P} \circ P$. If P is an isomorphism,

(7)We write the factors in a product of elements of an abstract semigroup from left to right. In case of some special semigroups this notation may be changed for another. E.g., we write the factors in a product of transformations from right to left. Thus P is a homomorphic representation of a semigroup S by transformations if $P(s_1 s_2) = P(s_2) \circ P(s_1)$ for all $s_1, s_2 \in S$.

then it is said to be *faithful.* It is obvious that in such a case $\epsilon_P = \Delta_S$, the identity relation on S.

We will now consider other relations on the semigroup S. By the *fundamental quasi-order relation* for the representation P we mean the binary relation

$$\zeta_P = \{(s_1, s_2) : P(s_1) \subset P(s_2)\}.$$

It can be shown that ζ_P is a stable quasi-ordered relation, i.e., $(s_1, s_2) \in \zeta_P \wedge (s_3, s_4) \in \zeta_P \longrightarrow (s_1 s_3, s_2 s_4) \in \zeta_P$. It is also clear that $\epsilon_P = \zeta_P \cap \overset{-1}{\zeta}_P$.

Next we define the relation

$$\underset{1}{\chi}_P = \{(s_1, s_2) : \mathrm{pr}_1 P(s_1) \subset \mathrm{pr}_1 P(s_2)\},$$

called the *first projection quasi-order relation* for P. The symmetric part of this relation, $\underset{1}{\pi}_P = \underset{1}{\chi}_P \cap \overset{-1}{\underset{1}{\chi}}_P$, is called the *first projection equivalence relation.* The *second projection relation* $\underset{2}{\chi}_P$ and its symmetric part $\underset{2}{\pi}_P$ are defined analogously.

Let $\{P_i\}_{i \in I}$ be a family of representations, and form their sum $P(s) = \Sigma_{i \in I} P_i(s)$. It is easy to prove that if ρ_P is one of the relations ϵ_P, ζ_P, $\underset{1}{\chi}_P$, $\underset{2}{\chi}_P$, $\underset{1}{\pi}_P$ or $\underset{2}{\pi}_P$, then $\rho_P = \bigcap_{i \in I} \rho_{P_i}$. An equivalence relation ϵ on a semigroup S is said to be *right regular* if $s_1 \equiv s_2\ (\epsilon) \longrightarrow s_1 s \equiv s_2 s\ (\epsilon)$ for all $s_1, s_2, s \in S$. A subset $W \subset S$ is called a *right ideal* of S if $WS \subset S$. We adjoin an element e to S such that $(\bigwedge x \in S^*)$ $(xe = ex = x)$. Then we obtain the semigroup with identity $S^* = S \cup \{e\}$.([8])

A pair (ϵ, W) is called a *determinative pair* for the semigroup S if ϵ is a right regular equivalence relation on S^* and either $W = \varnothing$ or W is a right ideal of S^* which is an equivalence class modulo ϵ.

Every equivalence relation induces a partition of S^*. Consider the set $(S^*/\epsilon)\backslash W$ of equivalence classes modulo ϵ which are different from W. Put the elements of this set into a one-to-one correspondence with some set A. Thus we obtain a family $(H_a)_{a \in A}$, where $H_{a_1} \neq H_{a_2} \longleftrightarrow a_1 \neq a_2$. It can be shown that ϵ is a right regular equivalence relation if and only if $(\bigwedge H)(\bigwedge s)(\bigvee H_1)\,[Hs \subset H_1]$, where H and H_1 are ϵ-classes.

Define the binary relation $P_{(\epsilon, W)} : S \longrightarrow F_A$ by

$$P_{(\epsilon, W)}(s) = \{(a_1, a_2) : H_{a_1} s \subset H_{a_2}\}.$$

Clearly $H_{a_1} s \subset W \longleftrightarrow a_1 \notin \mathrm{pr}_1 P_{(\epsilon, W)}(s)$. It is easily seen that $P_{(\epsilon, W)}$ is a single-valued. Prove that $P_{(\epsilon, W)}$ is a representation, i.e., that $P_{(\epsilon, W)}(s_2) \circ P_{(\epsilon, W)}(s_1) = P_{(\epsilon, W)}(s_1 s_2)$:

$$(a_1, a_2) \in P_{(\epsilon, W)}(s_2) \circ P_{(\epsilon, W)}(s_1) \longleftrightarrow (\bigvee a)[(a_1, a) \in P_{(\epsilon, W)}(s_1) \wedge (a, a_2) \in P_{(\epsilon, W)}(s_2)]$$

$$\longleftrightarrow (\bigvee a)[H_{a_1} s_1 \subset H_a \wedge H_a s_2 \subset H_{a_2}] \overset{1)}{\longleftrightarrow} H_{a_1} s_1 s_2 \subset H_{a_2} \longleftrightarrow (a_1, a_2) \in P_{(\epsilon, W)}(s_1, s_2).$$

([8]) Note that S^* does not coincide with S^1 if S already has an identity.

It remains only to justify the equivalence "1)". From the fact that $(\bigvee a)\,[H_{a_1}s_1 \subset H_a \wedge H_a s_2 \subset H_{a_2}]$ it follows immediately that $H_{a_1}s_1s_2 \subset H_{a_2}$. Conversely, assume that $H_{a_1}s_1s_2 \subset H_{a_2}$. Consider the product $H_{a_1}s_1$. Suppose that $H_{a_1}s_1 \subset W$. Then $H_{a_1}s_1s_2 \subset Ws_2 \subset W$. We have arrived at a contradiction, since $H_{a_2} \neq W$. In such a case $(\bigvee a)\,[H_{a_1}s_1 \subset H_a]$, whence it follows that $H_{a_1}s_1s_2 \subset H_a s_2$. But the hypothesis $H_{a_1}s_1s_2 \subset H_{a_2}$ implies that $H_{a_2} \cap H_a s_2 \neq \varnothing$ and $H_a s_2 \subset H_{a_2}$. Hence $P_{(\epsilon, W)}$ is a representation, called the *simplest* representation corresponding to the determinative pair (ϵ, W).

THEOREM. *Every representation of a semigroup by single-valued transformations is the join of a family of simplest representations.*

REMARK. In this case, and later on, by "transformation" we mean "partial transformation".

Let there be given a representation P by transformations of the set. A. Take $a \in A$, and define the relation $\epsilon^a \subset S^* \times S^*$ and the subset $W^a \subset S^*$ by

$$\epsilon^a = \{(s_1, s_2) \mid P(s_1)\langle a\rangle = P(s_2)\langle a\rangle\},\text{(}^9\text{)}$$

$$W^a = \{s \mid P(s)\langle a\rangle = \varnothing\}.$$

We will prove that (ϵ^a, W^a) is a determinative pair. The equivalence relation ϵ^a is right regular, for

$$s_1 \equiv s_2(\epsilon^a) \longleftrightarrow P(s_1)\langle a\rangle = P(s_2)\langle a\rangle$$

$$\longrightarrow P(s) \circ P(s_1)\,\langle a\rangle = P(s) \circ P(s_2)\langle a\rangle$$

$$\longleftrightarrow P(s_1 s)\langle a\rangle = P(s_2 s)\langle a\rangle \longleftrightarrow s_1 s \equiv s_2 s\ (\epsilon^a).$$

Now we show that if $W^a \neq \varnothing$ then W^a is a right ideal of S^*. For if $w \in W^a$ then

$$P(ws)\langle a\rangle = P(s) \circ P(w)\langle a\rangle = \overset{\smile}{\overline{P(s)}}(P(w)\langle a\rangle) = \overset{\smile}{\overline{P(s)}}(\varnothing) = \varnothing,$$

i.e., $ws \in W^a$. The fact that W^a is an ϵ^a-class is obvious.

Consider the following subset of S^*:

$$H_{a_2}^{a_1} = \{s : (a_1, a_2) \in P(s)\},\quad e \in H_{a_2}^{a_1} \longleftrightarrow a_1 = a_2.$$

Let $(H_{a_1}^a)_{a_1 \in \mathfrak{A}_a \subset A}$, where $\mathfrak{A}_a = \{a_1 : H_{a_1}^a \neq \varnothing\}$. The $H_{a_1}^a$ are ϵ^a-classes different from W^a. Then

(9) If $\rho \subset A \times B$ is a binary relation and $a \in A$, then $\rho\langle a\rangle = \{b \mid (a, b) \in \rho\}$. If $\mathfrak{A} \subset A$ then $\breve{\rho}(\mathfrak{A}) = \bigcup_{a \in \mathfrak{A}} \rho\langle a\rangle$.

$$s \in H^a_{a_1} \longleftrightarrow (a, a_1) \in P(s) \longleftrightarrow a_1 \in P(s)\langle a\rangle \longrightarrow P(s)\langle a\rangle \neq \varnothing \longrightarrow H^a_{a_1} \neq W^a,$$

$$s_1 \in H^a_{a_1} \wedge s_1 \equiv s_2\ (\epsilon^a) \longleftrightarrow P(s_1)\langle a\rangle = P(s_2)\langle a\rangle = \{a_1\} \longleftrightarrow s_1, s_2 \in H^a_{a_1}.$$

Hence $H^a_{a_1}$ is an ϵ^a-class different from W^a. If $s \notin W^a$ then $\{a_1\} = P(s)\langle a\rangle \neq \varnothing$, which means that $(a, a_1) \in P(s)$, i.e., $s \in H^a_{a_1}$. It remains to show that the subscripts are one-to-one. So assume that $H^a_{a_1} = H^a_{a_2}$ and take $s \in H^a_{a_1} = H^a_{a_2}$. This means that $(a, a_1), (a, a_2) \in P(s)$, whence $a_1 = a_2$ since $P(s)$ is a single-valued binary relation.

Form a family of simplest representations $(P_{(\epsilon^a, W^a)})_{a\in A}$. We will show that this system is joinable, and that its join is equal to P, i.e., $P(s) = \bigcup_{a\in A} P_{(\epsilon^a, W^a)}(s)$. Indeed,

$$(a_1, a_2) \in \bigcup_{a\in A} P_{(\epsilon^a, W^a)}(s) \longleftrightarrow (\bigvee a)\,[(a_1, a_2) \subset P_{(\epsilon^a, W^a)}(s)]$$

$$\longleftrightarrow (\bigvee a)\,[H^a_{a_1} s \subset H^a_{a_2}] \longleftrightarrow (a_1, a_2) \in P(s).$$

It remains to verify the last equivalence. Assume that $(\bigvee a)\,[H^a_{a_1} s \subset H^a_{a_2}]$. We have $s_1 \in H^a_{a_1} \longrightarrow s_1 s \in H^a_{a_2}$. But $s_1 \in H^a_{a_1} \longleftrightarrow (a, a_1) \in P(s_1)$ and $s_1 s \in H^a_{a_2} \longleftrightarrow (a, a_2) \in P(s_1 s) = P(s) \circ P(s_1)$. Consequently $a_2 = (P(s) \circ P(s_1))(a) = P(s)(P(s_1)(a)) = P(s)(a_1)$, i.e., $(a_1, a_2) \in P(s)$. Now let $(a_1, a_2) \in P(s)$. Take the class $H^{a_1}_{a_1}$. It is non-empty since $e \in H^{a_1}_{a_1}$. Thus $(a_1, a_2) \in P(s) \longleftrightarrow s \in H^{a_1}_{a_2} \longleftrightarrow es = s \in H^{a_2}_{a_1} \longrightarrow H^{a_1}_{a_1} s \subset H^{a_1}_{a_2}$, since ϵ^{a_1} is right regular. This completes the proof of the theorem.

Note that P can also be the join of other sets of simplest representations.

Consider a subset $H \subset S^*$ and the relation

$$\epsilon_H = \{(s_1, s_2) : (\bigwedge s \in S^*)(s_1 s \in H \longleftrightarrow s_2 s \in H)\}.$$

It is not hard to verify that ϵ_H is a right regular equivalence relation. Define the subset W_H by

$$W_H = \{s : (\bigwedge s_1 \in S^*)(s s_1 \notin H)\}.$$

If $W_H \neq \varnothing$, it can be shown that W_H is an ϵ_H-class and is a right ideal. Hence (ϵ_H, W_H) is a determinative pair. This pair corresponds to the representation $P_H = P_{(\epsilon_H, W_H)}$, which is called an *elementary* representation. This raises the question (analogous to that considered for simplest representations): is every representation the join of some family of elementary representations? The answer is negative. Conditions under which a representation will be a join of elementary representations are not known. We point out one more result: a representation P is a simplest representation if and only if there exists an element $a \in A$ such that $\tau_P\langle a\rangle \cup \{a\} = A$, i.e. if for every $a_1 \neq a$ there exists $s \in S$ such that $P(s)$ transforms a into a_1.

We will now touch upon representations of semigroups by one-to-one transformations. Of the results related to this question, we can state the following: if the join of a family of representations is a representation by one-to-one transformations, then so is each term in the family. In order for $P_{(\epsilon, W)}$ to be a representation of a semigroup by one-to-one (invertible) transformations, it is necessary and sufficient that the following condition hold:

$$s_1 s \equiv s_2 s\ (\epsilon) \wedge s_1 s \notin W \longrightarrow s_1 \equiv s_2\ (\epsilon).$$

Indeed,

$$((a_1, a), (a_2, a) \in P_{(\epsilon, W)}(s) \longrightarrow a_1 = a_2) \longleftrightarrow (H_{a_1} s \subset H_a \wedge H_{a_2} s \subset H_a \longrightarrow a_1 = a_2)$$

$$\longleftrightarrow (\bigwedge s_1, s_2)[s_1 s, s_2 s \in H_a \longrightarrow s_1 \equiv s_2\ (\epsilon)] \longleftrightarrow$$

$$\longleftrightarrow (\bigwedge s_1, s_2)[s_1 s \equiv s_2 s\ (\epsilon) \wedge s_1 s \notin W \longrightarrow s_1 \equiv s_2\ (\epsilon)].$$

A subset $H \subset S$ is said to be *strong* if for all $x, y, u, v \in S$

$$xv \in H \wedge uv \in H \wedge uy \in H \longrightarrow xy \in H.$$

An elementary representation P_H will be a representation by one-to-one transformations if and only if H is a strong subset of S^*. Every representation by one-to-one transformations is a join of elementary representations corresponding to a family of strong subsets of S^*. A more detailed account of the method of determinative pairs can be found in [18] and [28]. The latter paper also contains other methods for constructing all representations of a given semigroup. Representations by multi-valued transformations are described in [19].

§5. Representation of ordered semigroups

In this section we will find an abstract characterization of the fundamental order relation on semigroups of transformations, i.e., necessary and sufficient conditions insuring the existence of a faithful representation P of an abstract semigroup S with an ordering relation ζ into a semigroup of transformations such that $\zeta = \zeta_P$, where ζ_P denotes inclusion of transformations, i.e., $s_1 \prec s_2\ (\zeta_P) \longleftrightarrow P(s_1) \subset P(s_2)$. In other words, the given concrete relation algebra is of the genre $\sigma = (O_0, P_\zeta)$, where O_0 is a formula defining multiplication of binary relations and P_ζ defines inclusion of binary relations, and where by "binary relations" we mean single-valued binary relations (transformations). We are going to find an axiomatization of the corresponding class of relation algebras of the genre σ. By the fundamental theorem this class admits a system of universal axioms. This problem bears particular interest since it treats the relation of inclusion, i.e., the relation of extendability of transformations (if $\varphi \subset \psi$, then ψ extends φ), which plays an important role in many matters.

Some trivial necessary conditions for the existence of such a representation P present themselves immediately. For example, the multiplication must be associative and the relation ζ must be a stable ordering. Assume that the desired isomorphism P can be chosen as a sum of simplest representations (as we will see later, such a P does indeed exist). The desire to find a representation P which is not simply a join, but rather a sum, of simplest representations is motivated by the fact that the structure of sums of simplest representations is known far better than their join.

So let $P = \Sigma_{i \in I} P_{(\epsilon_i, W_i)}$. Then $\zeta_P = \bigcap_{i \in I} \zeta_{P_{(\epsilon_i, W_i)}}$. We require that $\zeta_P = \zeta$. As a result, we obtain some conditions which will be obviously sufficient for the desired representation P to exist. Clearly

$$\zeta \subset \zeta_{P_{(\epsilon_i, W_i)}} \quad \text{for each } i \in I. \tag{6}$$

Thus we must consider those $\zeta_{P_{(\epsilon_i, W_i)}}$ which contain ζ, and we have to find as many of them as satisfy the inclusion

$$\bigcap_{i \in I} \zeta_{P_{(\epsilon_i, W_i)}} \subset \zeta. \tag{7}$$

Of course, it is possible that the intersection of all $\zeta_{P_{(\epsilon_i, W_i)}}$ containing ζ does not satisfy (7). This would mean that the desired representation cannot be assumed to be a sum of simplest representations. To help us decide which simplest representations must be taken for (7) to hold, we will analyse the relation $\zeta_{P_{(\epsilon_i, W_i)}}$. To this end we will investigate (6).

Denote the elements in the semigroup S^* by $x, y, \ldots$, and the elements of S by $s_1, s_2, \ldots$. Thus

$$(s_1, s_2) \in \zeta_{P_{(\epsilon, W)}} \longleftrightarrow (P_{(\epsilon, W)}(s_1) \subset P_{(\epsilon, W)}(s_2))$$

$$\longleftrightarrow (\bigwedge a_1, a_2)[(a_1, a_2) \in P_{(\epsilon, W)}(s_1) \longrightarrow (a_1, a_2) \in P_{(\epsilon, W)}(s_2)]$$

$$\longleftrightarrow (\bigwedge a_1, a_2)[H_{a_1} s_1 \subset H_{a_2} \longrightarrow H_a s_2 \subset H_{a_2}]$$

$$\longleftrightarrow (\bigwedge a_1, a_2)(\bigwedge x \in H_{a_1})[xs_1 \in H_{a_2} \longrightarrow xs_1 \equiv xs_2 \ (\epsilon)]$$

$$\longleftrightarrow (\bigwedge x)[xs_1 \notin W \longrightarrow xs_1 \equiv xs_2 \ (\epsilon)].$$

We will prove the last equivalence. Let $xs_1 \in H_{a_2} \longrightarrow xs_1 \equiv xs_2$ (ϵ) and $xs_1 \notin W$. Then $xs_1 \in H_{a_2}$ for some a_2. Therefore $xs_1 \equiv xs_2$ (ϵ). Conversely, let $xs_1 \notin W \longrightarrow xs_1 \equiv xs_2$ (ϵ). Assume that $x \in H_{a_1}$ and $xs_1 \in H_{a_2}$. Then $xs_1 \notin W$, so $xs_1 \equiv xs_2$ (ϵ). Thus we have proved that

$$(s_1, s_2) \in \zeta_{P_{(\epsilon, W)}} \longleftrightarrow (\bigwedge x)[xs_1 \notin W \longrightarrow xs_1 \equiv xs_2 \ (\epsilon)].$$

We can write $\zeta \subset \zeta_{P_{(\epsilon_i, W_i)}}$ in the form $s_1 \prec s_2 \longrightarrow (s_1, s_2) \in \zeta_{P_{(\epsilon_i, W_i)}}$, or, taking the obtained result into account, in the form

$$s_1 \prec s_2 \longrightarrow (\bigwedge x)\ [xs_1 \notin W_i \longrightarrow xs_1 \equiv xs_2 \ (\epsilon_i),$$

where $\prec$ is the infix denotation of ζ, i.e., $s_1 \prec s_2$ means that $(s_1, s_2) \in \zeta$. The condition so obtained can be rewritten as follows:

$$s_1 \prec s_2 \wedge xs_1 \notin W_i \longrightarrow xs_1 \equiv xs_2(\epsilon_i).$$

Letting $x = e$, we obtain

$$s_1 \prec s_2 \wedge s_1 \notin W_i \longrightarrow s_1 \equiv s_2 \ (\epsilon_i). \tag{8}$$

If (8) holds and $s_1 \prec s_2 \wedge xs_1 \notin W_i$, then the left regularity of ζ implies that $xs_1 \prec xs_2$. By (8), $xs_1 \equiv xs_2$ (ϵ_i). Hence (8) means that $\zeta \subset \zeta_{P_{(\epsilon_i, W_i)}}$. Condition (8) means that each ϵ_i-class different from W_i is *majorantly saturated* with respect to ζ, which means $s \in H \wedge s < t \longrightarrow t \in H$ for every ϵ_i-class $H \neq W_i$. Now it behooves us to find all determinative pairs (ϵ, W) which satisfy (6). In our case it turns out that one must only consider those determinative pairs for which the representation P is a sum of elementary representations. In such a case, inclusion can be written as

$$\zeta \subset \zeta_{P_{(\epsilon_H, W_H)}} \longleftrightarrow (\bigwedge s_1, s_2)[s_1 < s_2 \wedge s_1 \notin W_H \longrightarrow s_1 \equiv s_2 \ (\epsilon_H)].$$

The right-hand side of this formula can be written in the following equivalent form:

$$s_1 \prec s_2 \wedge s_1 y \in H \longrightarrow (\bigwedge z)(s_1 z \in H \longleftrightarrow s_2 z \in H),$$

or

$$\bigwedge (y, z)[s_1 \prec s_2 \wedge s_1 y \in H \longrightarrow (s_1 z \in H \longleftrightarrow s_2 z \in H)].$$

Replacing the equivalence in the last formula by two implications, we obtain the conjunction

$$[s_1 \prec s_2 \wedge s_1 y \in H \longrightarrow (s_1 z \in H \longrightarrow s_2 z \in H)]$$

$$\wedge [s_1 \prec s_2 \wedge s_1 y \in H \longrightarrow (s_2 z \in H \longrightarrow s_1 z \in H)],$$

or, by the conjunction of the premises,

$$[s_1 \prec s_2 \wedge s_1 y \in H \wedge s_1 z \in H \longrightarrow s_2 z \in H]$$

$$\wedge [s_1 \prec s_2 \wedge s_1 y \in H \wedge s_2 z \in H \longrightarrow s_1 z \in H].$$

We will consider what information the first term of this conjunction reveals about H. We write it out separately:

$$s_1 \prec s_2 \wedge s_1 y \in H \wedge s_1 z \in H \longrightarrow s_2 z \in H.$$

It is obvious that without losing any information we can discard the second term of the premise. We obtain

$$s_1 \prec s_2 \wedge s_1 z \in H \longrightarrow s_2 z \in H.$$

Letting $z = e$, we obtain $s_1 \prec s_2 \quad s_1 \in H \longrightarrow s_2 \in H$, i.e., H is majorantly saturated. Conversely, if H is majorantly saturated and $s_1 \prec s_2 \wedge s_1 z \in H$, then $s_1 z \prec s_2 z$, so $s_2 z \in H$.

Thus the first term of the conjunction means that $\breve{\zeta}(H) \subset H$, where $\zeta(H) = \bigcup_{h \in H} \zeta\langle h \rangle$. The second term has the form

$$s_1 \prec s_2 \wedge s_1 y \in H \wedge s_2 z \in H \longrightarrow s_1 z \in H. \tag{9}$$

Hence the first and second terms are equivalent to the formula $\breve{\zeta}(H) \subset H$ and (9). Thus it is obvious that we must take subsets H satisfying these formulas, construct their elementary representations, take their sum, and require, moreover, that $\bigcap \zeta_{(\epsilon_H, w_H)} \subset \zeta$. Then we obtain sufficient conditions for the existence of a faithful representation P. These conditions will also turn out to be necessary.

However, the conditions thus obtained are not elementary axioms but axioms of the second degree, since the quantifier binds the subsets H. Thus we must continue our search for an elementary system of axioms.

Let us take for the majorantly saturated subsets H the subsets of the form $H = \zeta\langle s \rangle$. We require that they satisfy the second of the above conditions, and verify that this condition is necessary, i.e., we show that the transformations satisfy the condition

$$y \circ s_1,\ z \circ s_2 \in \zeta_P\langle s \rangle \wedge s_1 \subset s_2 \longrightarrow z \circ s_1 \in \zeta_P\langle s \rangle,$$

or equivalently

$$s \subset y \circ s_1 \wedge s \subset z \circ s_2 \wedge s_1 \subset s_2 \longrightarrow s \subset z \circ s_1.$$

We have

$$s \subset y \circ s_1 \longrightarrow \mathrm{pr}_1 s \subset \mathrm{pr}_1 s_1 \longrightarrow \Delta_{\mathrm{pr}_1 s} \subset \Delta_{\mathrm{pr}_1 s_1},$$

$$s \subset z \circ s_2 \wedge \Delta_{\mathrm{pr}_1 s} \subset \Delta_{\mathrm{pr}_1 s_1}, \longrightarrow s = s \circ \Delta_{\mathrm{pr}_1 s} \subset z \circ (s_2 \circ \Delta_{\mathrm{pr}_1 s}) = z \circ s_1.$$

Hence we have obtained that the relation ζ for $H = \zeta\langle s \rangle$ necessarily satisfies the second of our conditions. Thus if we take for H sets of the form $(\zeta\langle s \rangle)_{s \in S}$ and construct the elementary representation $P_{\zeta\langle s \rangle}$, then

$$(\bigwedge s \in S)\,[\zeta \subset \zeta_{P_{\zeta\langle s \rangle}}].$$

Next, we have

$$(s_1, s_2) \in \bigcap_{s \in S} \zeta_{P_{\zeta\langle s \rangle}} \longleftrightarrow (\bigwedge s)[(s_1, s_2) \in \zeta_{P_{\zeta\langle s \rangle}}]$$

$$\longleftrightarrow (\bigwedge s, x, y)[x s_1 y \in \zeta\langle s \rangle \longrightarrow (\bigwedge z)(x s_1 z \in \zeta\langle s \rangle \longleftrightarrow x s_2 z \in \zeta\langle s \rangle)].$$

Let $x = y = z = e$ and $s = s_1$. This yields the following implication:

$$s_1 \in \zeta\langle s_1\rangle \longrightarrow (s_1 \in \zeta\langle s_1\rangle \longleftrightarrow s_2 \in \zeta\langle s_1\rangle).$$

The premise always holds, and so the equivalence $s_1 \in \zeta\langle s_1\rangle \longleftrightarrow s_2 \in \zeta\langle s_1\rangle$ does also. Hence we obtain $s_2 \in \zeta\langle s_1\rangle$, or $(s_1, s_2) \in \zeta$. In this way we have shown that $\bigcap_{s\in S}\zeta_{P_{\zeta\langle s\rangle}} \subset \zeta$.

From the equality $\epsilon_P = \zeta_P \cap \overset{-1}{\zeta_P} = \Delta_S$ (recall that ζ_P is an ordering) we conclude that the constructed representation $P = \Sigma_{s\in S}P_{\zeta\langle s\rangle}$ is an isomorphism of the semigroup S with ordering ζ onto a semigroup of transformations with ordering $\subset$, mapping ζ onto $\subset$. Consequently we have proved the

THEOREM. *An ordered semigroup is isomorphic to a fundamentally ordered semigroup of transformations if and only if the ordering is stable and weakly steady.*

We point out that an ordering of a semigroup is called *weakly steady* if it satisfies

$$\left.\begin{array}{l} s \prec s_1x \\ s \prec s_2y \\ s_1 \prec s_2 \end{array}\right\} \longrightarrow s \prec s_1y. \tag{10}$$

Clearly (10) is merely (9) for $H = \zeta\langle s\rangle$.

In order to solve the analogous problem for invertible transformations it will be necessary to impose certain additional conditions on the subset H. We will determine them by using the requirement that the transformation $P_H(s)$ be invertible for each $s \in S$, i.e.,

$$(\bigwedge a_1, a_2, a)[(a_1, a) \in P_H(s) \wedge (a_2, a) \in P_H(s) \longrightarrow a_1 = a_2].$$

It is easy to prove that this is equivalent to the formula

$$\left.\begin{array}{l} xv \\ uv \\ uy \end{array}\right\} \in H \longrightarrow xy \in H. \tag{11}$$

This means that H must be a strong subset of the semigroup S^*. If for H we take the cuts $(\zeta\langle s\rangle)_{s\in S}$, not all of them will be strong. Sometimes it will be necessary to adjoin an identity to them. It is not hard to prove that the cut

$$\zeta^*\langle z\rangle = \begin{cases} \zeta^*\langle s\rangle, & \text{if } (s_1s^2) \notin \zeta, \\ \zeta\langle s\rangle \cup \{e\}, & \text{if } (s_1s^2) \in \zeta, \end{cases}$$

is a strong subset of S^*.

We can then write (11) in the form

$$\left.\begin{array}{l} s \prec xv \\ s \prec uv \\ s \prec uy \end{array}\right\} \longrightarrow s \prec xy, \tag{12}$$

which is called the *steadiness* of the order relation. We shall prove that this condition is necessary, i.e., we show that the condition

$$s \subset v \circ x \wedge s \subset v \circ u \wedge s \subset y \circ u \longrightarrow s \subset y \circ x$$

holds for invertible transformations. Indeed,

$$s = s \circ \overset{-1}{s} \circ s \subset y \circ u \circ \overset{-1}{u} \circ \overset{-1}{v} \circ v \circ x = y \circ \Delta_{\mathrm{pr}_2 u} \circ \Delta_{\mathrm{pr}_1 v} \circ x \subset y \circ x,$$

which was to be proved.

It is easy to show that steadiness is a stronger condition than weak steadiness, whence (10) follows from (12).

Therefore the theorem on the abstract characterization of the ordering on semigroups of invertible transformations can be formulated in this form:

THEOREM. *An ordered semigroup is isomorphic to a fundamentally ordered semigroup of invertible transformations if and only if the ordering is stable and steady.*

The results in this section can be found in [22].

*Note that if no restrictions are imposed on the binary relations considered, we have the following result [6]:

THEOREM. *An ordered semigroup is isomorphic to a fundamentally ordered semigroup of binary relations if and only if the order is stable.*

*Indeed, if S is a semigroup endowed with a stable order $\prec$, then one can easily verify that

$$P(s) = \{(x, y) \in S \times S^* : x \prec sy\}$$

produces the representation needed, i.e. P is an order-isomorphism of S onto an inclusion-ordered semigroup of binary relations.

*If 'a relation' means 'a reflexive relation', 'a transitive relation', 'a quasi-order relation' and so on, then characterizations of fundamentally ordered semigroups of relations can be found in [6] and [31].

§6. Representation of first projection semigroups

We will seek here a system of elementary axioms for an algebra of transformations of the genre $(O_0, P_{\underset{1}{\chi}})$, where O_0 is a formula defining multiplication of binary relations and $P_{\underset{1}{\chi}}$ defines the first projection quasi-order relation

$$(\varphi, \psi) \in \underset{1}{\chi}_P \leftrightarrow \mathrm{pr}_1\varphi \subset \mathrm{pr}_1\psi$$

(sometimes we denote this by $\varphi \underset{P}{\sqsubset} \psi$). In other words, we will find necessary and sufficient conditions for an algebraic system of the from $(S, \sqsubset)$, where S is a set with a given algebraic binary operation and $\sqsubset$ (or $\underset{1}{\chi}$) is a binary relation on S, to be isomorphic to a *first projection semigroup* of transformations (that is, to a semigroup of transformations with the first projection quasi-order relation).

Certain necessary conditions for that are fairly obvious. Clearly the operation on S must be associative, i.e., S must be a semigroup, and $\sqsubset$ must be a quasi-order (since $\underset{1}{\chi}$ is a quasi-order). Our plan of attack is analogous to the preceding one.

Thus, we must construct a representation P such that $\chi = \underset{1}{\chi}_P$ and $\epsilon_P = \Delta_S$, i.e., P should be a faithful representation (isomorphism).

We point out at once that we need not demand that P be an isomorphism. For let P be a representation and $\underset{1}{\chi} = \underset{1}{\chi}_P$. Let P_1 be a faithful representation of S by everywhere-defined transformations (for example, the representation of S by right translations of S^*). Then $P_0 = P + P_1$ is a faithful representation, since

$$\epsilon_{P_0} = \epsilon_P \cap \epsilon_{P_1} = \epsilon_P \cap \Delta_S = \Delta_S.$$

On the other hand, $\mathrm{pr}_1 P_1(s_1) = \mathrm{pr}_1 P_1(s_2)$ for any $s_1, s_2 \in S$, i.e., $s_1 \underset{P_1}{\sqsubset} s_2$ and $\underset{1}{\chi}_{P_1} = S \times S$. Therefore

$$\underset{1}{\chi}_{P_0} = \underset{1}{\chi}_P \cap \underset{1}{\chi}_{P_1} = \underset{1}{\chi}_P \cap S \times S = \underset{1}{\chi}_P.$$

As in the preceding section, we seek a representation P in the form of a sum of a family of simplest representations $(P_{(\epsilon_i, W_i)})_{i \in I}$. Then

$$\underset{1}{\chi}_P = \bigcap_{i \in I} \underset{1}{\chi}_{P_{(\epsilon_i, W_i)}}.$$

We require that $\underset{1}{\chi} \subset \underset{1}{\chi}_P$. Obviously then

$$\underset{1}{\chi} \subset \underset{1}{\chi}_{P_{(\epsilon_i, W_i)}} \quad \text{for } i \in I. \tag{13}$$

Consider the relation $\underset{1}{\chi}_{P_{(\epsilon_i, W_i)}}$. We have

$$(s_1, s_2) \in \underset{1}{\chi}{}_{P_{(\epsilon_i, W_i)}} \longleftrightarrow \mathrm{pr}_1 P_{(\epsilon_i, W_i)}(s_1) \subset \mathrm{pr}_1 P_{(\epsilon_i, W_i)}(s_2)$$

$$\longleftrightarrow (\bigwedge a)[a \in \mathrm{pr}_1 P_{(\epsilon_i, W_i)}(s_1) \longrightarrow a \in \mathrm{pr}_1 P_{(\epsilon_i, W_i)}(s_2)]$$

$$\longleftrightarrow (\bigwedge a)[(\bigvee a_1)[(a, a_1) \in P_{(\epsilon_i, W_i)}(s_1)] \longrightarrow (\bigvee a_2)[(a, a_2) \in P_{(\epsilon_i, W_i)}(s_2)]]$$

$$\longleftrightarrow (\bigwedge a)[(\bigvee a_1)[H_a s_1 \subset H_{a_1}] \longrightarrow (\bigvee a_2)[H_a s_2 \subset H_{a_2}]]$$

$$\longleftrightarrow (\bigwedge a)[H_a s_1 \not\subset W_i \longrightarrow H_a s_2 \not\subset W_i] \longleftrightarrow (\bigwedge a)(\bigwedge x \in H_a)[xs_1 \notin W_i \longrightarrow xs_2 \notin W_i]$$

$$\longleftrightarrow (\bigwedge x \notin W_i)[xs_1 \notin W_i \longrightarrow xs_2 \notin W_i] \longleftrightarrow (\bigwedge x)[xs_1 \notin W_i \longrightarrow xs_2 \notin W_i]$$

$$\longleftrightarrow (\bigwedge x)[xs_2 \in W_i \longrightarrow xs_1 \in W_i].$$

Here we have made use of the facts that ϵ_i is right regular and W_i is a right ideal. We draw the important conclusion that $\underset{1}{\chi}{}_{P_{(\epsilon_i, W_i)}}$ depends only on the second component W_i of the determinative pair (ϵ_i, W_i). Clearly (13) implies that

$$(s_1, s_2) \in \underset{1}{\chi} \longrightarrow (\bigwedge x)[xs_2 \in W_i \longrightarrow xs_1 \in W_i]$$

or that

$$(s_1, s_2) \in \underset{1}{\chi} \wedge xs_2 \in W_i \longrightarrow xs_1 \in W_i.$$

Putting $x = e$, we obtain $(s_1, s_2) \in \underset{1}{\chi} \wedge s_2 \in W_i \longrightarrow s_1 \in W_i$. On the other hand, if $(s_1, s_2) \in \underset{1}{\chi} \wedge s_2 \in W_i \longrightarrow s_1 \in W_i$, then it follows from $(s_1, s_2) \in \underset{1}{\chi}$ and and the left regularity of $\underset{1}{\chi}$ that $(xs_1, xs_2) \in \underset{1}{\chi}$. Therefore $(s_1, s_2) \in \underset{1}{\chi} \wedge xs_2 \in W_i$ implies that $xs_i \in W_i$. We prove now that the left regularity of $\underset{1}{\chi}$ is necessary. For,

$$(s_1, s_2) \in \underset{1}{\chi}{}_P \longleftrightarrow \mathrm{pr}_1 P(s_1) \subset \mathrm{pr}_1 P(s_2)$$

$$\longrightarrow \mathrm{pr}_1 P(ss_1) = \mathrm{pr}_1 P(s_1) \circ P(s) = \overset{1}{\overline{P(s)}}(\mathrm{pr}_1 P(s_1)) \subset \overset{-1}{\overline{P(s)}}(\mathrm{pr}_1 P(s_2))$$

$$= \mathrm{pr}_1 P(s_2) \circ P(s) = \mathrm{pr}_1 P(ss_2)$$

$$\longleftrightarrow (ss_1, ss_2) \in \underset{1}{\chi}{}_P.$$

The left regularity of $\underset{1}{\chi}$ could have been observed from the beginning, but we arrived at this condition naturally here.

Thus (13) indicates that $(s_1, s_2) \in \underset{1}{\chi} \wedge s_2 \in W_i \longrightarrow s_1 \in W_i$, i.e., $\underset{1}{\chi}^{-1}(W_i) \subset W_i$.

In other words, the subset W_i must be a right ideal, minorantly saturated with respect to the quasi-order $\underset{1}{\chi}$. In fact, the first of these conditions follows from the second. For let H be a subset of S^* minorantly saturated with respect to $\underset{1}{\chi}$, and let $s \in H$. For any $s_2 \in S$,

$$\mathrm{pr}_1 P(s_1 s_2) = \mathrm{pr}_1 P(s_2) \circ P(s_1) \subset \mathrm{pr}_1 P(s_1),$$

so $\underset{1}{\chi}$ is *right negative*, i.e., $s_1 s_2 \vdash s_1$. Hence it follows that $ss_2 \in H$, i.e., H is a right ideal of S^*.

Note that for any right ideal $W \subset S^*$ there exists a determinative pair (ϵ, W) whose second component is W. (For ϵ one can take, for instance, the binary relation $\Delta_{S^*} \cup (W \times W)$ or ϵ_W. It is easily verified that both are right regular equivalence relations on S^*, and if $W \neq \varnothing$ then W is an equivalence class.) The representations $P_{(\epsilon_i, W_i)}$ must also satisfy $\bigcap_{i \in I} \underset{1}{\chi}_{P_{(\epsilon_i, W_i)}} \subset \underset{1}{\chi}$. Since

$$(s_1, s_2) \in \bigcap_{i \in I} \underset{1}{\chi}_{P_{(\epsilon_i, W_i)}} \longleftrightarrow (\bigwedge i)[(s_1, s_2) \in \underset{1}{\chi}_{P(\epsilon_i, W_i)}]$$

$$\longleftrightarrow (\bigwedge i)(\bigwedge x)[xs_2 \in W_i \longrightarrow xs_1 \in W_i],$$

we obtain

$$(\bigwedge i)(\bigwedge x)[xs_2 \in W_i \longrightarrow xs_1 \in W_i] \longrightarrow (s_1, s_2) \in \underset{1}{\chi}.$$

This condition holds, in particular, for $I = S$ and $W_i = \overset{-1}{\underset{1}{\chi}} \langle i \rangle$. In this case

$$(\bigwedge i)(\bigwedge x)[xs_2 \in W_i \longrightarrow xs_1 \in W_i] \longrightarrow (\bigwedge i)[s_2 \in W_i \longrightarrow s_1 \in W_i]$$

$$\longleftrightarrow (\bigwedge i)[s_i \vdash i \longrightarrow s_1 \vdash i] \longrightarrow (s_2 \vdash s_2 \longrightarrow s_1 \vdash s_2) \longleftrightarrow s_1 \vdash s_2.$$

Another variant is $P = \Sigma_{s \in S} P_{\underset{1}{\chi}\langle s \rangle}$.

Consequently we have proved the

THEOREM. *A quasi-ordered semigroup S is isomorphic to a first projection semigroup of transformations if and only if the quasi-order is left regular* ($s_1 \vdash s_2 \longrightarrow ss_1 \vdash ss_2$) *and right negative* ($s_1 s_2 \vdash s_1$).

A more extensive account of the results in this section can be found in [21] and [22].

*Let $R = \{(s_1, s_2) \in S \times S : (\bigvee x \in S^*)[s_1 = s_2 x]\}$ be the left divisibility relation on S. In other words, $(s_1, s_2) \in R$ if and only if $(s_1)_r \subset (s_2)_r$, where $(s)_r$ denotes the principal right ideal of S generated by s. If Q_1 is the set of all left regular quasi-orders on S, then $R \in Q$. On the other hand, a quasi-order $\zeta \in Q_1$ is right negative if and only if $R \in \zeta$. Thus ζ can be represented as a first projection quasi-order of S if and only if it belongs to the segment $[R, \infty)$ of Q_1.

*Clearly, the first projection quasi-order on semigroups of arbitrary binary relations is also left regular and right negative. Thus we arrive at the following theorem first proved in [24]:

*THEOREM. *A quasi-ordered semigroup is isomorphic to a first projection semigroup of binary relations if and only if the quasi-order is left regular and right negative.*

§7. Representation of ordered first projection semigroups

We will now find a system of elementary axioms for an algebra of transformations of the genre $(O_0, P_\zeta, P_{\underset{1}{\chi}})$ where O_0 is a formula defining multiplication, P_ζ defines inclusion, and $P_{\underset{1}{\chi}}$ defines the first projection quasi-order. In other words, we must find necessary and sufficient conditions for an algebraic system of the form $(S, \prec, \vdash)$, where S is a set with a given operation and binary relations $\prec$ and $\vdash$, to be isomorphic to a relation algebra of the genre $(O_0, P_\zeta, P_{\underset{1}{\chi}})$; It is natural to call such a system a *fundamentally ordered first projection semigroup of transformations.*

Using the results on the abstract characterizations of ζ and $\underset{1}{\chi}$ obtained in the preceding sections, we can observe beforehand that ζ must be a stable and weakly steady ordering and $\underset{1}{\chi}$ a left regular and right negative quasi-order.

It is necessary for us to construct an isomorphism P of an abstract relation algebra of genre $(O_0, P_\zeta, P_{\underset{1}{\chi}})$ into a concrete relation algebra of the same genre such that $\zeta = \zeta_P$ and $\underset{1}{\chi} = \underset{1}{\chi}_P$. We will assume that the desired isomorphism P is a sum of simplest representations (mentioning at once that, in contrast to the preceding problems, there may be no P which is the sum of elementary representations).

Let $P = \Sigma_{i \in I} P_{(\epsilon_i, W_i)}$. Then $\zeta = \zeta_P = \bigcap_{i \in I} \zeta_{P_{(\epsilon_i, W_i)}}$ and $\underset{1}{\chi} = \underset{1}{\chi}_P = \bigcap_{i \in I} \underset{1}{\chi}_{P_{(\epsilon_i, W_i)}}$. Consequently $\zeta \subset \zeta_{P_{(\epsilon_i, W_i)}}$ and $\underset{1}{\chi} \subset \underset{1}{\chi}_{P_{(\epsilon_i, W_i)}}$ for all $i \in I$. By virtue of (8), the first inclusion means that

$$s_1 \prec s_2 \wedge s_1 \notin W_i \longrightarrow s_1 \equiv s_2 \ (\epsilon_i). \tag{14}$$

As was proved in the preceding section, the second inclusion means that W_i is minorantly saturated with respect to $\underset{1}{\chi}$. The determinative pair $(\epsilon_{\zeta\langle s\rangle}, W_{\zeta\langle s\rangle})$ satisfies (14), and the determinative pair $(\epsilon_{\underset{1}{\chi}\langle s\rangle}, W_{\underset{1}{\chi}\langle s\rangle})$ satisfies the second condition. It is easy to see that $\zeta \subset \underset{1}{\chi}$. For,

$$s_1 \prec_P s_2 \longleftrightarrow P(s_1) \subset P(s_2) \longrightarrow \mathrm{pr}_1 P(s_1) \subset \mathrm{pr}_1 P(s_2) \longleftrightarrow s_1 \underset{P}{\vdash} s_2.$$

Thus it follows without difficulty that $\zeta\langle s\rangle \subset \underset{1}{\chi}\langle s\rangle$ and $W_{\underset{1}{\chi}\langle s\rangle} \subset W_{\zeta\langle s\rangle}$. However, the pair $(\epsilon_{\underset{1}{\chi}\langle s\rangle}, W_{\underset{1}{\chi}\langle s\rangle})$ does not satisfy (14), and generally speaking, $W_{\zeta\langle s\rangle}$ is not minorantly saturated with respect to $\underset{1}{\chi}$. We will attempt to construct a determinative pair of the form $(\epsilon_s, W_{\underset{1}{\chi}\langle s\rangle})$, where ϵ_s "resembles" $\epsilon_{\zeta\langle s\rangle}$ so much that ϵ_s satisfies (14). Since $\underset{1}{\chi}$ is right negative, it follows that

$$s_1 \in W_{\underset{1}{\chi}\langle s\rangle} \longleftrightarrow (\bigwedge x)[s_1 x \notin \underset{1}{\chi}\langle s\rangle] \longleftrightarrow (\bigwedge x)\ [s \nvdash s_1 x] \longleftrightarrow s \nvdash s_1$$

$$\longleftrightarrow s_1 \notin \underset{1}{\chi}\langle s\rangle;$$

that is, $W_{\underset{1}{\chi}\langle s\rangle} = \underset{1}{\chi}'\langle s\rangle$, where $\underset{1}{\chi}'$ is the complement of $\underset{1}{\chi}$. We will assume that if $W_{\underset{1}{\chi}\langle s\rangle} \neq \varnothing$, then $W_{\underset{1}{\chi}\langle s\rangle}$ is an ϵ_s-class, $s_1 \equiv s_2\ (\epsilon_s)$, and $s_1 x \notin W_{\underset{1}{\chi}\langle s\rangle}$. From the right regularity of ϵ_s it follows that $s_2 x \in W'_{\underset{1}{\chi}\langle s\rangle}$. Conversely, $s_2 x \in W'_{\underset{1}{\chi}\langle s\rangle} \longrightarrow s_1 x \in W'_{\underset{1}{\chi}\langle s\rangle}$, i.e., $s_1 \equiv s_2\ (\epsilon_{W'_{\underset{1}{\chi}\langle s\rangle}})$. Therefore $\epsilon_s \subset \epsilon_{W'_{\underset{1}{\chi}\langle s\rangle}} = \epsilon_{\underset{1}{\chi}\langle s\rangle}$. Put $\epsilon_s = \epsilon_{\zeta\langle s\rangle} \cap \epsilon_{\underset{1}{\chi}\langle s\rangle}$. If $W_{\underset{1}{\chi}\langle s\rangle} \neq \varnothing$, then $W_{\underset{1}{\chi}\langle s\rangle}$ is an $\epsilon_{\underset{1}{\chi}\langle s\rangle}$-class and $W_{\underset{1}{\chi}\langle s\rangle}$ is included in the $\epsilon_{\zeta\langle s\rangle}$-class $W_{\zeta\langle s\rangle}$. Therefore $W_{\underset{1}{\chi}\langle s\rangle}$ is an ϵ_s-class.

It is clear that ϵ_s is a right regular equivalence relation, whence $(\epsilon_s, W_{\underset{1}{\chi}\langle s\rangle})$ is a determinative pair. For this pair, (14) assumes the form

$$s_1 \prec s_2 \wedge s \vdash s_1 \longrightarrow s_1 \equiv s_2\ (\epsilon_{\zeta\langle s\rangle} \cap \epsilon_{\underset{1}{\chi}\langle s\rangle}).$$

Applying the definitions of $\epsilon_{\zeta\langle s\rangle}$ and $\epsilon_{\underset{1}{\chi}\langle s\rangle}$, we write this as the conjunction of the four terms:

$$\begin{array}{ll} \text{(I)} & s_1 \prec s_2 \wedge s \vdash s_1 \wedge s \prec s_1 y \longrightarrow s \prec s_2 y, \\ \text{(II)} & s_1 \prec s_2 \wedge s \vdash s_1 \wedge s \prec s_2 y \longrightarrow s \prec s_1 y, \\ \text{(III)} & s_1 \prec s_2 \wedge s \vdash s_1 \wedge s \prec s_1 y \longrightarrow s \vdash s_2 y, \\ \text{(IV)} & s_1 \prec s_2 \wedge s \vdash s_1 \wedge s \prec s_2 y \longrightarrow s \vdash s_1 y. \end{array} \qquad (15)$$

These four conditions together with those which we have already assumed, namely that the operation on S is associative, ζ is a stable ordering, $\underset{1}{\chi}$ is a left regular and right negative quasi-order, and $\zeta \subset \underset{1}{\chi}$, are sufficient conditions for representability, since

$$\bigcap_{s\in S} \zeta_{(\epsilon_s, W_{\underset{1}{\chi}\langle s\rangle})} \subset \zeta \quad \text{and} \quad \bigcap_{s\in S} \underset{1}{\chi}_{(\epsilon_s, W_{\underset{1}{\chi}\langle s\rangle})} \subset \underset{1}{\chi}.$$

Indeed, let

$$(s_1, s_2) \in \bigcap_{s\in S} \xi'_{(\epsilon_s, W_{\underset{1}{\chi}\langle s\rangle})};$$

then

$$(s_1, s_2) \in \zeta_{(\epsilon_{s_1}, W_{\underset{1}{\chi}<s_1>})},$$

which implies that $xs_1 \notin W_{\underset{1}{\chi}\langle s\rangle} \longrightarrow xs_1 \equiv xs_2\ (\epsilon_{s_1})$. Putting $x = e$ yields $s_1 \in \underset{1}{\chi}\langle s_1\rangle \longrightarrow s_1 \equiv s_2\ (\epsilon_{s_1})$, i.e., $s_1 \equiv s_2\ (\epsilon_{s_1})$. Therefore $s_1 \equiv s_2\ (\epsilon_{\zeta\langle s_1\rangle})$, i.e., $s_1 y \in \zeta\langle s_1\rangle \longleftrightarrow s_2 y \in \zeta\langle s_1\rangle$. Putting $y = e$ yields $s_1 \in \zeta\langle s_1\rangle \longleftrightarrow s_2 \in \zeta\langle s_1\rangle$, i.e., $s_2 \subset \zeta\langle s_1\rangle$ and $(s_1, s_2) \in \zeta$. The inclusion

$$\bigcap_{s\in S} \underset{1}{\chi}_{(\epsilon_s, W_{\underset{1}{\chi}\langle s\rangle})} \subset \underset{1}{\chi}$$

is proved similarly.

It remains to prove that formulas (15) are necessary.

We shall show that 15(I) and 15(III) yield no new information, and so can be discarded.

First we prove this for 15(I). Let the premise hold, i.e., $s_1 \prec s_2 \wedge s \vdash s_1 \wedge s \prec s_1 y$. Since ζ is right regular and $\zeta \subset \underset{1}{\chi}$, $s_1 y \prec s_2 y \wedge s \vdash s_1 y \wedge s \prec s_1 y$ holds, and since ζ is transitive, $s \prec s_2 y$. Thus 15(I) follows from the right regularity and transitivity of ζ and the inclusion $\zeta \subset \underset{1}{\chi}$. Similarly, 15(III) follows from the right regularity of ζ, transitivity of $\underset{1}{\chi}$, and the fact that $\zeta \in \underset{1}{\chi}$.

Thus only conditions 15(II) and 15(IV) remain. As it turns out, these conditions provide new information on our algebra of transformations. We prove the necessity of 15(II), which takes the following form for transformations:

$$s_1 \subset s_2 \wedge \mathrm{pr}_1 s \subset \mathrm{pr}_1 s_1 \wedge s \subset y \circ s_2 \longrightarrow s \subset y \circ s_1.$$

If the premise is valid, then $\Delta_{\mathrm{pr}_1 s} \subset \Delta_{\mathrm{pr}_1 s_1}$ and $s = s \circ \Delta_{\mathrm{pr}_1 s} \subset y \circ s_2 \circ \Delta_{\mathrm{pr}_1 s_1}$. Since $s_1 \subset s_2$ and $s_2 \circ \Delta_{\mathrm{pr}_1 s_1} = s_1$, we obtain $s \subset y \circ s_1$, thus proving our assertion.

We prove further that when 15(IV) holds, 15(II) is equivalent to

$$s_1 \prec s_2 \wedge s \vdash s_1 \wedge s \prec s_2 \longrightarrow s \prec s_1. \tag{16}$$

It is obvious that by putting $y = e$ we have 15(II) $\longrightarrow$ (16). Conversely, suppose that (16) and 15(IV) hold, and that $s_1 \prec s_2 \wedge s \vdash s_1 \wedge s \prec s_2 y$ (the premise of 15(II)). We will show that $s \prec s_1 y$. Since $\zeta \subset \underset{1}{\chi}$, we have $s \prec s_2 y \longrightarrow s \vdash s_2 y$, so the premise of 15(IV) holds. Therefore the conclusion holds too, so $s \vdash s_1 y$. The right regularity of ζ implies that $s_1 \prec s_2 \longrightarrow s_1y \prec s_2y$. Thus $s_1y \prec s_2y \wedge s \sqsubset s_1y \wedge s \prec s_2y$, which is the premise of (16) obtained by substituting s_1y for s_1 and s_2y for s_2. Hence $s \prec s_1y$, which was to be proved.

We will write (16) in the following equivalent form:

$$(s_1, s_2) \in \zeta \wedge (s, s_2) \in \zeta \wedge (s, s_1) \in \underset{1}{\chi} \longrightarrow (s, s_1) \in \zeta,$$

or, what is the same,

$$(s, s_1) \in \overset{-1}{\zeta} \circ \zeta \wedge (s, s_1) \in \underset{1}{\chi} \longrightarrow (s, s_1) \in \zeta,$$

or

$$(s, s_1) \in (\overset{-1}{\zeta} \circ \zeta \cap \underset{1}{\chi}) \longrightarrow (s, s_1) \in \zeta,$$

which in turn can be written $\overset{-1}{\zeta} \circ \zeta \cap \underset{1}{\chi} \subset \zeta$. Since $\Delta_S \subset \zeta$ and $\zeta \subset \underset{1}{\chi}$, we finally have $\zeta = \overset{-1}{\zeta} \circ \zeta \cap \underset{1}{\chi}$.

It remains to prove that 15(IV) is necessary. Without loss of generality this condition can be rewritten as

$$s_1 \prec s_2 \wedge s \sqsubset s_1 \wedge s \sqsubset s_2s_3 \longrightarrow s \sqsubset s_1s_3 \tag{17}$$

(since the case $y = e$ holds trivially). For transformations, (17) can be rewritten:

$$s_1 \subset s_2 \wedge \mathrm{pr}_1 s \subset \mathrm{pr}_1 s_1 \wedge \mathrm{pr}_1 s \subset \mathrm{pr}_1 s_3 \circ s_2 \longrightarrow \mathrm{pr}_1 s \subset \mathrm{pr}_1 s_3 \circ s_1.$$

Let the premise hold. Then $s_1 \circ \Delta_{\mathrm{pr}_1 s} \subset s_2 \circ \Delta_{\mathrm{pr}_1 s}$. But

$$\mathrm{pr}_1(s_1 \circ \Delta_{\mathrm{pr}_1 s}) = \mathrm{pr}_1 s \text{ since } \mathrm{pr}_1 s \subset \mathrm{pr}_1 s_1,$$

and

$$\mathrm{pr}_1(s_2 \circ \Delta_{\mathrm{pr}_1 s}) = \mathrm{pr}_1 s \text{ since } \mathrm{pr}_1 s \subset \mathrm{pr}_1 s_2.$$

Therefore $s_1 \circ \Delta_{\mathrm{pr}_1 s} = s_2 \circ \Delta_{\mathrm{pr}_1 s}$. Hence we have

$$(s_3 \circ s_1) \circ \Delta_{\mathrm{pr}_1 s} = s_3 \circ s_2 \circ \Delta_{\mathrm{pr}_1 s}.$$

Since $\mathrm{pr}_1 s \subset \mathrm{pr}_1 s_3 \circ s_2$, it follows that

$$\mathrm{pr}_1((s_3 \circ s_1) \circ \Delta_{\mathrm{pr}_1 s}) = \mathrm{pr}_1 s.$$

It obviously follows that $\mathrm{pr}_2 \circ \Delta_{\mathrm{pr}_1 s} \subset \mathrm{pr}_1(s_3 \circ s_1)$, or equivalently, $\mathrm{pr}_1 s \subset \mathrm{pr}_1(s_3 \circ s_1)$, which completes our proof.

Thus we have found a complete system of axioms characterizing a relation algebra of the genre $(O_0, P_\zeta, P_{\underset{1}{\chi}})$. Note that ζ is necessarily weakly steady. Since we assert that we have obtained a system of axioms, it must yield all elementary properties of the relations ζ and $\underset{1}{\chi}$; in particular, the weak steadiness of ζ. As an example, we will show that the axioms we have obtained imply that ζ is weakly steady.

Thus, assume the premise for weak steadiness holds, i.e., $x \prec u \wedge z \prec xv \wedge z \prec uy$. We will show that $z \prec xy$. To this end, we will use of (17). Let $s = z$, $s_1 = x$, $s_2 = u$ and $s_3 = y$. We obtain $x \prec u \wedge z \vdash x \wedge z \prec uy \longrightarrow z \prec xy$. Now $\zeta \subset \underset{1}{\chi}$, whence $z \prec xv \longrightarrow z \vdash xv \vdash x$. We thus see that the premise $x \prec u \wedge z \vdash x \wedge z \prec uy$ of (17) holds. Consequently $z \prec xy$, as desired. We have proved the

THEOREM. *An ordered and quasi-ordered semigroup S is isomorphic to a fundamentally ordered first projection semigroup of transformations if and only if*

1) *$\underset{1}{\chi}$ is a left regular, right negative quasi-order on S,*
2) *ζ is a stable ordering, and*
3) *ζ and $\underset{1}{\chi}$ are connected by the two conditions*

$$\zeta = \underset{1}{\chi} \cap \overset{-1}{\zeta} \circ \zeta \text{ and } s \vdash s_1 \wedge s \vdash s_2 s_3 \wedge s_1 \prec s_2 \longrightarrow s \vdash s_1 s_3.$$

The proof of this result can also be found in [25] or [19]. For its applications to automata theory, see [27].

*The following result is proved in [24]:

THEOREM. *An ordered and quasi-ordered semigroup is isomorphic to a fundamentally ordered first projection semigroup of binary relations if and only if the conditions* 1) *and* 2) *of the preceding theorem hold and $\zeta \subset \underset{1}{\chi}$.*

*OUTLINE OF THE PROOF. The necessity of these conditions must be obvious by now. To prove sufficiency suppose that

$$P_1(s) = \{(x, y) \in S \times S^*\colon x \vdash sy\}.$$

It is easy to verify that P_1 is a representation of S by binary relations and $\underset{1}{\chi} = \underset{1}{\chi}_{P_1}$. If $(s_1, s_2) \in \zeta$ and $(x, y) \in P_1(s_1)$, then $x \vdash s_1 y$. By right regularity of ζ we have $s_1 y \prec s_2 y$. By $\zeta \subset \underset{1}{\chi}$ we have $s_1 y \vdash s_2 y$. By transitivity of $\vdash$ we have $x \vdash s_2 y$, i.e. $(x, y) \in P_1(s_2)$. Thus $P(s_1) \subset P(s_2)$ and $(s_1, s_2) \vdash \zeta_{P_1}$. We have proved that $\zeta \subset \zeta_{P_1}$.

*Now let

$$P_2(s) = \{(x, y) \in S \times S^*\colon sy \prec x\}.$$

It is easy to verify that P_2 is a representation of S by binary relations and $\zeta_{P_2} = \overset{-1}{\zeta}$. Thus P_2 is faithful, for

$$\epsilon_{P_2} = \zeta_{P_2} \cap \overset{-1}{\zeta}_{P_2} = \zeta \cap \overset{-1}{\zeta} = \Delta_S,$$

the diagonal. Define a binary relation $P_0(s)$ between the subsets of S^* as follows: if $U, V \subset S^*$, then

$$(U, V) \in P_0(s) \longleftrightarrow P_2(s)(U) \subset V,$$

where

$$P_2(s)(U) = \{v: (\forall u \in U)[(u, v) \in P_2(s)]\}.$$

It is not difficult to verify that P_0 is a representation of S by binary relations and

$$P_2(s_1) \subset P_2(s_2) \longleftrightarrow P_0(s_2) \subset P_0(s_1).$$

It follows that $\zeta_{P_0} = \overset{-1}{\zeta_{P_2}} = \zeta$. Now $\mathrm{pr}_1 P_0(s)$ is the set of all subsets of S^*; thus $\mathrm{pr}_1 P_0(s_1) \subset \mathrm{pr}_1 P_0(s_2)$ for all $s_1, s_2 \in S$, i.e. $\underset{1}{\chi}_{P_0} = S \times S$.

*Let $P = P_0 + P_1$. Then P is faithful representation, since P_0 is. Now

$$\zeta_P = \zeta_{P_0} \cap \zeta_{P_1} = \zeta \cap \zeta_{P_1} = \zeta \qquad \text{and} \qquad \underset{1}{\chi}_P = \underset{1}{\chi}_{P_0} \cap \underset{1}{\chi}_{P_1} = (S \times S) \cap \underset{1}{\chi} = \underset{1}{\chi}.$$

Thus P is the representation needed. It effectuates an isomorphism of $(S, \prec, \vdash)$ onto a fundamentally ordered first projection semigroup of binary relations.

§8. Representation of second projection semigroups

Our next task is to find necessary and sufficient conditions for the existence of a faithful representation P of an abstract semigroup S, endowed with a quasi-order $\underset{2}{\chi}(\llcorner\!\!-)$, by transformations which map $\underset{2}{\chi}$ onto the second projection quasi-order $\underset{2}{\chi}_P(\llcorner\!\!-_P)$, i.e., such that $\underset{2}{\chi} = \underset{2}{\chi}_P$. In other words, we seek an abstract characterization of the algebras of transformations of genre $\sigma = (O_0, P_{\underset{2}{\chi}})$, where O_0 is a formula defining multiplication of transformations and $P_{\underset{2}{\chi}}$ defines inclusion of the second projections of transformations (i.e., a characterization of second semigroups of transformations). Necessary conditions for the existence of a faithful representation P can be indicated by a manner similar to that for $\underset{1}{\chi}$, namely, $\underset{2}{\chi}$ must be a right regular left negative quasi-order, i.e., $s_1 \llcorner\!\!- s_2 \longrightarrow s_1 s \llcorner\!\!- s_2 s$ and $s_1 s_2 \llcorner\!\!- s_2$.

That these conditions are necessary follows from the fact that the formulas expressing them for transformations, namely

$$\mathrm{pr}_2 s_1 \subset \mathrm{pr}_2 s_2 \longrightarrow \mathrm{pr}_2(s \circ s_1) \subset \mathrm{pr}_2(s \circ s_2) \qquad \text{and} \qquad \mathrm{pr}_2(s_2 \circ s_1) \subset \mathrm{pr}_2 s_2,$$

obviously hold not only for transformations, but for all binary relations. *These conditions are sufficient for characterization of second projection semigroups of arbitrary binary relations [24]. However, they are not sufficient for the existence of a representation of a given quasi-ordered semigroup by transformations. Previously, in solving analogous problems we succeeded in finding such a P which was a sum of simplest, and sometimes even elementary, representations. But for the problem at hand it turns out that the desired representation P, even if it exists, sometimes cannot be a sum of simplest representations. Nevertheless, in order to construct P as a sum of representations, we define certain representations by means of left translations extended to subsets of the semigroup S^*.

Let H be a subset of S^*. The quotient $H \mathbin{.\cdot} s \subset S^*$ is defined by

$$H \mathbin{.\cdot} s = \{x \in S^* : sx \in H\},$$

where $s \in S$ and $x \in S^*$. One of the basic properties of this division operation is expressed by the equality

$$(H \mathbin{.\cdot} s) \mathbin{.\cdot} t = H \mathbin{.\cdot} st, \tag{18}$$

which follows from

$$x \in (H \mathbin{.\cdot} s) \mathbin{.\cdot} t \longleftrightarrow tx \in H \mathbin{.\cdot} s \longleftrightarrow s(tx) \in H \longleftrightarrow (st)x \in H \longleftrightarrow x \in H \mathbin{.\cdot} st.$$

Now we will construct a representation P of S by the following transformations of the set $\mathfrak{P}(S^*)$ of all subsets of S^*:

$$(\bigwedge s \in S)(\bigwedge H \subset S^*)\ [P(s)(H) = H \mathbin{.\cdot} s].$$

Obviously this representation is everywhere defined. We prove that P is a homomorphism:

$$\begin{aligned}(H_1, H_2) \in P(s_2) \circ P(s_1) &\longleftrightarrow (\bigvee H)\,[(H_1, H) \in P(s_1) \wedge (H, H_2) \in P(s_2)]\\ &\longleftrightarrow (\bigvee H)\,[H_1 \mathbin{.\cdot} s_1 = H \wedge H \mathbin{.\cdot} s_2 = H_2]\\ &\longleftrightarrow (\bigvee H)\,[H_1 \mathbin{.\cdot} s_1 = H \wedge H_1 \mathbin{.\cdot} s_1 s_2 = H_2]\\ &\longleftrightarrow H_1 \mathbin{.\cdot} s_1 s_2 = H_2 \longleftrightarrow (H_1, H_2) \in P(s_1 s_2).\end{aligned}$$

Note that we made use of property (10) of division, and the fact that since the expression $(\bigvee H)\,[H_1 \mathbin{.\cdot} s_1 = H]$ always holds, it can be omitted. It is easy to prove that P is an isomorphism, i.e., a faithful representation. For if $P(s_1) = P(s_2)$ then obviously $P(s_1)(\{s_1\}) = P(s_2)(\{s_1\})$, or $\{s_1\} \mathbin{.\cdot} s_1 = \{s_1\} \mathbin{.\cdot} s_2$. Clearly $e \in \{s_1\} \mathbin{.\cdot} s_1 = \{s_1\} \mathbin{.\cdot} s_2$, i.e., $s_2 = s_2 e \in \{s_1\}$. This means that $s_1 = s_2$, which proves our assertion.

We shall restrict P. First of all we extend the quasi-order $\underset{2}{\chi}$ to S^*. We shall assume that

$$(\bigwedge s \in S)[\underset{2}{\chi}^*\langle s\rangle = \underset{2}{\chi}\langle s\rangle \cup \{e\}],$$

where $\underset{2}{\chi}^*\langle s\rangle \subset S^*$, and consider subsets of the form $\mathfrak{P}(\underset{2}{\chi}^*\langle s\rangle) \subset \mathfrak{P}(S^*)$. We restrict P to such subsets, i.e., we construct the restriction P_s of P in the following manner:

$$(H_1, H_2) \in P_s(s_1) \longleftrightarrow H_1 \mathbin{.\cdot} s_1 = H_2 \wedge H_1, H_2 \subset \underset{2}{\chi}^*\langle s\rangle.$$

Using the fact that $\underset{2}{\chi}^*$ is left negative, it is easy to see that $H \subset \underset{2}{\chi}^*\langle s\rangle \longrightarrow H \mathbin{.\cdot} s_1 \subset \underset{2}{\chi}^*\langle s\rangle$. Then P_s can be defined in a simpler way:

$$(H_1, H_2) \in P_s(s_1) \longleftrightarrow H_1 \mathbin{.\!\cdot} s_1 = H_2 \wedge H_1 \subset \underset{2}{\chi}^*\langle s\rangle.$$

That P_s is indeed a representation can be seen from the following:

$$(H_1, H_2) \in P_s(s_2) \circ P_s(s_1) \longleftrightarrow (\bigvee H)[(H_1, H) \in P_s(s_1) \wedge (H, H_2) \in P_s(s_2)]$$

$$\longleftrightarrow (\bigvee H)[H_1 \mathbin{.\!\cdot} s_1 = H \wedge H_1 \subset \underset{2}{\chi}^*\langle s\rangle \wedge H \mathbin{.\!\cdot} s_2 = H_2 \wedge H \subset \underset{2}{\chi}^*\langle s\rangle]$$

$$\longleftrightarrow ((\bigvee H)[H_1 \mathbin{.\!\cdot} s_1 = H] \wedge H_1 \in \underset{2}{\chi}^*\langle s\rangle \wedge H_1 \mathbin{.\!\cdot} s_1 s_2 = H_2 \wedge H_1 \mathbin{.\!\cdot} s_2 \subset \underset{2}{\chi}^*\langle s\rangle)$$

$$\longleftrightarrow (H_1 \subset \underset{2}{\chi}^*\langle s\rangle \wedge H_1 \mathbin{.\!\cdot} s_1 s_2 = H_2) \longleftrightarrow (H_1, H_2) \in P_s(s_1 s_2),$$

i.e., $P_s(s_2) \circ P_s(s_1) = P_s(s_1 s_2)$.

For each element $s \in S$ we thus construct a representation P_s. Now we shall try to write the required representation P as a sum of simplest representations, i.e., $P = \Sigma_{s\in S} P_s$. First of all, we show that $\epsilon_P = \bigcap_{s\in S} \epsilon_{P_s} = \Delta_S$. Let $s_1 \equiv s_2\ (\epsilon_P)$, or, equivalently,

$$(\bigwedge s)[P_s(s_1) = P_s(s_2)].$$

Then $P_{s_1}(s_1) = P_{s_1}(s_2)$, whence

$$P_{s_1}(s_1)(\{s_1\}) = P_{s_1}(s_2)(\{s_1\}).$$

Since $\underset{2}{\chi}$ is reflexive, we have $\{s_1\} \subset \underset{2}{\chi}^*\langle s_1\rangle$, so $\{s_1\} \mathbin{.\!\cdot} s_1 = \{s_1\} \mathbin{.\!\cdot} s_2$. Since $e \in \{s_1\} \mathbin{.\!\cdot} s_1$, it follows that $s_2 = s_2 e \in \{s_1\}$, i.e., $s_1 = s_2$, which was to be proved.

It remains to clarify under what conditions we have $\underset{2}{\chi}_P = \bigcap_{s\in S} \underset{2}{\chi}_{P_s} = \underset{2}{\chi}$, or

$$\bigcap_{s\in S} (\underset{2}{\chi}_{P_s}) \subset \underset{2}{\chi} \wedge \underset{2}{\chi} \subset \bigcap_{s\in S} \underset{2}{\chi}_{P_s}. \tag{19}$$

We prove the first part of (19). If $(s_1, s_2) \in \bigcap_{s\in S} \underset{2}{\chi}_{P_s}$, then $(s_1, s_2) \in \underset{2}{\chi}_{P_{s_1}}$. We have already seen that $\{s_1\} \mathbin{.\!\cdot} s_1 \in \mathrm{pr}_2 P_{s_1}(s_1)$. Since $\mathrm{pr}_2 P_{s_1}(s_1) \subset \mathrm{pr}_2 P_{s_1}(s_2)$, we obtain $(H, \{s_1\} \mathbin{.\!\cdot} s_1) \in P_{s_1}(s_2)$ for some $H \subset \mathfrak{P}(\underset{2}{\chi}^*\langle s_1\rangle)$. Therefore $e \in \{s_1\} \mathbin{.\!\cdot} s_1 = H \mathbin{.\!\cdot} s_2$, i.e., $s_2 = s_2 e \in H$. Consequently $s_2 \in \underset{2}{\chi}^*\langle s_1\rangle$, i.e., $(s_1, s_2) \in \underset{2}{\chi}$.

Now we will find out under what conditions the second term in (19) holds. This term means that

$$\bigwedge (s_1, s_2, s)\ [(s_1, s_2) \in \underset{2}{\chi} \longrightarrow (\bigwedge H_1)(\bigvee H_2)[H_1 \subset \underset{2}{\chi}^*\langle s\rangle \longrightarrow$$

$$H_1 \mathbin{.\!\cdot} s_1 = H_2 \mathbin{.\!\cdot} s_2 \wedge H_2 \subset \underset{2}{\chi}^*\langle s\rangle]\,],$$

or, equivalently,

$$(\bigwedge H_1)(\bigvee H_2)[s_1 \mathrel{\llcorner\!\!-} s_2 \wedge H_1 \subset \underset{2}{\chi}^*\langle s\rangle \longrightarrow H_1 \mathbin{.\cdot} s_1 = H_2 \mathbin{.\cdot} s_2 \wedge H_2 \subset \underset{2}{\chi}^*\langle s\rangle] \tag{20}$$

(the quantifiers $(\bigwedge s_1)$, $(\bigwedge s_2)$ and $(\bigwedge s)$ have been omitted). Condition (20) is sufficient for the existence of the desired representation P. But it is not satisfactory, since, in the first place, it is not a universal axiom (the existential quantifier $(\bigvee H_2)$ is present), and secondly, it is an axiom of the second degree (the quantifiers bind set variables). Therefore first of all we try to get rid of the existential quantifier $(\bigvee H_2)$. By the condition we know that for each H_1 there exists at least one H_2. We will attempt to construct some concrete subset H_2.

From the equality $H_1 \mathbin{.\cdot} s_1 = H_2 \mathbin{.\cdot} s_2$, which appears in (20), it follows that

$$H_1 \mathbin{.\cdot} s_1 \subset H_2 \mathbin{.\cdot} s_2 \wedge H_2 \mathbin{.\cdot} s_2 \subset H_1 \mathbin{.\cdot} s_1.$$

By the definition of the division operation, the inclusion $H_1 \mathbin{.\cdot} s_1 \subset H_2 \mathbin{.\cdot} s_2$ can also be written $s_2(H_1 \mathbin{.\cdot} s_1) \subset H_2$. Hence (fiom the obvious fact that the relation of inclusion is stable with respect to division) we obtain $s_2(H_1 \mathbin{.\cdot} s_1) \mathbin{.\cdot} s_2 \subset H_2 \mathbin{.\cdot} s_2$. Since $H_2 \mathbin{.\cdot} s_2 \subset H_1 \mathbin{.\cdot} s_1$, it follows that $s_2(H_1 \mathbin{.\cdot} s_1) \mathbin{.\cdot} s_2 \subset H_1 \mathbin{.\cdot} s_1$. On the other hand, it follows again from the definition of division that $H_1 \mathbin{.\cdot} s_1 \subset s_2(H_1 \mathbin{.\cdot} s_1) \mathbin{.\cdot} s_2$. Thus $H_1 \mathbin{.\cdot} s_1 = s_2(H_1 \mathbin{.\cdot} s_1) \mathbin{.\cdot} s_2$.

We proved above that $s_2(H_1 \mathbin{.\cdot} s_1) \subset H_2$, but by (20), $H_2 \subset \underset{2}{\chi}^*\langle s\rangle$, whence $s_2(H_1 \mathbin{.\cdot} s_1) \subset \underset{2}{\chi}^*\langle s\rangle$. Thus for H_2 we can take the subset $s_2(H_1 \mathbin{.\cdot} s_1)$. Then (20) is equivalent to the formula

$$(\bigwedge H)[s_1 \mathrel{\llcorner\!\!-} s_2 \wedge H \subset \underset{2}{\chi}^*\langle s\rangle \longrightarrow H \mathbin{.\cdot} s_1 = s_2(H \mathbin{.\cdot} s_1) \mathbin{.\cdot} s_2 \wedge s_2(H \mathbin{.\cdot} s_1) \subset \underset{2}{\chi}^*\langle s\rangle], \tag{21}$$

which is universal. Now it is necessary to simplify it in order that it become elementary. Since $H \mathbin{.\cdot} s_1 \subset s_2(H \mathbin{.\cdot} s_1) \mathbin{.\cdot} s_2$ (from the definition of division), (21) can be written as

$$(\bigwedge H)[s_1 \mathrel{\llcorner\!\!-} s_2 \wedge H \subset \underset{2}{\chi}^*\langle s\rangle \longrightarrow s_2(H \mathbin{.\cdot} s_1) \mathbin{.\cdot} s_2 \subset H \mathbin{.\cdot} s_1 \wedge s_2(H \mathbin{.\cdot} s_1) \subset \underset{2}{\chi}^*\langle s\rangle].$$

Thus we obtain the following equivalent formula:

$$\begin{aligned}&(\bigwedge H)[s_1 \mathrel{\llcorner\!\!-} s_2 \wedge H \subset \underset{2}{\chi}^*\langle s\rangle \longrightarrow s_2(H \mathbin{.\cdot} s_1) \mathbin{.\cdot} s_2 \subset H \mathbin{.\cdot} s_1].\\ &\wedge (\bigwedge H)[s_1 \mathrel{\llcorner\!\!-} s_2 \wedge H \subset \underset{2}{\chi}^*\langle s\rangle \longrightarrow s_2(H \mathbin{.\cdot} s_1) \subset \underset{2}{\chi}^*\langle s\rangle].\end{aligned} \tag{22}$$

We will consider the implications in these conjunctive terms. We have

$$s_2(H \mathbin{.\cdot} s_1) \mathbin{.\cdot} s_2 \subset H \mathbin{.\cdot} s_1 \longleftrightarrow (\bigwedge x)\,[x \in s_2(H \mathbin{.\cdot} s_1) \mathbin{.\cdot} s_2 \longrightarrow x \in H \mathbin{.\cdot} s_1]$$

$$\longleftrightarrow (\bigwedge x)\,[s_2 x \in s_2(H \mathbin{.\cdot} s_1) \longrightarrow s_1 x \in H]$$

$$\longleftrightarrow (\bigwedge x)[(\bigvee y)[y \in H \mathbin{.\cdot} s_1 \wedge s_2 x = s_2 y] \longrightarrow s_1 x \in H]$$

$$\longleftrightarrow (\bigwedge x)(\bigwedge y)[s_1 y \in H \wedge s_2 x = s_2 y \longrightarrow s_1 x \in H],$$

$$s_2(H \mathbin{.\cdot} s_1) \subset \underset{2}{\chi}{}^{*}\langle s\rangle \longleftrightarrow (\bigwedge t)[t \in s_2(H \mathbin{.\cdot} s_1) \longrightarrow t \in \underset{2}{\chi}{}^{*}\langle s\rangle]$$

$$\longleftrightarrow (\bigwedge t)[(\bigvee z)[z \in H \mathbin{.\cdot} s_1 \wedge t = s_2 z] \longrightarrow t \in \underset{2}{\chi}{}^{*}\langle s\rangle]$$

$$\longleftrightarrow (\bigwedge t)(\bigwedge z)[t = s_2 z \wedge s_1 z \in H \longrightarrow s_2 z \in \underset{2}{\chi}{}^{*}\langle s\rangle]$$

$$\longleftrightarrow (\bigwedge z)\,[s_1 z \in H \longrightarrow s_2 z \in \underset{2}{\chi}{}^{*}\langle s\rangle].$$

Then (22) assumes the form

$$(\bigwedge H)\,[s_1 \mathrel{\llcorner\!\!-} s_2 \wedge H \subset \underset{2}{\chi}{}^{*}\langle s\rangle \longrightarrow (s_1 y \in H \wedge s_2 x = s_2 y \longrightarrow s_1 x \in H)]$$

$$\wedge (\bigwedge H)[s_1 \mathrel{\llcorner\!\!-} s_2 \wedge H \subset \underset{2}{\chi}{}^{*}\langle s\rangle \longrightarrow (s_1 z \in H \longrightarrow s_2 z \in \underset{2}{\chi}{}^{*}\langle s\rangle)],$$

or, by the conjunction of the premises,

$$(\bigwedge H)[s_1 \mathrel{\llcorner\!\!-} s_2 \wedge H \subset \underset{2}{\chi}{}^{*}\langle s\rangle \wedge s_1 y \in H \wedge s_2 x = s_2 y \longrightarrow s_1 x \in H]$$

$$\wedge (\bigwedge H)\,[s_1 \mathrel{\llcorner\!\!-} s_2 \wedge H \subset \underset{2}{\chi}{}^{*}\langle s\rangle \wedge s_1 z \in H \longrightarrow s_2 z \in \underset{2}{\chi}{}^{*}\langle s\rangle]. \qquad (23)$$

We shall see what information the second term in (23) furnishes. By simplifying it, we obtain

$$(\bigwedge H)\,[s_1 \mathrel{\llcorner\!\!-} s_2 \wedge H \subset \underset{2}{\chi}{}^{*}\langle s\rangle \wedge s_1 z \in H \longrightarrow s_2 z \in \underset{2}{\chi}{}^{*}\langle s\rangle]$$

$$\longleftrightarrow [s_1 \mathrel{\llcorner\!\!-} s_2 \wedge (\bigvee H)(H \subset \underset{2}{\chi}{}^{*}\langle s\rangle \wedge s_1 z \in H) \longrightarrow s_2 z \in \underset{2}{\chi}{}^{*}\langle s\rangle]$$

$$\longleftrightarrow [s_1 \mathrel{\llcorner\!\!-} s_2 \wedge s_1 z \in \underset{2}{\chi}{}^{*}\langle s\rangle \longrightarrow s_2 z \in \underset{2}{\chi}{}^{*}\langle s\rangle] \longleftrightarrow [s_1 \mathrel{\llcorner\!\!-} s_2 \longrightarrow s_1 z \mathrel{\llcorner\!\!-} s_2 z].$$

The last equivalence follows from the property

$$(s_1, s_2) \in \varphi \longleftrightarrow (\bigwedge s)[(s, s_1) \in \varphi \longrightarrow (s, s_2) \in \varphi]$$

which characterizes quasi-orders φ and is easy to verify. By considering the last term in the chain of equivalences, we conclude that the second part of (23) expresses the right regularity of $\underset{2}{\chi}$.

Next we investigate the first term in (23). It means that

$$s_1 \mathrel{\llcorner\!-} s_2 \wedge s_2x = s_2y \longrightarrow (\bigwedge H)[H \subset \underset{2}{\chi}^*\langle s\rangle \wedge s_1y \in H \longrightarrow s_1x \in H]. \tag{24}$$

We will simplify the second implication in this expression:

$$\begin{aligned}
&(\bigwedge H)[H \subset \underset{2}{\chi}^*\langle s\rangle \wedge s_1y \in H \longrightarrow s_1x \in H] \\
&\longleftrightarrow (\bigwedge H)[H \subset \underset{2}{\chi}^*\langle s\rangle \wedge s_1y \in H \wedge s_1y \in \underset{2}{\chi}^*\langle s\rangle \longrightarrow s_1x \in H] \\
&\longleftrightarrow (\bigwedge H)[H \subset \underset{2}{\chi}^*\langle s\rangle \wedge s_1y \in \underset{2}{\chi}^*\langle s\rangle \longrightarrow (s_1y \in H \longrightarrow s_1x \in H)] \\
&\longleftrightarrow [s_1y \in \underset{2}{\chi}^*\langle s\rangle \longrightarrow s_1y = s_1x].
\end{aligned}$$

We shall prove the last equivalence. Let

$$(\bigwedge H)[H \subset \underset{2}{\chi}^*\langle s\rangle \wedge s_1y \in \underset{2}{\chi}^*\langle s\rangle \longrightarrow (s_1y \in H \longrightarrow s_1x \in H)].$$

Take $H = \{s_1y\}$. Then

$$\{s_1y\} \subset \underset{2}{\chi}^*\langle s\rangle \wedge s_1y \in \underset{2}{\chi}^*\langle s\rangle \longrightarrow (s_1y \in \{s_1y\} \longrightarrow s_1x \in \{s_1y\}),$$

or, equivalently, $s_1y \in \underset{2}{\chi}^*\langle s\rangle \longrightarrow s_1x = s_1y$, so we have proved that

$$\begin{aligned}
&(\bigwedge H)[H \subset \underset{2}{\chi}^*\langle s\rangle \wedge s_1y \in \underset{2}{\chi}^*\langle s\rangle \longrightarrow (s_1y \in H \longrightarrow s_1x \in H)] \\
&\qquad \longrightarrow [s_1y \in \underset{2}{\chi}^*\langle s\rangle \longrightarrow s_1y = s_1x].
\end{aligned}$$

The converse is trivial. Hence

$$\begin{aligned}
&(\bigwedge H)[H \subset \underset{2}{\chi}^*\langle s\rangle \wedge s_1y \in \underset{2}{\chi}^*\langle s\rangle \longrightarrow (s_1y \in H \longrightarrow s_1x \in H)] \\
&\qquad \longleftrightarrow [s_1y \in \underset{2}{\chi}^*\langle s\rangle \longrightarrow s_1y = s_1x].
\end{aligned}$$

By substituting this result into (24), we obtain that

$$s_1 \mathrel{\llcorner\!-} s_2 \wedge s_2x = s_2y \longrightarrow (s_1y \in \underset{2}{\chi}^*\langle s\rangle \longrightarrow s_1x = s_1y),$$

or, equivalently,

$$s_1 \mathrel{\llcorner\!-} s_2 \wedge s_2x = s_2y \wedge s \mathrel{\llcorner\!-} s_1y \longrightarrow s_1x = s_1y.$$

Since the expression $(\bigvee s)(s \mathrel{\llcorner\!-} s_1y)$ always holds, the last formula can be written as

$$s_1 \mathrel{\llcorner\!-} s_2 \wedge s_2x = s_2y \longrightarrow s_1x = s_1y. \tag{25}$$

We have obtained an elementary formula which is a sufficient condition for the representability of our semigroup.

If $x = y = e$, then (25) holds trivially. Therefore we can rewrite it

$$s_1 \sqsubseteq s_2 \wedge s_2 x = s_2 s \longrightarrow s_1 x = s_1 s. \tag{26}$$

We will prove that (26) is necessary, i.e., that

$$\mathrm{pr}_2 s_1 \subset \mathrm{pr}_2 s_2 \wedge x \circ s_2 = s \circ s_2 \longrightarrow x \circ s_1 = s \circ s_1,$$

where s, s_1, s_2 and x are transformations. Assume that the premise holds. Then $x \circ s_2 \circ \overset{-1}{s_2} = s \circ s_2 \circ \overset{-1}{s_2}$. Starting from the fact that the formula $s \circ \overset{-1}{s} = \Delta_{\mathrm{pr}_2 s}$ holds for all transformations, the last equality can be written $x \circ \Delta_{\mathrm{pr}_2 s_2} = s \circ \Delta_{\mathrm{pr}_2 s_2}$, whence

$$x \circ \Delta_{\mathrm{pr}_2 s_2} \circ \Delta_{\mathrm{pr}_2 s_1} = s \circ \Delta_{\mathrm{pr}_2 s_2} \circ \Delta_{\mathrm{pr}_2 s_1},$$

or, granting that $\mathrm{pr}_2 s_1 \subset \mathrm{pr}_2 s_2$,

$$x \circ s_1 \circ \overset{-1}{s_1} = s \circ s_1 \circ \overset{-1}{s_1}.$$

Multiplying both sides by s_1, we have

$$x \circ s_1 \circ \overset{-1}{s_1} \circ s_1 = s \circ s_1 \circ \overset{-1}{s_1} \circ s_1,$$

or $x \circ s_1 = s \circ s_1$, which was to be proved.

Thus we have proved the following

THEOREM. *A quasi-ordered semigroup is isomorphic to a second projection semigroup of transformations if and only if its quasi-order is right regular, left negative, and satisfies the condition*

$$s_1 \sqsubseteq s_2 \wedge s_2 x = s_2 s \longrightarrow s_1 x = s_1 s.$$

The proof of this result can be found in [25] (cf. also [22] and [29]). The problems considered in the three preceding sections were solved for semigroups of invertible transformations by V. S. Garvac'kiĭ [3]. For a discussion of the method of construction of representations used here (which forms the rudiments for the method of vector-complexes), see [28].

*Let $L = \{(s_1, s_2) \in S \times S\colon (\bigvee x \in S^*)[s_1 = xs_2]\}$ be the right divisibility relation on S. In other words, $(s_1, s_2) \in L$ if and only if $(s_1)_1 \subset (s_2)_1$, where $(s)_1$ is the principal left ideal generated by an element $s \in S$. If Q_r is the set of all right regular quasi-orders on S, then $L \in Q_r$. A quasi-order $\zeta \in Q_r$ is left negative if and only if $L \subset \zeta$.

*Now consider the following binary relation σ on S:

$$\sigma = \{(s_1, s_2) : (\bigwedge x \in S^*, s \in S)[s_2 x = s_2 s \longrightarrow s_1 x = s_1 s]\}.$$

It is clear that $\sigma \in Q_r$ and $L \subset \sigma$. Now condition (26) means that $\underset{2}{\chi} \subset \sigma$. Thus ζ can be represented as a second projection quasi-order on S if and only if it belongs to the segment $[L, \sigma]$ of Q_r.

*We will give another solution to the same problem (axioms for second projection semigroups) in order to illustrate still another method of constructing representations of semigroups by transformations. First we outline some heuristic foundations of this method.

*The idea is to embed S in a larger semigroup T and then consider the representation of S by (inner) right translations ρ_s of T^* corresponding the elements $s \in S$. Clearly we thus obtain a faithful representation P of S. We would like to have $\underset{2}{\chi} = \underset{2}{\chi_P}$. It is rather obvious that

$$(s_1, s_2) \in \underset{2}{\chi_P} \longleftrightarrow (\bigvee t \in T^*)[s_1 = ts_2],$$

so we need the formula

$$s_1 \llcorner\!\!- s_2 \longleftrightarrow (\bigvee t \in T^*)[s_1 = ts_2]$$

to hold.

*This formula is obviously equivalent to the conjunction of two formulas:

$$s_1 \llcorner\!\!- s_2 \longrightarrow (\bigvee t \in T^*)[s_1 = ts_2] \tag{27}$$

and

$$s_1 = ts_2 \longrightarrow s_1 \llcorner\!\!- s_2. \tag{28}$$

Our aim is to find such a T. Suppose some kind of inverse s_2^{-1} exists in T for s_2, and $s_1 s_2^{-1} s_2 = s_1$, if $s_1 \llcorner\!\!- s_2$. Then we could take $t = s_1 s_2^{-1}$ in (27) thus satisfying (27). Now, if $t \in S^*$ in (28), then $s_1 \llcorner\!\!- s_2$, for $\llcorner\!\!-$ is left negative. So we must define T in such a way that $s_1 = ts_2$ is impossible for $t \notin S^*$, unless $s_1 \llcorner\!\!- s_2$. We may try to construct T as a semigroup generated by the set S and $S^{-1} = \{s^{-1} : s \in S\}$ subject to the defining relations given by the multiplication table of S and $s_2^{-1} s_2 = s_1$ if $s_1 \llcorner\!\!- s_2$. This way really leads us to success. In what follows we skip the remaining 'heuristic' steps and give the final construction.

*So suppose that a semigroup S is endowed with a right regular and left negative quasi-order relation $\llcorner\!\!-$. Let $\bar{A}$ be the set of all words over S^* of the form $a_1 b_1 a_2 b_2 \cdots$ such that $a_i \llcorner\!\!- b_i$ for all i (here we suppose that $s \llcorner\!\!- e$ for all $s \in S^*$). Such a word is called *reduced* if $(b_{i+1}, a_i) \notin R$ for all i. In our above 'heuristic' pictures such a word corresponds to the element $a_1 b_1^{-1} a_2 b_2^{-1} \cdots$ of T.

*Now define multiplication of reduced words by elements of S. Let A be the set of all reduced words, and let $a \in A$ and $s \in S$. Consider two cases, according to whether a is of odd or even length.

**Case* 1. Let $a = a_1 b_1 \cdots a_n b_n$ have even length. If $(s, b_n) \notin R$, then as is the reduced word $a_1b_1 \cdots a_nb_ns$. Now suppose that $(s, b_n) \in R$, i.e. $s = b_nx$ for some $x \in S^*$. In this case *as* is equal to the product of $a_1b_1 \cdots a_n$ and x. Whether such a definition leads us to a uniquely defined *as* still remains to be seen.

**Case* 2. Let $a = a_1b_1 \cdots a_nb_na_{n+1}$. Then *as* equals the product of $a_1b_1 \cdots a_nb_n$ and the element $a_{n+1} \cdot s$ of S.

*Thus we have given an inductive definition of *as* in all cases. Now we show that *as* does not depend on the choice of x in Case 1. Indeed, suppose that $a = a_1b_1 \cdots a_nb_n$ and $s = b_n \cdot x = b_n \cdot y$ for some $x, y \in S^*$. Then, by our definition, *as* equals the product of $a_1b_1 \cdots a_n$ and x (or y). The latter one, again by our definition, equals the product of $a_1b_1 \cdots a_{n-1}b_{n-1}$ and $a \cdot x$ (or $a \cdot y$), where $a_n \cdot x$ and $a_n \cdot y$ are considered as products in S^*. However, by (26)

$$b_n \cdot x = b_n \cdot y \longrightarrow a_n \cdot x = a_n \cdot y,$$

since $a_n \llcorner b_n$. Had we not known the axiom (26) we would have found it now.

*Suppose that there exists a chain $x_0, x_1, \ldots, x_{n-k}$ of elements of S^* such that

$$\begin{aligned} a_{n+1} \cdot s &= b_n \cdot x_0, \\ a_n \cdot x_0 &= b_{n-1} \cdot x_1, \\ a_{n-1} \cdot x_1 &= b_{n-2} \cdot x_2, \\ &\cdots \\ a_{k+1} \cdot x_{n-k-1} &= b_k \cdot x_{n-k} \end{aligned} \tag{29}$$

and $(a_k \cdot x_{n-k}, b_{k-1}) \notin R$. Then, applying our definition to $a = a_1b_1 \cdots a_nb_na_{n+1}$ (if a has even length we may suppose $a_{n+1} = e$ to comprise this case as well) we obtain that *as* equals $(a_1b_1 \cdots a_nb_n)(a_{n+1} \cdot s)$, which, in turn, equals

$$(a_1b_1 \cdots a_{n-1}b_{n-1})(a_nx_0),$$

which equals $(a_1b_1 \cdots a_{n-2}b_{n-2})(a_{n-1} \cdot x_1), \cdots$, which equals

$$a_1b_1 \cdots a_{k-1}b_{k-1}(a_k \cdot x_{n-k})$$

which is a reduced word not depending on the choice of $x_i \in S^*$.

*For every $s \in S$ define a transformation ρ_s of the set A of all reduced words as follows: $\rho_s(a) = as$ for all $a \in A$.

*Next we show that for all $a \in A$ and $s, t \in S$

$$(as)t = a(s \cdot t). \tag{30}$$

Indeed, suppose that (29) hold. Then

$$as = a_1b_1 \cdots b_{k-1}(a_k \cdot x_{n-k})$$

and so $(as)t$ equals the product of $a_1b_1 \cdots a_{k-1}b_{k-1}$ and $a_k \cdot x_{n-k} \cdot t$. Now multiplying all the equalities (29) by t on the right we obtain

$$a_{n+1} \cdot s \cdot t = b_n \cdot x_0 \cdot t,$$

$$a_n \cdot x_0 \cdot t = b_{n-1} \cdot x_1 \cdot t,$$

$$\cdots\cdots\cdots\cdots\cdots$$

$$a_{k+1} \cdot x_{n-k-1} = b_k \cdot x_{n-k} \cdot t$$

which show that $a(s \cdot t)$ equals the product of $a_1b_1 \cdots a_{k-1}b_{k-1}$ and $a_k \cdot x_{n-k} \cdot t$. Thus $(as)t = a(s \cdot t)$.

*It follows from (30) that the mapping $P : S \longrightarrow \mathcal{F}_A$, $P(s) = \rho_s$ is a representation of S by transformations. Now $e \in A$ and $\rho_s(e) = e \cdot s$ for all $s \in S$. Thus $P(s) = P(t)$ implies $\rho_s(e) = \rho_t(e)$ and $s = t$. So P is a faithful representation.

*Suppose now that $\mathrm{pr}_2P(s) \subset \mathrm{pr}_2P(t)$. Since, as we have just seen, $s \in \mathrm{pr}_2P(s)$, we obtain $s \in \mathrm{pr}_2P(t)$, i.e. $s = at$ for some $a \in A$. Thus, if $a = a_1b_1 \cdots a_{n+1}$, there exist $x_0, x_1, \ldots, x_{n-1} \in S^*$ such that

$$a_{n+1} \cdot t = b_n \cdot x_0, \quad a_n \cdot x_0 = b_{n-1} \cdot x_1, \quad \cdots, a_2 \cdot x_{n-2} = b_1 \cdot x_{n-1}$$

and so $at = a_1 \cdot x_{n-1} = s$.

*Since $\mathrel{\llcorner\!-}$ is right regular, we have $a_i \cdot x_{n-i} \mathrel{\llcorner\!-} b_i \cdot x_{n-i}$ for all i. Thus

$$s = a_1 \cdot x_{n-1} \mathrel{\llcorner\!-} b_1 \cdot x_{n-1} = a_2 \cdot x_{n-2} \mathrel{\llcorner\!-} b_2 \cdot x_{n-2} = \cdots = a_n \cdot x_0$$

$$\mathrel{\llcorner\!-} b_n \cdot x_0 = a_{n+1} \cdot t \mathrel{\llcorner\!-} t,$$

and $s \mathrel{\llcorner\!-} t$.

*Conversely, suppose that $s \mathrel{\llcorner\!-} t$ and $b \in \mathrm{pr}_2P(s)$, i.e. $b = as$ for some $a \in A$. Suppose that (29) hold. Then $b = a_1b_1 \cdots b_{k-1}(a_k \cdot x_{n-k})$. Now

$$a_k \cdot x_{n-k} \mathrel{\llcorner\!-} b_k \cdot x_{n-k} = a_{k+1} \cdot x_{n-k-1} \mathrel{\llcorner\!-} \cdots \mathrel{\llcorner\!-} b_n \cdot x_0 = a_{n+1}$$

$$\cdot s \mathrel{\llcorner\!-} s \mathrel{\llcorner\!-} t.$$

Therefore the word $a_1b_1 \cdots b_{k-1}(a_k \cdot x_{n-k})t$ is reduced and belongs to A. Now $a_1b_1 \cdots b_{k-1}(a_k \cdot x_{n-k})t$ multiplied by t gives

$$a_1b_1 \cdots b_{k-1}(a_k \cdot x_{n-k}) = b.$$

So $b \in \mathrm{pr}_2P(t)$ and $\mathrm{pr}_2P(s) \subset \mathrm{pr}_2P(t)$. We proved that $\mathrel{\llcorner\!-}$ coincides with $\chi_{\underset{2}{P}}$ if $\mathrel{\llcorner\!-}$ is a left negative, right regular quasi-order satisfying (26).

*Thus we have obtained a new proof of the main theorem of this section. An analogous proof was used in [13].

§9. Representation of projection semigroups of transformations

We turn to the solution of the following problem: find an abstract characterization of *projection semigroups of transformations* (that is, semigroups of transformations which are both first and second projection semigroups with respect to the first and second projection quasi-orders). Thus we must find necessary and sufficient conditions for the existence of a faithful representation P of an abstract semigroup S, endowed with two quasi-orders $\underset{1}{\chi}$ ($\ulcorner\!\!-$) and $\underset{2}{\chi}$ ($\llcorner\!\!-$), into a semigroup of transformations such that $\underset{1}{\chi} = \underset{1}{\chi}_P$ and $\underset{2}{\chi} = \underset{2}{\chi}_P$. In other words, we seek axioms for the class of abstract algebras of transformations of the genre $\sigma = (O_0, P_{\underset{1}{\chi}}, P_{\underset{2}{\chi}})$, where O_0 defines multiplication of transformations, and $P_{\underset{1}{\chi}}$ and $P_{\underset{2}{\chi}}$ define inclusion of the first and second projections of the transformations, respectively.

Obviously the quasi-orders $\underset{1}{\chi}$ and $\underset{2}{\chi}$ must satisfy all of the foregoing conditions giving abstract characterizations of first and second projection semigroups of transformations; namely, $\underset{1}{\chi}$ must be left regular and right negative, and $\underset{2}{\chi}$ right regular and left negative satisfying the condition

$$s_1 \llcorner\!\!- s_2 \wedge s_2 x = s_2 s \longrightarrow s_1 x = s_1 s.$$

Clearly these conditions need not be sufficient for finding a faithful representation P. It is possible that further conditions exist connecting $\ulcorner\!\!-$ and $\llcorner\!\!-$.

Now recall that, when finding an abstract characterization of the second projection quasi-order, we constructed a proper representation P_1 by everywhere defined transformations, where P_1 obviously satisfied the conditions $\epsilon_{P_1} = \Delta_S$, $\underset{1}{\chi}_{P_1} = S \times S$ and $\underset{2}{\chi}_{P_1} = \underset{2}{\chi}$. Taking this into account, we try to select a representation P_2 for which $P = P_1 + P_2$. Then P will be faithful (by virtue of the formula $\epsilon_P = \epsilon_{P_1} \cap \epsilon_{P_2} = \Delta_S$), and we will have $\underset{1}{\chi}_P = \underset{1}{\chi}_{P_1} \cap \underset{1}{\chi}_{P_2} = \underset{1}{\chi}_{P_2}$ and $\underset{2}{\chi}_P = \underset{2}{\chi}_{P_1} \cap \underset{2}{\chi}_{P_2} = \underset{2}{\chi} \cap \underset{2}{\chi}_{P_2}$. Thus our problem reduces to one of searching for such a representation P_2 (not necessarily faithful) which satisfies the conditions $\underset{1}{\chi} = \underset{1}{\chi}_{P_2}$ and $\underset{2}{\chi} \subset \underset{2}{\chi}_{P_2}$. Note that the representation constructed when finding an abstract characterization of $\underset{1}{\chi}$ cannot be used in this case, since it does not always satisfy the latter condition.

To construct P_2 we first extend the quasi-order $\underset{1}{\chi}$, defined on S, to S^*, i.e., we introduce $\underset{1}{\chi}^* = \underset{1}{\chi} \cup (S^* \times \{e\})$. We collect into one set $\mathfrak{M}$ all majorantly saturated (relative to the extended quasi-order $\underset{1}{\chi}^*$) nonempty subsets H of S^* relative to the extended quasi-order $\underset{1}{\chi}^*$, i.e., $H \in \mathfrak{M} \longleftrightarrow \underset{1}{\chi}^*(H) \subset H \wedge H \neq \varnothing$. We construct P_2 in the following way:

$$(H_1, H_2) \in P_2(s) \longleftrightarrow H_1 \mathbin{.\!\cdot} s = H_2 \wedge H_1 \in \mathfrak{M} \wedge H_2 \in \mathfrak{M}.$$

This definition can be written in the form

$$(H_1, H_2) \in P_2(s) \longleftrightarrow H_1 \mathbin{.\!\cdot} s = H_2 \wedge H_2 \neq \varnothing \wedge H_1 \in \mathfrak{M}.$$

The equivalence of these definitions follows from the fact that $\underset{1}{\chi}^*$ is left regular. Indeed, the second definition follows immediately from the first. Conversely, letting $H_1 \in \mathfrak{M} \wedge H_1 \mathrel{.\cdot} s \neq \varnothing$, we will show that $H_1 \mathrel{.\cdot} s \in \mathfrak{M}$. Since $H_1 \mathrel{.\cdot} s \neq \varnothing$, consider $x \in H_1 \mathrel{.\cdot} s$; that is, $sx \in H_1$. Let $x \ulcorner\!\!- y$. Since $\underset{1}{\chi}^*$ is left regular, it follows that $sx \ulcorner\!\!- sy$. Now $H_1 \in \mathfrak{M}$ implies $sy \in H_1$, i.e., $y \in H_1 \mathrel{.\cdot} s$, or $H_1 \mathrel{.\cdot} s \in \mathfrak{M}$, which completes the proof. The proof that P_2 is indeed a representation is analogous to the proof in the preceding section for P_s.

We will prove that our representation P_2 satisfies the formula $\underset{1}{\chi}_{P_2} = \underset{1}{\chi}$. In fact, we have

$$(s_1, s_2) \in \underset{1}{\chi}_{P_2} \longleftrightarrow \mathrm{pr}_1 P_2(s_1) \subset \mathrm{pr}_1 P_2(s_2)$$

$$\longleftrightarrow (\bigwedge H \in \mathfrak{M})\,[H \in \mathrm{pr}_1 P_2(s_1) \longrightarrow H \in \mathrm{pr}_1 P_2(s_2)]$$

$$\longleftrightarrow (\bigwedge H \in \mathfrak{M})\,[H \mathrel{.\cdot} s_1 \neq \varnothing \longrightarrow H \mathrel{.\cdot} s_2 \neq \varnothing]$$

$$\longleftrightarrow (\bigwedge H \in \mathfrak{M})\,[(\bigvee x)[s_1 x \in H] \longrightarrow (\bigvee y)[s_2 y \in H]\,]$$

$$\longleftrightarrow (\bigwedge H \in \mathfrak{M})\,[s_1 \in H \longrightarrow s_2 \in H] \longleftrightarrow (s_1, s_2) \in \underset{1}{\chi}.$$

We will verify the last two equivalences. First we prove that $(\bigvee x)\,[s_1 x \in H] \longleftrightarrow s_1 \in H$. Let $s_1 \in H$. Then, taking $x = e$, we have $(\bigvee x)[s_1 x \in H]$. Conversely, let $(\bigvee x)[s_1 x \in H]$; $\underset{1}{\chi}^*$ is right negative, so $s_1 x \ulcorner\!\!- s_1$; but H is majorantly saturated with respect to $\underset{1}{\chi}^*$, so $s_1 \in H$, which completes the proof. Incidentally, we thus see why we had to take the set $\mathfrak{M}$ consisting of the majorantly saturated subsets with respect to $\underset{1}{\chi}^*$.

Now we prove the last equivalence. We have $(\bigwedge H)[s_1 \in H \longrightarrow s_2 \in H]$. This holds for any majorantly saturated nonempty subset H. Then by setting $H = \underset{1}{\chi}^*\langle s_2 \rangle$, we obtain $s_1 \ulcorner\!\!- s_1 \longrightarrow s_1 \ulcorner\!\!- s_2$, which gives us $(s_1, s_2) \in \underset{1}{\chi}$. Conversely, let $s_1 \ulcorner\!\!- s_2$ and suppose $s_1 \in H$. Then $s_2 \in H$, since H is majorantly saturated with respect to $\underset{1}{\chi}^*$; therefore $(\bigwedge H)[s_1 \in H \longrightarrow s_2 \in H]$. Thus we have proved that

$$(\bigwedge H)\,[s_1 \in H \longrightarrow s_2 \in H] \longleftrightarrow s_1 \ulcorner\!\!- s_2.$$

It remains for us to determine what conditions insure that $\underset{2}{\chi} \subset \underset{2}{\chi}_{P_2}$. We have

$$\underset{2}{\chi} \subset \underset{2}{\chi}_{P_2} \longleftrightarrow (\bigwedge s_1, s_2)\,[s_1 \llcorner\!\!- s_2 \longrightarrow (s_1, s_2) \in \underset{2}{\chi}_{P_2}] \longleftrightarrow$$

$$\leftrightarrow (\bigwedge s_1, s_2)[s_1 \mathrel{\llcorner\!\!-} s_2 \rightarrow \mathrm{pr}_2 P_2(s_1) \subset \mathrm{pr}_2 P_2(s_2)]$$

$$\leftrightarrow (\bigwedge s_1, s_2)[s_1 \mathrel{\llcorner\!\!-} s_2 \rightarrow (\bigwedge H_2 \neq \varnothing)[(\bigvee H \in \mathfrak{M})[(H, H_2) \in P_2(s_1)]$$
$$\rightarrow (\bigvee H_1 \in \mathfrak{M})[(H_1, H_2) \in P_2(s_2)]]]$$

$$\leftrightarrow (\bigwedge s_1, s_2)[s_1 \mathrel{\llcorner\!\!-} s_2$$
$$\rightarrow (\bigwedge H_2 \neq \varnothing)(\bigwedge H \in \mathfrak{M})(\bigvee H_1 \in \mathfrak{M})[H_2 = H \mathbin{.\cdot} s_1 \rightarrow H_2 = H_1 \mathbin{.\cdot} s_2]]$$

$$\leftrightarrow (\bigwedge s_1, s_2)(\bigwedge H \in \mathfrak{M})(\bigvee H_1 \in \mathfrak{M})[s_1 \mathrel{\llcorner\!\!-} s_2 \wedge H \mathbin{.\cdot} s_1 \neq \varnothing \rightarrow H \mathbin{.\cdot} s_1 = H_1 \mathbin{.\cdot} s_2].$$

By omitting the quantifier $(\bigwedge s_1, s_2)$ and the expression $H \mathbin{.\cdot} s_1 \neq \varnothing$, we can rewrite the last term in the above chain of equivalences in the form

$$(\bigwedge H \in \mathfrak{M})(\bigvee H_1 \in \mathfrak{M})[s_1 \mathrel{\llcorner\!\!-} s_2 \rightarrow H \mathbin{.\cdot} s_1 = H_1 \mathbin{.\cdot} s_2]. \tag{31}$$

That these expressions are equivalent follows from the rules of logic and the fact that when $H \mathbin{.\cdot} s_1 = \varnothing$ the given formula holds trivially (for we may take $H_1 = \{e\}$). Condition (31) is sufficient for constructing P_2, and hence also P. But it is neither a universal axiom (the quantifier $(\bigvee H_1 \in \mathfrak{M})$ appears) nor an axiom of the first order (quantifiers binding sets). Therefore we must find a simplification of (31).

First of all we attempt to get rid of the existential quantifier $(\bigvee H_1 \in \mathfrak{M})$. By an argument similar to that in the previous section, we see that for H_1 we can take the subset $\underset{1}{\breve{\chi}}{}^*(s_2(H \mathbin{.\cdot} s_1))$. Then (31) is equivalent to the condition

$$(\bigwedge H \in \mathfrak{M})[s_1 \mathrel{\llcorner\!\!-} s_2 \rightarrow H \mathbin{.\cdot} s_1 = \underset{1}{\breve{\chi}}{}^*(s_2(H \mathbin{.\cdot} s_1)) \mathbin{.\cdot} s_2], \tag{32}$$

which is a universal formula. We must simplify it so that it becomes an elementary axiom. Since (by the definition of division and starting with the fact that $\underset{1}{\chi}{}^*$ is reflexive) $H \mathbin{.\cdot} s_1 \subset \underset{1}{\breve{\chi}}{}^*(s_2(H \mathbin{.\cdot} s_1)) \mathbin{.\cdot} s_2$, (32) can be written as

$$(\bigwedge H \in \mathfrak{M})[s_1 \mathrel{\llcorner\!\!-} s_2 \rightarrow \underset{1}{\breve{\chi}}{}^*(s_2(H \mathbin{.\cdot} s_1)) \mathbin{.\cdot} s_2 \subset H \mathbin{.\cdot} s_1].$$

Reducing this formula to an elementary form, we have

$$(\bigwedge H \in \mathfrak{M})[s_1 \mathrel{\llcorner\!\!-} s_2 \rightarrow (\bigwedge y)[y \in \underset{1}{\breve{\chi}}{}^*(s_2(H \mathbin{.\cdot} s_1)) \mathbin{.\cdot} s_2 \rightarrow y \in H \mathbin{.\cdot} s_1]]$$

$$\leftrightarrow (\bigwedge H \in \mathfrak{M})(\bigwedge y)[s_1 \mathrel{\llcorner\!\!-} s_2 \rightarrow (s_2 y \in \underset{1}{\breve{\chi}}{}^*(s_2(H \mathbin{.\cdot} s_1)) \rightarrow s_1 y \in H)]$$

$$\leftrightarrow (\bigwedge H \in \mathfrak{M})(\bigwedge y)[s_1 \mathrel{\llcorner\!\!-} s_2 \rightarrow (\bigwedge x)[s_1 x \in H \wedge s_2 x \mathrel{\ulcorner\!\!-} s_2 y \rightarrow s_1 y \in H]]$$

$$\leftrightarrow (\bigwedge H \in \mathfrak{M})(\bigwedge x, y)[s_1 \mathrel{\llcorner\!\!-} s_2 \wedge s_2 x \mathrel{\ulcorner\!\!-} s_2 y \rightarrow (s_1 x \in H \rightarrow s_1 y \in H)]$$

$$\leftrightarrow (\bigwedge x, y)[s_1 \mathrel{\llcorner\!\!-} s_2 \wedge s_2 x \mathrel{\ulcorner\!\!-} s_2 y \rightarrow s_1 x \mathrel{\ulcorner\!\!-} s_1 y].$$

Thus (31) is equivalent to

$$s_1 \llcorner\!- s_2 \wedge s_2 x \ulcorner\!- s_2 y \longrightarrow s_1 x \ulcorner\!- s_1 y, \tag{33}$$

which has an elementary form and is a sufficient condition for the representability of our semigroup. If $y = e$, then (33) is satisfied trivially ($s_1 x \ulcorner\!- s_1$ always holds). Therefore it can be written also as

$$s_1 \llcorner\!- s_2 \wedge s_2 x \ulcorner\!- s_2 s \longrightarrow s_1 x \ulcorner\!- s_1 s. \tag{34}$$

We will prove that (34) is necessary, i.e., that

$$\mathrm{pr}_2 s_1 \subset \mathrm{pr}_2 s_2 \wedge \mathrm{pr}_1(x \circ s_2) \subset \mathrm{pr}_1(s \circ s_2) \longrightarrow \mathrm{pr}_1(x \circ s_1) \subset \mathrm{pr}_1(s \circ s_1),$$

where s, s_1, s_2 and x are transformations. Assume that the premise holds. Then (since $\mathrm{pr}_1(s_2 \circ s_1) = \breve{\overset{-1}{s_1}}(\mathrm{pr}_1 s_2)$) we have that

$$\breve{\overset{-1}{s_2}}(\mathrm{pr}_1 x) \subset \breve{\overset{-1}{s_2}}(\mathrm{pr}_1 s),$$

whence $\breve{s}_2(\breve{\overset{-1}{s_2}}(\mathrm{pr}_1 x)) \subset \breve{s}_2(\breve{\overset{-1}{s_2}}(\mathrm{pr}_1 s))$, or, equivalently,

$$\breve{\overline{s_2 \circ \overset{-1}{s_2}}}(\mathrm{pr}_1 x) \subset \breve{\overline{s_2 \circ \overset{-1}{s_2}}}(\mathrm{pr}_1 s).$$

From the fact that $s \circ \overset{-1}{s} = \Delta_{\mathrm{pr}_2 s}$ for any transformation, the last equality can be rewritten as

$$\breve{\Delta}_{\mathrm{pr}_2 s_2}(\mathrm{pr}_1 x) \subset \breve{\Delta}_{\mathrm{pr}_2 s_2}(\mathrm{pr}_1 s),$$

whence

$$\breve{\Delta}_{\mathrm{pr}_2 s_1}(\breve{\Delta}_{\mathrm{pr}_2 s_2}(\mathrm{pr}_1 x)) \subset (\breve{\Delta}_{\mathrm{pr}_2 s_1}(\breve{\Delta}_{\mathrm{pr}_2 s_2}(\mathrm{pr}_1 s)),$$

or, granting that $\mathrm{pr}_2 s_1 \subset \mathrm{pr}_2 s_2$,

$$\breve{\Delta}_{\mathrm{pr}_2 s_1}(\mathrm{pr}_1 x) \subset \breve{\Delta}_{\mathrm{pr}_2 s_1}(\mathrm{pr}_1 s),$$

and so, finally,

$$\breve{\overline{s_1 \circ \overset{-1}{s_1}}}(\mathrm{pr}_1 x) \subset \breve{\overline{s_1 \circ \overset{-1}{s_1}}}(\mathrm{pr}_1 s).$$

Multiplying both sides of the inclusion by $\overset{-1}{s_1}$ and using the equality $\overset{-1}{s_1} \circ s_1 \circ \overset{-1}{s_1} = \overset{-1}{s_1}$, we obtain

$$\breve{\overline{\overset{-1}{s_1} \circ s_1 \circ \overset{-1}{s_1}}}(\mathrm{pr}_1 x) \subset \breve{\overline{\overset{-1}{s_1} \circ s_1 \circ \overset{-1}{s_1}}}(\mathrm{pr}_1 s),$$

and hence $\breve{\overset{-1}{s_1}}(\mathrm{pr}_1 x) \subset \breve{\overset{-1}{s_1}}(\mathrm{pr}_1 s)$, or finally, $\mathrm{pr}_1(x \circ s_1) \subset \mathrm{pr}_1(s \circ s_1)$, which completes the proof of the

THEOREM. *A semigroup with two quasi-orders* $\ulcorner\!-$ *and* $\llcorner\!-$ *is isomorphic to a projection semigroup of transformations if and only if* $\ulcorner\!-$ *is left regular and right negative and* $\llcorner\!\!-$ *is right regular and left negative, and*

$$s_1 \llcorner\!\!- s_2 \wedge s_2x = s_2s \longrightarrow s_1x = s_1s,$$

$$s_1 \llcorner\!\!- s_2 \wedge s_2x \ulcorner\!\!- s_2s \longrightarrow s_1x \ulcorner\!\!- s_1s.$$

We should point out that the analogous problem for invertible transformations has not been solved. The proof of the above theorem can be found in [25] (cf. also [29]). Another proof can be found in [13].

*Let $\mu = \{(s_1, s_2): (\bigwedge s \in S, x \in S^*)[s_2x \ulcorner\!\!- s_2s \longrightarrow s_1x \ulcorner\!\!- s_1s]\}$. Then μ is a right regular and left negative quasi-order, i.e. $\mu \in Q_r$ and $L \subset \mu$. Therefore $\sigma \cap \mu \in [L, \sigma]$. Now the latter two conditions of the above theorem mean that $\underset{2}{\chi} \subset \sigma \cap \mu$. Thus the pair $(\ulcorner\!\!-, \llcorner\!\!-)$ is representable by transformations if and only if $\ulcorner\!\!- \in [R, \infty) \subset Q_1$ and $\llcorner\!\!- \in [L, \sigma \cap \mu] \in Q_r$. In particular, given $\ulcorner\!\!-$, $\sigma \cap \mu$ is the largest of all quasi-orders $\llcorner\!\!-$ such that $(\ulcorner\!\!-, \llcorner\!\!-)$ is representable.

*It is proved in [24] that *a semigroup with two quasi-orders* $\ulcorner\!\!-$ *and* $\llcorner\!\!-$ *is isomorphic to a projection semigroup of binary relations if and only if* $\ulcorner\!\!-$ *is left regular and right negative and* $\llcorner\!\!-$ *is right regular and left negative.* Thus, if $\ulcorner\!\!-$ and $\llcorner\!\!-$ have representations by semigroups of binary relations, then $\ulcorner\!\!-$ and $\llcorner\!\!-$ have a join representation on the same semigroup of binary relations.

§10. Representation of ordered projection semigroups(10)

Now we are going to find an abstract characterization of ordered projection semigroups of transformations; that is, semigroups of transformations with the set-theoretical inclusion order and two projection quasi-orders of the first and second kinds. Thus we must find out when there exists a faithful representation P of an abstract semigroup S endowed with an order relation ς $(\prec)$ and two quasi-order relations $\underset{1}{\chi}$ $(\ulcorner\!\!-)$ and $\underset{2}{\chi}$ $(\llcorner\!\!-)$ by transformations so that

$$\varsigma = \varsigma_P, \qquad \underset{1}{\chi} = \underset{1}{\chi}_P, \qquad \underset{2}{\chi} = \underset{2}{\chi}_P. \tag{35}$$

In other words, we are looking for axioms for the class of abstract algebras of transformations of the genre $\sigma = (O_0, P_\varsigma, P_{\underset{1}{\chi}}, P_{\underset{2}{\chi}})$, where O_0 is a formula defining the composition of transformations, while $P_\varsigma, P_{\underset{1}{\chi}}, P_{\underset{2}{\chi}}$ are formulas defining the inclusion of transformations and the inclusion of their first (second) projections respectively. Clearly, to this end the relations $\prec$, $\ulcorner\!\!-$ and $\llcorner\!\!-$ must satisfy all those conditions we have found earlier when looking for abstract characterizations of ordered first projection semigroups of transformations and projection semigroups of transformations. It is clear also that these conditions need not be sufficient: some new conditions connecting $\prec$, $\ulcorner\!\!-$ and $\llcorner\!\!-$ may emerge, and we shall see that this is the case indeed.

To solve our problem we develop a method first used at the end of §8 to provide an alternative solution to the problem of characterizing second projection semigroups.

(10) §§10 and 11 have been added in translation.

We are going to construct a set A and a partial mapping $p\colon A \times S \longrightarrow A$, where $(a, s) \in \mathrm{pr}_1 p$ will be designated by $a \vdash s$ and $p(a, s)$ will be designated by as. A and p must satisfy the following conditions:

(1) $(as)t = a(st)$ for all $s, t \in S$, $a \in A$; that is, both sides of the equality are defined or undefined simultaneously, and whenever both sides are defined they are equal.

(2) $s \prec t$ if and only if for all $a \vdash s$ we have both $a \vdash t$ and $as = at$.

(3) $s \mathrel{\ulcorner} t$ if and only if $a \vdash s \longrightarrow a \vdash t$ for all $a \in A$.

(4) $s \mathrel{\llcorner} t$ if and only if $(\exists b \in A)[bs = a] \longrightarrow (\exists c \in A)[ct = a]$ for all $a \in A$.

Define a mapping $P\colon S \longrightarrow \mathfrak{F}_A$ as follows:

$$(a, b) \in P(s) \longleftrightarrow as = b.$$

Then (1) means that P is a representation, (2) means that $\zeta = \zeta_P$, (3) means that $\underset{1}{\chi} = \underset{1}{\chi}_P$, and (4) means that $\underset{2}{\chi} = \underset{2}{\chi}_P$. (2) implies that P is faithful. Conversely, if P is a representation of S, $P\colon S \longrightarrow \mathfrak{F}_A$, and if (35) are satisfied, then we may define

$$a \vdash s \longleftrightarrow a \in \mathrm{pr}_1 P(s),\ as = b \longleftrightarrow (a, b) \in P(s).$$

We obtain a partial mapping $A \times S \longrightarrow A$ satisfying (1)–(4) above. E.g., the faithful representation of S by right translations of S^1 amounts to S acting on the right on $A = S^1$. This acting does not satisfy our condition (2). Indeed, suppose that $s \prec t$, $s \neq t$. By (2) we must have $us = ut$ for all $u \in S^1$, which surely is not the case. To amend this we may try to restrict the domain of $p\colon S^1 \times S \longrightarrow S$. Namely, we shall make p a partial mapping such that $us = ut$ for $s \prec t$ and all $u \in S^1$ such that $u \vdash s$. To this end it suffices to suppose $u \vdash s \longleftrightarrow u \mathrel{\ulcorner} us$.

Indeed, let $u \mathrel{\ulcorner} us$. It follows from $s \prec t$ that $us \prec ut$ (by left regularity of $\prec$). Now $\prec \subset \mathrel{\ulcorner}$, so that $us \mathrel{\ulcorner} ut$. Since $u \mathrel{\ulcorner} us$ and $\ulcorner$ is transitive, we have $u \mathrel{\ulcorner} ut$. So $u \vdash t$. Now we use (16) in which $s_1 = us$, $s_2 = s = ut$. Then $s_1 \prec s_2$ and $s \prec s_2$. To see that $s \mathrel{\ulcorner} s_1$ we note that $s = ut \mathrel{\ulcorner} u \mathrel{\ulcorner} us = s_1$. By (16), $s \prec s_1$, i.e. $ut \prec us$. By antisymmetry of $\prec$, $us = ut$.

So we may suppose that

$$u \vdash s \longleftrightarrow u \mathrel{\ulcorner} us \tag{36}$$

for all $u \in S^1$ and $s \in S$. If S is a semigroup without identity, we can extend $\prec$, $\ulcorner$ and $\llcorner$ to S^1 as follows: $x \prec 1 \longleftrightarrow 1 \prec x \longleftrightarrow x = 1$ and $x \mathrel{\ulcorner} 1$, $x \mathrel{\llcorner} 1$ for all $x \in S^1$.

Nevertheless, (2) need not hold, for if $us = ut$ for all $u \in S^1$, $u \vdash s$, we need not have $s \prec t$. To circumvent this difficulty we try to extend S^1 to a larger set A for which (1)–(4) hold.

Consider (4). Suppose that $bs = a$ for some $a, b \in A$ and $s \mathrel{\llcorner} t$. Then $ct = a = bs$ for some $c \in A$. Choose one such c and denote it by $c = a_t$. Then $a_t \vdash t$ and $a_t t = a$. Of course, $a_t u$ may make sense for $u \neq t$. E. g., if $t \mathrel{\ulcorner} u$, then $a_t \vdash u$ by (3), and so $a_t u$ is defined.

Thus some elements of A can be represented either in the form

$$s^{(1)}_{t_1} s^{(2)}_{t_2} \cdots s^{(n)}_{t_n} \tag{37}$$

or in the form

$$s^{(1)}_{t_1} s^{(2)}_{t_2} \cdots s^{(n)}_{t_n} s^{(n+1)}, \tag{38}$$

where the superscripts are just indices (not degrees), $s^{(i)}$, $t_j \in S$, $s^{(i)} \mathrel{\llcorner\!\!-} t_i$ and $s^{(1)}_{t_1} \cdots s^{(i)}_{t_i} \vdash s^{(i+1)}$ for all i. To make this situation clearer, note that $s^{(1)} \in A$, and if $s^{(1)} \mathrel{\llcorner\!\!-} t_1$, then $(s^{(1)})_{t_1} = s^{(1)}_{t_1} \in A$ by the above. Now, if $s^{(1)}_{t_1} \vdash s^{(2)}$, then $s^{(1)}_{t_1} s^{(2)} \in A$. If $s^{(2)} \mathrel{\llcorner\!\!-} t_2$, then by (4) $(s^{(1)}_{t_1} s^{(2)})_{t_2} = s^{(1)}_{t_1} s^{(2)}_{t_2} \in A$, etc. Here the last two equalities are merely definitions, a short-hand form of longer expressions (some parentheses are omitted).

We may suppose that $a1 = a$ for all $a \in A$. Then

$$s^{(1)}_{t_1} \cdots s^{(n)}_{t_n} = s^{(1)}_{t_1} \cdots s^{(n)}_{t_n} 1$$

and so we may consider elements of the form (38) only in which $s^{(n+1)} \in S^1$, for elements of the form (37) are expressible in the form (38) as well. For typographical reasons we will write down elements of the form (38) as follows:

$$(s_1, t_1, s_2, t_2, \ldots, s_n, t_n, s_{n+1}). \tag{39}$$

If $n = 1$, we write s or (s) interchangeably. It will turn out that A need not contain any further elements.

Now we must decide what $s^{(1)}_{t_1} \cdots s^{(i)}_{t_i} \vdash s^{(i+1)}$ should mean. Using the form (39) and substituting n for i, we must give a meaning to $(s_1, t_1, \ldots, s_n, t_n, 1) \vdash s_{n+1}$. More generally, we must define when

$$(s_1, t_1, \ldots, s_n, t_n, s_{n+1}) \vdash u. \tag{40}$$

If $n = 0$, then $s \vdash u$ means $s \mathrel{\ulcorner\!\!-} su$, as we have already seen. Now $(s_1, t_1, \ldots, s_n, t_n, s_{n+1}) \in A$ implies that

$$(s_1, t_1, \ldots, s_n, t_n, 1) \vdash s_{n+1}$$

and

$$(s_1, t_1, \ldots, s_n, t_n, 1)s_{n+1} = (s_1, t_1, \ldots, s_n, t_n, s_{n+1}).$$

Therefore $(s_1, t_1, \ldots, s_n, t_n, 1) \vdash s_{n+1}u$ implies by (1)

$$(s_1, t_1, \ldots, s_n, t_n, 1)s_{n+1}u = (s_1, t_1, \ldots, s_n, t_n, s_{n+1})u,$$

so that $(s_1, t_1, \ldots, s_n, t_n, s_{n+1}) \vdash u$. This argument may be considered backwards so that (40) is equivalent to

$$(s_1, t_1, \ldots, s_n, t_n) \vdash s_{n+1}u.$$

Thus we must decide when

$$(s_1, t_1, \ldots, s_n, t_n, 1) \vdash s. \tag{41}$$

By definition, $a_n = (s_1, t_1, \ldots, s_n, t_n, 1) \vdash t_n$. Suppose there exists $x_n \in S^1$ such that

$$a_n \vdash t_n x_n \tag{42}$$

and $t_n x_n \ulcorner\!\!- s$. Then (41) holds by (3). Now (1) and (41) imply

$$a(t_n s) = (at_n)s = (s_1, t_1, \ldots, s_{n-1}, t_{n-1}, s_n)s = (a_{n-1}s_n)s = a_{n-1}(s_n s).$$

So (41) is equivalent to

$$a_{n-1} \vdash s_n x_n. \tag{43}$$

Suppose there exists $x_{n-1} \in S^1$ such that

$$a_{n-1} \vdash t_{n-1}x_{n-1} \tag{44}$$

and $t_{n-1}x_{n-1} \ulcorner\!\!- s_n x_n$. By (3) formula (43) holds. Now (1) and (44) imply

$$\begin{aligned} a_{n-1}(t_{n-1}x_{n-1}) &= (a_{n-1}t_{n-1})x_{n-1} = (s_1, t_1, \ldots, s_{n-2}, t_{n-2}, s_{n-1})x_{n-1} \\ &= (a_{n-2}s_{n-1})x_{n-1} = a_{n-2}(s_{n-2}x_{n-2}), \end{aligned}$$

so that (44) is equivalent to

$$a_{n-2} \vdash s_{n-1}x_{n-1}. \tag{45}$$

Proceeding as above we see that, provided there exist $x_1, \ldots, x_n \in S^1$ such that

$$t_n x_n \ulcorner\!\!- s,\ t_{n-1}x_{n-1} \ulcorner\!\!- s_n x_n, \ldots, t_1 x_1 \ulcorner\!\!- s_2 x_2, \tag{46}$$

(41) is equivalent to $(s_1, t_1, 1) = a_1 \vdash t_1 x_1$. Now $a_1(t_1 x_1) = (a_1 t_1)x_1 = (s_1)x_1 = s_1 x_1$, so that $a_1 \vdash t_1 x_1$ is equivalent to $s_1 \vdash x_1$, i.e. to $s_1 \ulcorner\!\!- s_1 x_1$. Thus we may try to define (41) as follows: there exist $x_1, \ldots, x_n \in S^1$ such that

$$\begin{aligned} &s_1 \ulcorner\!\!- s_1 x_1 \wedge t_1 x_1 \ulcorner\!\!- s_2 x_2 \wedge s_1 \llcorner\!\!- t_1 \wedge t_2 x_2 \ulcorner\!\!- s_3 x_3 \wedge s_2 \llcorner\!\!- t_2 \\ &\wedge \cdots \wedge t_n x_n \ulcorner\!\!- s \wedge s_n \llcorner\!\!- t_n. \end{aligned} \tag{47}$$

Representation of elements of A in the form (39) is not unique. E.g.

$$(s_1, t_1, \ldots, s_n, t_n, t_n) = a_n t_n = (s_1, t_1, \ldots, s_{n-1}, t_{n-1}, s_n).$$

So the same element of A may be represented by different sequences of the form (39). To circumvent this difficulty we first introduce some definitions.

Finite sequences of the form (39) with $s_i, t_i \in S$ and $s_{n+1} \in S^1$ are called *odd words.* Indeed, all these sequences are of odd length. An odd word is called *permissible* whenever there exist $x_1, \ldots, x_n \in S^1$ such that (47) holds, where $s = s_{n+1}$.

Next we try to pick out certain representatives in classes of pairwise equal permissible words (here equality has no precise meaning; we consider two permissible words as equal if they represent the same element of A).

Suppose there exist $y_k, y_{k+1}, \ldots, y_n \in S^1$ such that

$$s_k \ulcorner s_ky_k \wedge t_ky_k \prec s_{k+1}y_{k+1} \wedge t_{k+1}y_{k+1} \prec s_{k+2}y_{k+2} \wedge \cdots \wedge t_{n-1}y_{n-1} \prec s_ny_n \wedge t_ny_n \prec s_{n+1}. \tag{48}$$

Then $(s_1, t_1, \ldots, s_{k-1}, t_{k-1}, s_ky_k) = a_{k-1}(s_ky_k)$. By right negativity of $\ulcorner$ we have $s_ky_k \ulcorner s_k$, and by (48) $s_k \ulcorner s_ky_k$. So by (3) $a_{k-1} \vdash s_ky_k \longleftrightarrow a_{k-1} \vdash s_k$. Clearly, every odd beginning of a permissible word is permissible, so $(s_1, t_1, \ldots, s_{k-1}, t_{k-1}, s_k)$ is permissible. Therefore $a_{k-1} \vdash s_k$. We conclude that $a_{k-1} \vdash s_ky_k$, and so $(s_1, t_1, \ldots, s_{k-1}, t_{k-1}, s_ky_k)$ is a permissible word. Now

$$(s_1, t_1, \ldots, s_{k-1}, t_{k-1}, s_ky_k) = (s_1, t_1, \ldots, s_{k-1}, t_{k-1}, s_k)y_k$$
$$= (a_kt_k)y_k = a_k(t_ky_k) = a_k(s_{k+1}y_{k+1}).$$

The last equality follows from (2) and $t_ky_k \prec s_{k+1}y_{k+1}$ in (48). Now

$$a_k(s_{k+1}y_{k+1}) = (s_1, t_1, \ldots, s_k, t_k, s_{k+1})y_{k+1}.$$

Repeating the above argument we see that

$$(s_1, t_1, \ldots, s_k, t_k, s_{k+1})y_{k+1} = (s_1, t_1, \ldots, s_{k+1}, t_{k+1}s_{k+2}y_{k+2})$$
$$= \cdots = (s_1, t_1, \ldots, s_{n-1}, t_{n-1}, s_n)y_n = (a_nt_n)y_n = a_n(t_ny_n) = a_ns_{n+1}$$
$$= (s_1, t_1, \ldots, s_n, t_n, s_{n+1}).$$

The odd word $b_k = (s_1, t_1, \ldots, s_{k-1}, t_{k-1}, s_ky_k)$ is called a *contraction* of the permissible word $a = (s_1, t_1, \ldots, s_n, t_n, s_{n+1})$. It is easy to see that b_k is permissible. Indeed,

$$s_1 \ulcorner s_1x_1 \wedge t_1x_1 \prec s_2x_2 \wedge s_1 \llcorner t_1 \wedge \cdots \wedge t_{k-2}x_{k-2} \prec s_{k-1}x_{k-1} \wedge s_{k-2} \llcorner t_{k-2}$$

follows from (47). Now $t_{k-1}x_{k-1} \ulcorner s_kt_k \ulcorner s_k \ulcorner s_ky_k$ and $s_{k-1} \llcorner t_{k-1}$ follow from (47) and the right negativity of $\ulcorner$. Thus b_k is permissible.

A contraction of a permissible word which cannot be contracted any more is called a *convolution* of the permissible word. If a permissible word and all its beginnings of odd length possess no proper contractions, then the original word is called *reduced.*

Our plan of attack is as follows: we take the set of all reduced words for A. If $a = (s_1, t_1, \ldots, s_n, t_n, s_{n+1}) \in A$ and $u \in S$, then we define $a \vdash u$ if and only if $(s_1, t_1, \ldots, s_n, t_n, s_{n+1}u)$ is permissible. This being the case, we define $a \cdot u$ to be a convolution of the word $au = (s_1, t_1, \ldots, s_n, t_n, s_{n+1}u)$. At this stage we need a convolution of a permissible word to be uniquely determined, to be *the* convolution of the word. The unicity of convolution means that certain properties of $\prec$, $\mathrel{\ulcorner\!\!-}$ and $\mathrel{\llcorner\!\!-}$ must hold in S. These properties turn out to be new axioms for ordered projection semigroups. Then we investigate whether conditions (1)–(4) hold. It turns out that they do hold, provided another series of axioms are satisfied. Thus we arrive at the conditions which characterize ordered projection semigroups.

So let b_k be a convolution of the above word a along the formulas (48) and let

$$s_l \mathrel{\ulcorner\!\!-} s_l x_l \wedge t_l x_l \prec s_{l+1} x_{l+1} \wedge \cdots \wedge t_{n-1} x_{n-1} \prec s_n x_n \wedge t_n x_n \prec s_{n+1} \tag{49}$$

for some $x_1, \ldots, x_n \in S^1$ define another convolution

$$c = (s_1, t_1, \ldots, s_{l-1}, t_{l-1}, s_l x_l)$$

of a. Without loss of generality we may suppose that $l \leqslant k$. Then $s_k x_k \mathrel{\ulcorner\!\!-} s_k \mathrel{\ulcorner\!\!-} s_k y_k$. Using this, (48) and the last $n - k + 1$ lines of (49), we may write

$$\begin{aligned}
& s_k x_k \prec s_k x_k \wedge s_k y_k \prec s_k y_k \wedge t_k x_k \prec s_{k+1} x_{k+1} \wedge t_k y_k \prec s_{k+1} y_{k+1} \wedge s_k \mathrel{\llcorner\!\!-} t_k \\
& \wedge \ldots \\
& \wedge t_{n-1} x_{n-1} \prec s_n x_n \wedge t_{n-1} y_{n-1} \prec s_n y_n \wedge s_{n-1} \mathrel{\llcorner\!\!-} t_{n-1} \\
& \wedge t_n x_n \prec s_{n+1} \wedge t_n y_n \prec s_{n+1} \wedge s_n \mathrel{\llcorner\!\!-} t_n \\
& \qquad \wedge s_k x_k \mathrel{\ulcorner\!\!-} s_k y_k .
\end{aligned} \tag{50}$$

Suppose that (50) implies $s_k x_k \prec s_k y_k$. If $l < k$, then b_k can be contracted along the first $k - l + 1$ lines of (49) and $s_k x_k \prec s_k y_k$. Since b_k cannot be contracted any more, $l = k$. Now interchanging (48) and (49) we obtain $s_k y_k \prec s_k x_k$. So $s_k y_k = s_k x_k$, and b_k and c coincide. We conclude that every permissible word has a uniquely determined convolution.

Now we must look into our supposition concerning (50). Consider the formula A_n:

$$\begin{aligned}
& t \prec s_1 x_1 \wedge t_0 \prec s_1 y_1 \\
& \wedge t_1 x_1 \prec s_2 x_2 \wedge t_1 y_1 \prec s_2 y_2 \wedge s_1 \mathrel{\llcorner\!\!-} t_1 \\
& \wedge \ldots \\
& \wedge t_{n-1} x_{n-1} \prec s_n x_n \wedge t_{n-1} y_{n-1} \prec s_n y_n \wedge s_{n-1} \mathrel{\llcorner\!\!-} t_{n-1} \\
& \wedge t_n x_n \prec s_{n+1} \wedge t_n y_n \prec s_{n+1} \wedge s_n \mathrel{\llcorner\!\!-} t_n \\
& \wedge t \mathrel{\ulcorner\!\!-} t_0 \rightarrow t \prec t_0 ,
\end{aligned} \tag{51}$$

where $s_i, t, t_j \in S$ and $x_i, y_i \in S^1$.

Suppose that A_n holds for all $n \geqslant 0$. Then $(50) \longrightarrow s_k x_k \prec s_k y_k$ is just a particular case of A_{n-k+1}.

Now we show that every ordered projection semigroup satisfies A_n for all $n \geqslant 0$. Without loss of generality we may suppose that S is a semigroup of transformations of a set A and, if S has no identity, then 1 is the identity transformation of A.

Suppose that the antecedent of A_n holds for some transformations s_i, t, t_j, where $st = t \circ s$ and $\prec, \sqsubset, \llcorner$ have their usual meaning for transformations. To prove (51) we first introduce the compatibility relation for transformations. Two transformations s and t are called *compatible* (written as $s \Downarrow t$), if $s \cup t$ is a transformation, i.e. if $s(a) = t(a)$ for all $a \in \mathrm{pr}_1 s \cap \mathrm{pr}_1 t$. The following condition holds for transformations s, t, x, y:

$$s \llcorner t \wedge tx \Downarrow ty \longrightarrow sx \Downarrow sy. \tag{52}$$

Indeed, let $s \llcorner t$ and $tx \Downarrow ty$. Since $\Downarrow$ is obviously left regular, $st^{-1}tx \Downarrow st^{-1}ty$. Now $st^{-1}t = s\Delta_{\mathrm{pr}_2 t} = s$, so that $sx \Downarrow sy$. Recall that $\Delta_{\mathrm{pr}_2 t}$ is the identity transformation on $\mathrm{pr}_2 t$.

Turning to the antecedent of (51), we see that $t_n x_n \prec s_{n+1}$ and $t_n y_n \prec s_{n+1}$ implies $t_n x_n \Downarrow t_n y_n$. Since $t_n \llcorner s_n$, we obtain by (52) $s_n x_n \Downarrow s_n y_n$. Now $t_{n-1} x_{n-1} \prec s_n x_n$ and $t_{n-1} y_{n-1} \prec s_n y_n$, so that $t_{n-1} x_{n-1} \Downarrow t_{n-1} y_{n-1}$. Repeating this argument several times, we obtain $t \Downarrow t_0$. This and $t \sqsubset t_0$ obviously imply $t \prec t_0$, i.e. (51) holds.

Now let $a = (s_1, t_1, \ldots, s_n, t_n, s_{n+1}) \in A$, i.e. let a be a reduced word. Let $a \vdash u$, i.e., $au = (s_1, t_1, \ldots, s_n, t_n, s_{n+1}u)$ is a permissible word. Clearly, the convolution $a \cdot u$ of au is a reduced word, so $a \cdot u \in A$.

Next we prove (1). Let $a \vdash u$ and let $a \cdot u \vdash v$ for some $a \in A$ and $u, v \in S$. Let the convolution $a \cdot u$ correspond to the formulas

$$s_k \sqsubset s_k x_k \wedge t_k x_k \prec s_{k+1} x_{k+1} \wedge \cdots \wedge t_{n-1} x_{n-1} \prec s_n x_n \wedge t_n x_n \prec s_{n+1} u, \tag{53}$$

i.e.

$$a \cdot u = (s_1, t_1, \ldots, s_{k-1}, t_{k-1}, s_k x_k).$$

Now $a \cdot u \vdash v$ means that $(s_1, t_1, \ldots, s_{k-1}, t_{k-1}, s_k x_k v)$ is a permissible word, i.e. there exist $y_1, \ldots, y_{k-1} \in S^1$ such that

$$s_1 \sqsubset s_1 y_1 \wedge t_1 y_1 \sqsubset s_2 y_2 \wedge \cdots \wedge t_{k-1} y_{k-1} \sqsubset s_k x_k v. \tag{54}$$

Using right regularity of $\prec$ and the fact that $\prec \subset \sqsubset$, we obtain (multiplying the last line of (48) by v at the right and then replacing $\prec$ by $\sqsubset$)

$$t_k x_k v \sqsubset s_{k+1} y_{k+1} v \wedge \cdots \wedge t_n x_n v \sqsubset s_{n+1} uv.$$

Adding these formulas to (54), we obtain formulas of the type (47) showing that $a \vdash uv$.

Conversely, suppose that $a \vdash uv$, i.e. there exist $z_1, \ldots, z_n \in S^1$ such that

$$s_1 \sqsubset s_1 z_1,\ t_1 z_1 \sqsubset s_2 z_2, \ldots, t_n z_n \sqsubset s_{n+1} uv.$$

From these formulas and $s_{n+1} uv \sqsubset s_{n+1} u$ we deduce that $a \vdash u$. Now suppose that the convolution $a \cdot u$ corresponds to (53), i.e.

$$a \cdot u = (s_1, t_1, \ldots, s_{k-1}, t_{k-1}, s_k x_k).$$

Using (53) and the formulas $t_k z_k \sqsubset s_{k+1} z_{k+1}, \ldots, t_n z_n \sqsubset s_{n+1} uv$ and $s_k z_k \sqsubset s_k \sqsubset s_k x_k$, we may write

$$\begin{aligned} & s_k z_k \sqsubset s_k z_k \wedge s_k x_k \prec s_k x_k \\ & \wedge t_k z_k \sqsubset s_{k+1} z_{k+1} \wedge t_k x_k \prec s_{k+1} x_{k+1} \wedge t_k \llcorner s_k \\ & \quad \wedge \ldots \ldots \ldots \ldots \ldots \ldots \\ & \wedge t_{n-1} z_{n-1} \sqsubset s_n z_n \wedge t_{n-1} x_{n-1} \prec s_n x_n \wedge t_{n-1} \llcorner s_{n-1} \\ & \wedge t_n z_n \sqsubset s_{n+1} uv \wedge t_n x_n \prec s_{n+1} u \wedge t_n \llcorner s_n \\ & \quad \wedge s_k z_k \sqsubset s_k x_k. \end{aligned} \tag{55}$$

Suppose that (55) implies

$$s_k z_k \sqsubset s_k x_k v. \tag{56}$$

Then

$$(a \cdot u) v = (s_1, t_1, \ldots, s_{k-1}, t_{k-1}, s_k x_k v)$$

is a permissible word, i.e. $a \cdot u \vdash v$. Thus $(a \cdot u) \cdot v$ and $a \cdot (uv)$ are defined simultaneously. It remains to prove that $(a \cdot u) \cdot v = a \cdot (uv)$.

Suppose that the convolution $(a \cdot u) \cdot v$ corresponds to the formulas

$$s_l \sqsubset s_l y_l \wedge t_l y_l \prec s_{l+1} y_{l+1} \wedge \cdots \wedge t_{k-1} y_{k-1} \prec s_k x_k v. \tag{57}$$

Multiplying the formulas (53) by v at the right, we obtain

$$t_k x_k v \prec s_{k+1} x_{k+1} v \wedge \cdots \wedge t_n x_n v \prec s_{n+1} uv.$$

Together with (57) these formulas show that $(a \cdot u) \cdot v$ is a contraction of $a(uv)$. Since $a \cdot (uv)$ is not contractible, $(a \cdot u) \cdot v = a \cdot (uv)$.

Next we consider the implication (55) $\longrightarrow$ (56) which we have used above. Let B_n denote the formula

$$
\begin{aligned}
& t \sqsubset s_1x_1 \wedge t_0 \prec s_1y_1 \\
& \wedge t_1x_1 \sqsubset s_2x_2 \wedge t_1y_1 \prec s_2y_2 \wedge t_1 \llcorner s_1 \\
& \wedge \; . \; . \; . \; . \; . \; . \; . \; . \; . \; . \\
& \wedge t_{n-1}x_{n-1} \sqsubset s_nx_n \wedge t_{n-1}y_{n-1} \prec s_ny_n \wedge t_{n-1} \llcorner s_{n-1} \\
& \wedge t_nx_n \sqsubset s_{n+1}x_{n+1} \wedge t_ny_n \prec s_{n+1} \wedge t_n \llcorner s_n \\
& \wedge t \sqsubset t_0 \longrightarrow t \sqsubset t_0x_{n+1},
\end{aligned}
\tag{B_n}
$$

where s_i, t, $t_j \in S$ and $x_i, y_j \in S^1$

Clearly, the implication (55) $\longrightarrow$ (56) is just a particular case of B_{n-k+1}.

Now we prove that B_n holds for every ordered projection semigroup of transformations for all $n \geqslant 0$. Indeed, let S be a subsemigroup of $\mathfrak{F}_A$, and let $\prec, \sqsubset$ and $\llcorner$ have their usual meaning for transformations. Suppose that the antecedent of B_n holds and $a_1 \in \mathrm{pr}_1 t$. Then $a_1 \in \mathrm{pr}_1 s_1x_1$, i.e. $(a_1, b_1) \in s_1$ and $b_1 \in \mathrm{pr}_1 x_1$ for some b_1. Since $s_1 \llcorner t_1$, we have $(a_2, b_1) \in t_1$ for some a_2. It follows that $a_2 \in \mathrm{pr}_1 t_1x_1 \subset \mathrm{pr}_1 s_2x_2$, i.e. $(a_2, b_2) \in s_2$ and $b_2 \in \mathrm{pr}_1 x_2$ for some b_2. Repeating this argument and using the first and third columns of formulas in the antecedent of B_n, we have $(a_i, b_i) \in s_i$, $b_i \in \mathrm{pr}_1 x_i$, and $(a_{i+1}, b_i) \in t_i$ for all $i = 1, 2, \ldots, n+1$. Now $t \sqsubset t_0$ implies $a_1 \in \mathrm{pr}_1 t_0$. Therefore $(a_1, c) \in t_0$ for some c. It follows that $(a_1, c) \in s_1y_1$. Since $(a_1, b_1) \in s_1$, we have $(b_1, c) \in Y_1$. Thus $(a_2, c) \in t_1y_1 \prec s_2y_2$; hence $(b_2, c) \in y_2$. Proceeding as above we obtain $(a_{n+1}, c) \in t_ny_n \prec s_{n+1}$, i.e. $c = b_{n+1}$. Thus $(a_1, c) \in t_0$ and $c \in \mathrm{pr}_1 x_{n+1}$; hence $a_1 \in \mathrm{pr}_1 t_0x_{n+1}$. Consequently, $t \sqsubset t_0x_{n+1}$ and B_n holds.

Next we consider (2).

Let $u \prec v$. If $a \cdot u = b$ for some $a, b \in A$ and (53) holds, then (53) and left regularity of $\prec$ imply $s_{n+1}u \prec s_{n+1}v$ and $b = a \cdot u$ is a contraction of av. Since $a \cdot u$ cannot be contracted, $a \cdot v = b$. Now $a \vdash v$ follows from $\prec \subset \sqsubset$, which gives $u \sqsubset v$, and from $a \vdash u$. Thus $a \cdot u = a \cdot v$.

Conversely, suppose that $a \cdot u = a \cdot v$ for all $a \vdash u$. The odd word $a = (u, u, 1)$ belongs to A because $u \llcorner u$, the words ending with 1 being noncontractible and the word (u) being obviously noncontractible. Now $au = (u, u, u)$. Since $u \sqsubset u1 \wedge u1 \prec u$, we have $a \cdot u = (u)$. Thus $a \cdot v = (u)$, i.e. (u) is a contraction of $av = (u, u, v)$. This means that there exists $x \in S^1$ such that $u \sqsubset ux$, $ux \prec v$ and $(ux) = (u)$, i.e. $ux = u$. Therefore $u \prec v$. Thus (2) holds.

Now we prove (3). Let $u \sqsubset v$. If $a \vdash u$, then $a \vdash v$ follows obviously from the definition of $\vdash$. Now let $a \vdash u \longrightarrow a \vdash v$ for all $a \in A$. When proving (2) we saw that $a = (u, u, 1) \vdash u$. Therefore $a \vdash v$. Thus $av = (u, u, v)$ is a permissible word, i.e. $u \sqsubset ux \wedge ux \sqsubset v$ for some $x \in S^1$. It follows that $u \sqsubset v$.

Next we prove (4). Let $u \llcorner v$. If $b = a \cdot u$ for some $a \in A$, then consider the word $c = (b, v, 1)$ which is a concatenation of b and two letters, v and 1. Let

(53) hold. Right regularity of $\mathrel{\llcorner\!\!-}$ implies

$$s_k x_k \mathrel{\llcorner\!\!-} t_k x_k, \ldots, s_n x_n \mathrel{\llcorner\!\!-} t_n x_n.$$

Now $\prec \subset \mathrel{\llcorner\!\!-}$ produces

$$t_k x_k \mathrel{\llcorner\!\!-} s_{k+1} x_{k+1}, \ldots, t_n x_n \mathrel{\llcorner\!\!-} s_{n+1} u.$$

These and the above formulas give $s_k x_k \mathrel{\llcorner\!\!-} s_{n+1} u$. Using left negativity of $\mathrel{\llcorner\!\!-}$, we obtain $s_{n+1} u \mathrel{\llcorner\!\!-} u$. Since $u \mathrel{\llcorner\!\!-} v$, we have $s_k x_k \mathrel{\llcorner\!\!-} v$. The conditions of the type (47) hold for c obviously, since they hold for b. Thus $c \in A$. Now $cv = (b, v, v)$. Since $s_k x_k 1 \mathrel{\ulcorner\!\!-} s_k x_k \wedge v1 \prec v$, we see that b is a contraction of cv. Therefore $c \cdot v = b$.

Conversely, let $b \cdot u = a$ imply $c \cdot v = a$ for some $c \in A$. As we saw when proving (2), $(u) = b \cdot u$ for some $b \in A$. Therefore $(u) = c \cdot v$ for some $c \in A$. Thus there exist $x_i \in S^1$ such that

$$s_1 \mathrel{\ulcorner\!\!-} s_1 x_1 \wedge t_1 x_1 \prec s_2 x_2 \wedge \cdots \wedge t_n x_n \prec s_{n+1} v$$

and $c \cdot v = (s_1 x_1) = (u)$, i.e. $s_1 x_1 = u$. Therefore $t_i x_i \mathrel{\llcorner\!\!-} s_{i+1} x_{i+1}$ and $s_n x_n \mathrel{\llcorner\!\!-} s_{n+1} v$, since $\prec \subset \mathrel{\llcorner\!\!-}$. Right regularity of $\mathrel{\llcorner\!\!-}$ and $s_i \mathrel{\llcorner\!\!-} t_i$ produce $s_i x_i \mathrel{\llcorner\!\!-} t_i x_i$. Using left negativity of $\mathrel{\llcorner\!\!-}$ we obtain

$$u = s_1 x_1 \mathrel{\llcorner\!\!-} t_1 x_1 \mathrel{\llcorner\!\!-} s_2 x_2 \mathrel{\llcorner\!\!-} \cdots \mathrel{\llcorner\!\!-} t_n x_n \mathrel{\llcorner\!\!-} s_{n+1} v \mathrel{\llcorner\!\!-} v.$$

Hence $u \mathrel{\llcorner\!\!-} v$, and (4) holds.

We have proved the following

THEOREM. *An algebraic system* $(S, \prec, \mathrel{\ulcorner\!\!-}, \mathrel{\llcorner\!\!-})$ *is isomorphic to a fundamentally ordered projection semigroup of transformations if and only if S is a semigroup, $\prec$ is a stable order relation, $\mathrel{\ulcorner\!\!-}$ is a left regular and right negative quasi-order relation, $\mathrel{\llcorner\!\!-}$ is a right regular and left negative quasi-order relation, $\prec$ is included both in $\mathrel{\ulcorner\!\!-}$ and $\mathrel{\llcorner\!\!-}$ (i.e. $s \prec t \rightarrow s \mathrel{\ulcorner\!\!-} t$, $s \prec t \rightarrow s \mathrel{\llcorner\!\!-} t$), and the axioms* A_n *and* B_n *hold for all* $n \geqslant 0$.

Note that A_0 is the following condition:

$$t \prec s_1 \wedge t_0 \prec s_1 \wedge t \mathrel{\ulcorner\!\!-} t_0 \rightarrow t \prec t_0,$$

i.e. A_0 coincides with (16), while B_0 is

$$t \mathrel{\ulcorner\!\!-} s_1 x_1 \wedge t_0 \prec s_1 \wedge t \mathrel{\ulcorner\!\!-} t_0 \rightarrow t \mathrel{\ulcorner\!\!-} t_0 x_1,$$

i.e. B_0 coincides with (17).

So ordered projection semigroups are characterized by an infinite system of elementary axioms. They cannot be characterized by any finite set of elementary axioms. We omit the proof of the latter statement. It is based on the fact that no finite subsystem of the system of axioms found in the above theorem is equivalent to the whole system of axioms (to prove this one needs to construct algebraic systems in which any finite number of the above axioms are satisfied, while the systems do not satisfy all the axioms). The next step in the proof used the Gödel Completeness Theorem for the lower predicate calculus; it follows from the theorem that if an infinite system of elementary axioms is not equivalent to any of its finite subsystems, the whole system of axioms is not equivalent to any finite set of elementary axioms whatever.

It is easy to see that if $m > n$, then A_m follows from A_n and B_m follows from B_n.

The above theorem was first proved in [30]. All other theorems on semigroups of transformations endowed with one or two relations $\prec$, $\sqsubset$ or $\llcorner$ can be deduced as corollaries of the above theorem (see [30]).

If one considers ordered projection semigroups of binary relations, then the situation is much simpler, for interrelations between $\prec$, $\sqsubset$ and $\llcorner$ are much weaker in this case (the only such interrelation is $\prec \subset \sqsubset$, $\prec \subset \llcorner$). The following result was first proved in [24]:

THEOREM. *A semigroup S endowed with an order $\prec$ and two quasi-orders $\sqsubset$ and $\llcorner$ is isomorphic to a fundamentally ordered projection semigroup of binary relations if and only if the order $\prec$ is stable, the quasi-order $\sqsubset$ is left regular and right negative, the quasi-order $\llcorner$ is right regular and left negative, and $\prec$ is included both in $\sqsubset$ and $\llcorner$.*

§11. Representation of ordered second projection semigroups

In this section we give axioms for semigroups S endowed with an order relation $\prec$ and quasi-order relation $\llcorner$ and isomorphic to fundamentally ordered and second projection quasi-ordered semigroups of transformations. The only method of arriving at this result known to the author is deducing it from the main theorem of §10 (or repeating the argument leading to the main theorem of §10 in a not too much simpler situation when only two relations $\prec$ and $\llcorner$ are given on S).

Clearly, $(S, \prec, \llcorner)$ is an ordered second projection semigroup if and only if there exists a quasi-order $\sqsubset$ on S such that $(S, \prec, \sqsubset, \llcorner)$ is an ordered projection semigroup. Now $\prec \subset \sqsubset$ and $(st, s) \in \sqsubset$ for all $s, t \in S$. It follows that $s \prec tx$ for some $x \in S^1$ implies $s \sqsubset t$. Let us try

$$\rho = \{(s, t) : (\exists x \in S^1)[s \prec tx]\}$$

for $\sqsubset$ on S.

Suppose that $(S, \prec, \llcorner)$ is an ordered second projection semigroup, i.e. a suitable $\sqsubset$ exists on S such that $(S, \prec, \sqsubset, \llcorner)$ is an ordered projection semigroup. Thus $(S, \prec, \sqsubset, \llcorner)$ satisfies the axioms from the main theorem of §10. We prove that $(S, \prec, \rho, \llcorner)$ also satisfies these axioms. Indeed, $\prec$ is a stable order, and $\llcorner$ is a right regular and left negative quasi-order. Left regularity of ρ follows from left regularity of $\prec$ and the definition of ρ. For all $s, t \in S$ we have $st \prec st$, so $(st, s) \in \rho$, i.e. ρ is right negative. To verify A_n suppose that the antecedent of A_n holds for some elements of S^1 with ρ standing for $\mathrel{\ulcorner\!\!-}$. Clearly, $\rho \subset \mathrel{\ulcorner\!\!-}$, as we have already mentioned above. Thus we may replace ρ by $\mathrel{\ulcorner\!\!-}$ in the antecedent of A_n. Since A_n holds in $(S, \prec, \mathrel{\ulcorner\!\!-}, \mathrel{\llcorner\!\!-})$, we obtain the consequent $t \prec t_0$ which holds in $(S, \prec, \mathrel{\ulcorner\!\!-}, \mathrel{\llcorner\!\!-})$. So it holds in $(S, \prec, \rho, \mathrel{\llcorner\!\!-})$ as well, i.e. A_n is true for $(S, \prec, \rho, \mathrel{\llcorner\!\!-})$.

Now suppose the antecedent of B_n holds for some elements of S^1 with ρ standing for $\mathrel{\ulcorner\!\!-}$. Replacing ρ by its definition (in which we use z_i instead of x), we obtain

$$
\begin{array}{l}
t \prec s_1x_1z_1 \wedge t_0 \prec s_1y_1 \\
\wedge\, t_1x_1 \prec s_2x_2z_2 \wedge t_1y_1 \prec s_2y_2 \wedge s_1 \mathrel{\llcorner\!\!-} t_1 \\
\wedge \;.\;.\;.\;.\;.\;.\;.\;.\;. \\
\wedge\, t_{n-1}x_{n-1} \prec s_nx_nz_n \wedge t_{n-1}y_{n-1} \prec s_ny_n \wedge s_{n-1} \mathrel{\llcorner\!\!-} t_{n-1} \\
\wedge\, t_nx_n \prec s_{n+1}x_{n+1}z_{n+1} \wedge t_ny_n \prec s_{n+1} \wedge s_n \mathrel{\llcorner\!\!-} t_n \\
\wedge\, t \prec t_0z_0
\end{array}
\tag{58}
$$

for some $z_i \in S^1$. Using right regularity of $\prec$ we multiply the inequalities in the first column of (58) at the right: the second inequality is multiplied by z_1, the third one by $z_2z_1, \ldots$, the nth one by $z_{n-1}z_{n-2} \cdots z_1$, and the $(n+1)$th one by $z_nz_{n-1} \cdots z_1$. Let $u_0 = z_0$ and $u_i = x_iz_iz_{i-1} \cdots z_2z_1$ for $i = 1, \ldots, n+1$. Then we have

$$
\begin{array}{l}
t \prec s_1u_1 \wedge t_0 \prec s_1y_1 \\
\wedge\, t_1u_1 \prec s_2u_2 \wedge t_1y_1 \prec s_2y_2 \wedge t_1 \mathrel{\llcorner\!\!-} s_1 \\
\wedge \;.\;.\;.\;.\;.\;.\;.\;.\;.\;. \\
\wedge\, t_{n-1}u_{n-1} \prec s_nu_n \wedge t_{n-1}y_{n-1} \prec s_ny_n \wedge t_{n-1} \mathrel{\llcorner\!\!-} s_{n-1} \\
\wedge\, t_nu_n \prec s_{n+1}u_{n+1} \wedge t_ny_n \prec s_{n+1} \wedge t_n \mathrel{\llcorner\!\!-} s_n \\
\wedge\, t \prec t_0u_0.
\end{array}
\tag{59}
$$

Since $\prec \subset \mathrel{\ulcorner\!\!-}$, we may replace $\prec$ by $\mathrel{\ulcorner\!\!-}$ in the first column of (59), and since $\rho \subset \mathrel{\ulcorner\!\!-}$, we may replace $t \prec t_0u_0$ by $t \mathrel{\ulcorner\!\!-} t_0$ in the last line of (59). Then we obtain the antecedent of B_n in $(S, \prec, \mathrel{\ulcorner\!\!-}, \mathrel{\llcorner\!\!-})$. Since B_n holds in $(S, \prec, \mathrel{\ulcorner\!\!-}, \mathrel{\llcorner\!\!-})$, it follows that $t \mathrel{\ulcorner\!\!-} t_0u_{n+1}$. Now multiply by u_{n+1} at the right all the inequalities in the second column of (59). Then we obtain the antecedent of A_n with x_i replaced by u_i, t_0 replaced by t_0u_{n+1}, y_i replaced by y_iu_{n+1} and s_{n+1} replaced by $s_{n+1}u_{n+1}$. By A_n, $t \prec t_0u_{n+1}$. Therefore $(t, t_0) \in \rho$, i.e. B_n holds in $(S, \prec, \rho, \mathrel{\llcorner\!\!-})$. Simultaneously we have proved that $(S, \prec, \mathrel{\llcorner\!\!-})$ satisfies the following axiom C_n:

$$
\begin{aligned}
& t \prec s_1x_1 \wedge t_0 \prec s_1y_1 \\
& \wedge t_1u_1 \prec s_2x_2 \wedge t_1y_1 \prec s_2y_2 \wedge t_1 \llcorner\!\!- s_1 \\
& \wedge \ldots\ldots\ldots\ldots \\
& \wedge t_{n-1}u_{n-1} \prec s_nx_n \wedge t_{n-1}y_{n-1} \prec s_ny_n \wedge t_{n-1} \llcorner\!\!- s_{n-1} \\
& \wedge t_nu_n \prec s_{n+1}x_{n+1} \wedge t_ny_n \prec s_{n+1} \wedge t_n \llcorner\!\!- s_n \\
& \wedge t \prec t_0x_0 \rightarrow t \prec t_0x_{n+1}.
\end{aligned} \tag{C_n}
$$

(Here x_i stand for u_i.)

Thus $(S, \prec, \rho, \llcorner\!\!-)$ is an ordered projection semigroup.

Now, if $(S, \prec, \rho, \llcorner\!\!-)$ is an ordered projection semigroup, then $(S, \prec, \llcorner\!\!-)$ is an ordered second projection semigroup. Thus $(S, \prec, \llcorner\!\!-)$ is an ordered second projection semigroup if and only if $(S, \prec, \rho, \llcorner\!\!-)$ is an ordered projection semigroup. As we have just seen, if A_n and B_n hold for $(S, \prec, \rho, \llcorner\!\!-)$, then C_n holds for $(S, \prec, \llcorner\!\!-)$. Conversely, suppose that C_n holds for $(S, \prec, \llcorner\!\!-)$. A_n is a particular case of C_n with $x_{n+1} = 1$ (where $t \,{}_{r}\!\!\ulcorner\!\!- t_0$ is replaced by $t \prec t_0x_0$ according to the definition of ρ). Thus A_n holds in $(S, \prec, \rho, \llcorner\!\!-)$. If $(S, \prec, \rho, \llcorner\!\!-)$ satisfies the antecedent of B_n, i.e. if (58) holds, we may transform (58) into (59) using the right regularity of $\prec$. Applying C_n, we obtain $t \prec t_0x_{n+1}$. Since $\prec \subset \rho$, we have $(t, t_0x_{n+1}) \in \rho$, i.e. B_n holds for $(S, \prec, \rho, \llcorner\!\!-)$.

Thus we have proved the following

THEOREM. *A semigroup S endowed with an order relation $\prec$ and a quasi-order relation $\llcorner\!\!-$ is isomorphic to a fundamentally ordered and second projection quasi-ordered semigroup of transformations if and only if $\prec$ is stable, $\llcorner\!\!-$ is right regular and left negative, $\prec$ is included in $\llcorner\!\!-$, and C_n holds for all $n \geqslant 0$.*

Note that C_0 is

$$t \prec s_1x_1 \wedge t \prec t_0x_0 \wedge t_0 \prec s_1 \rightarrow t \prec t_0x_1,$$

which is precisely the weak steadiness of $\prec$.

This theorem was first proved in [30].

For semigroups of binary relations we have the following result from [24]:

THEOREM. *A semigroup S endowed with an order relation $\prec$ and a quasi-order relation $\llcorner\!\!-$ is isomorphic to a fundamentally ordered and second projection quasi-ordered semigroup of binary relations if and only if $\prec$ is stable, $\llcorner\!\!-$ is right regular and left negative, and $\prec$ is included in $\llcorner\!\!-$.*

BIBLIOGRAPHY

1. Garrett Birkhoff, *Lattice theory*, 2nd rev. ed., Amer. Math. Soc., Providence, R. I., 1948.

2. V. V. Vagner, *The theory of generalized grouds and generalized groups*, Mat. Sb. 32 (74)(1953), 545–632. (Russian)

3. V. S. Garvac'kiĭ, *Projection semigroups of invertible transformations*, Studies in Algebra, Izdat. Saratov. Univ., Saratov, 1969, pp. 11–22. (Russian)

4. D. Hilbert and W. Ackermann, *Grundzuge der theoretischen Logik,* 2nd ed., Springer-Verlag, Berlin, 1938; English transl., Chelsea, New York, 1950.

5. George Grätzer, *Universal algebra,* Van Nostrand, Princeton, N. J., 1968.

6. K. A. Zareckiĭ, *The representation of ordered semigroups by binary relations,* Izv. Vysš. Učebn. Zaved. Matematika 1959, no. 6 (13), 48–50. (Russian)

7. Bjarni Jónsson, *Representation of modular lattices and of relation algebras,* Trans. Amer. Math. Soc. 92 (1959), 449–464.

8. P. M. Cohn, *Universal algebra,* Harper & Row, New York, 1965.

9. Roger C. Lyndon, *The representation of relation algebras.* II, Ann. of Math. (2) 63 (1956), 294–307.

10. ———, *Notes on logic,* Van Nostrand, Princeton, N. J., 1966.

11. A. I. Mal'cev, *Über die Einbettung von assoziativen Systemen in Gruppen,* Mat. Sb. 6 (48) (1939), 331–336. (Russian; German summary)

12. Abraham Robinson, *Introduction to model theory and to the metamathematics of algebra,* North-Holland, Amsterdam, 1963.

13. V. N. Saliĭ, *Projective semigroups,* Izv. Vysš. Učebn. Zaved Matematika 1966, no. 3 (52), 144–149. (Russian)

14. M. H. Stone, *The theory of representations for Boolean algebras,* Trans. Amer. Math. Soc. 40 (1936), 37–111.

15. Alfred Tarski, *Contributions to the theory of models.* III, Nederl. Akad. Wetensch. Proc. Ser. A 58 = Indag. Math. 17 (1955), 56–64.

16. Alonzo Church, *Introduction to mathematical logic.* Vol. I, 2nd rev. ed., Princeton Univ. Press, Princeton, N. J., 1966.

17. B. M. Šaĭn, *A system of axioms for semigroups embeddable in generalized groups,* Dokl. Akad. Nauk SSSR 134(1960), 1030–1033; English transl. in Soviet Math. Dokl. 1 (1960).

18. ———, *Embedding of semigroups in generalized groups,* Mat. Sb. 55(97)(1961), 379–400. (Russian)

19. ———, *Representation of semigroups by means of binary relations,* Mat. Sb. 60(102) (1963), 293–303. (Russian)

20. ———, *On the theory of generalized groups,* Dokl. Akad. Nauk SSSR 153(1963), 296–299; English transl. in Soviet Math. Dokl. 4 (1963).

21. ———, *On the theory of transformation semigroups,* Works of Young Scientists, Izdat. Saratov. Univ. Saratov, 1964, pp. 120–122. (Russian) MR 36 #3904.

22. ———, *The representation of ordered semigroups,* Mat. Sb. 65(107)(1964), 188–197. (Russian)

23. ———, *Relation algebras,* Bull. Acad. Polon. Sci. Sér. Sci. Math. Astronom. Phys. 13 (1965), 1–5.

24. ———, *On certain comitants of semigroups with binary relations,* Ural. Gos. Univ. Mat. Zap. 5 (1965), Tetrad' 1, 73–91. (Russian)

25. ———, *Relation algebras,* Doctoral Dissertation, Saratov Univ., Saratov, 1965. (Russian)

26. ———, *Algebras of relations,* Inter-University Scientific Sympos. General Algebra, Tartu Gos. Univ., Tartu, 1966, pp. 130–168. (Russian) MR 34 #2753.

27. ———, *On some problems in the theory of partial automata,* Kybernetika (Prague) 5 (1969), 44–49.

28. ———, *Representation of semigroups,* First All-Union Sympos. Theory of Semigroups, Izdat. Ural. Gos. Univ., Sverdlovsk, 1969, pp. 79–89. (Russian) MR 41 #8548.

29. ———, *Relation algebras and function semigroups,* Semigroup Forum 1 (1970), 1–62.

30. ———, *Representation of ordered projection semigroups,* Mat. Sb. 84(126) (1971), 559–574; English transl. in Math. USSR Sb. 13 (1971).

31. ———, *On certain classes of semigroups of binary relations,* Sibirsk. Mat. Ž. 6 (1965), 616–635. (Russian)

32. ———, *Representation of involuted semigroups by binary relations,* Fund. Math. 82 (1974/75), 121–141.

33. ———, Colloq. Math. 39 (1978), no. 1 (to appear).

34. V. S. Garvac'kiĭ, *Stationary relations,* Izv. Vysš. Učebn. Zaved. Matematika 1971, no. 1 (104), 26–34. (Russian)

Translated by D. E. ZITARELLI
Translation edited by the author

Amer. Math. Soc. Transl.
(2) Vol. 113, 1979

IRREDUCIBLE REPRESENTATIONS OF THE GROUP $SU(n)$ OF CLASS I RELATIVE TO $SU(n-1)$*

R. L. ŠAPIRO

The aim of this note is the description of all (up to equivalence) class I representations of $SU(n)$ relative to $SU(n-1)$. The realization that we shall construct is a generalization to the case of complex space of the realization of class I representations of $SO(n)$ proposed by N. Ja. Vilenkin [1].

1. *Terminology and notation.*

1°. $\mathbf{z} = (z_1, \ldots, z_n)$ is a vector in n-dimensional complex space C_n, and $|\mathbf{z}|^2 = Z = z_1\bar{z}_1 + \cdots + z_n\bar{z}_n$. The sphere $Z = 1$ in C_n is denoted by S; for vectors in S we use the symbol $\omega = (\omega_1, \ldots, \omega_n) \in C_n$, $|\omega| = 1$. Finally, the Laplace operator $\Delta = 4\Sigma_1^n \partial^2/\partial z_i \partial\bar{z}_i$.

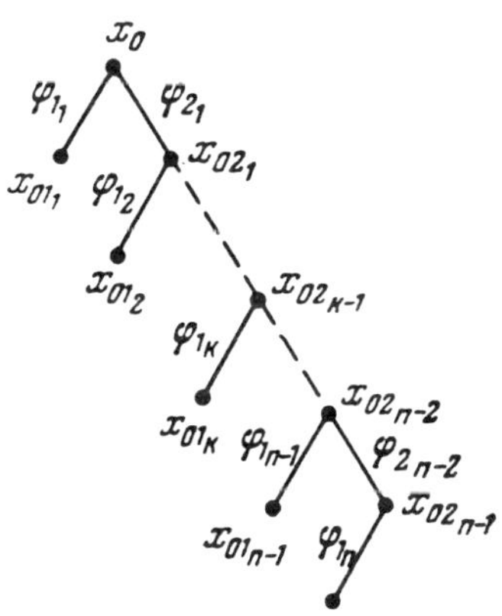

2°. $u_{p,p+1}(\varphi, \theta, \psi)$ is a unitary transformation in the (z_p, z_{p+1}) plane, defined for $0 \leqslant \theta \leqslant \pi/2$ and $0 \leqslant \varphi, \psi < 2\pi \pmod{2\pi}$ by the formulas

$$z'_p = z_p \cos\theta \cdot e^{-i\varphi} + z_{p+1}\sin\theta \cdot e^{i\psi},$$

$$z'_{p+1} = -z_p \sin\theta \cdot e^{-i\psi} + z_{p+1}\cos\theta \cdot e^{i\varphi}.$$

3°. $\mathfrak{R}^{n(k,l)}$ is the space of bihomogeneous polynomials of degree (kl) in the variables $\mathbf{z}$ and $\bar{\mathbf{z}}$, and $\mathfrak{H}^{n(k,l)}$ is the subspace of bihomogeneous harmonic polynomials in $\mathfrak{R}^{n(k,l)}$.

4°. All terms connected with representation theory are understood in the sense of [1].

2. *Coordinates in C_n and parametrization of $SU(n)$.* In C_n we define a coordinate system by means of a tree [1] (see the figure). Here

$$x_{0j_p} = x_{0,\underbrace{2\ldots 2}_{p}j},\ \varphi_{j_p} = \varphi_{\underbrace{2\ldots 2}_{p}j},\ j = 1, 2,$$

$$\left.\begin{aligned} &0 \leqslant \varphi_{1_p} < 2\pi \pmod{2\pi},\ 0 \leqslant \varphi_{2_p} \leqslant \frac{\pi}{2}, \\ &z_p = x_{02_{n-p}} + ix_{01_{n-p+1}},\ z_n = x_0 + ix_{01_1}. \end{aligned}\right\} \qquad (1)$$

AMS (MOS) subject classifications (1970). Primary 22E45.

*Translation of Uspehi Mat. Nauk 22 (1967), no. 1 (133), 180–182. MR 35 #1707.

The following measure on the sphere S is invariant under the mapping $\mathbf{z} \longrightarrow u\mathbf{z}$ ($u \in SU(n)$):

$$d\omega^{(n)} = \frac{\Gamma(n)}{2\pi^n} \prod_{m=1}^{n-1} (\sin \varphi_{2m})^{2(n-m)-1} \cos \varphi_{2_m} d\varphi_{2_m} \prod_{m=1}^{n} d\varphi_{1_m}. \tag{2}$$

Applying the method of introducing parameters on $SU(n)$ described in [3], we get

THEOREM 1. *Any unitary unimodular matrix $u \in SU(n)$ can be represented in the form $u = u^{(n)} \cdot u^{(n-1)} \cdot \cdot \cdot u^{(2)}$, where*

$$u^{(k)} = u_{12}(\varphi_{1_{n-1}}^{(k)}, \varphi_{2_{n-1}}^{(k)}, \varphi_{1_n}^{(k)}) \times u_{23}(\varphi_{1_{n-2}}^{(k)}, \varphi_{2_{n-2}}^{(k)}, 0) \dots u_{k-1\,k}(\varphi_{1_{n-k+1}}^{(k)}, \varphi_{2_{n-k+1}}^{(k)}, 0) \tag{3}$$

$(0 \leqslant \varphi_{1_p}^{(k)} < 2\pi \pmod{2\pi};\ 0 \leqslant \varphi_{2_p}^{(k)} \leqslant \pi/2)$, *and for almost all u the parameters $\varphi_{j_p}^{(k)}$ are defined uniquely within the indicated bounds.*

REMARK. For $\mathbf{z} = u \cdot \omega_n$, where $\omega_n = (0, \dots, 0, 1)$, in the coordinates (1) we have $\varphi_{j_p} = \varphi_{j_p}^{(n)}$, while the coordinates (1) of the vector $[u^{(n)} \cdots u^{(k+1)}]u\omega_n$ coincide with $\varphi_{j_p}^{(k)}$, and therefore the normalized invariant measure on $SU(n)$ in the parameters (3) can be computed from the formula

$$du(n) = d\omega^{(n)} d\omega^{(n-1)} \cdots d\omega^{(2)}. \tag{4}$$

3. *Some properties of bihomogeneous polynomials.*

THEOREM 2. *The space $\mathfrak{R}^{n(k,l)}$ can be decomposed into a direct sum of subspaces $Z^s \mathfrak{H}^{n(k-s,l-s)}$, $s = 0, 1, \dots, q$, where $q = \min(k, l)$. The projection operators $P_s \mathfrak{R}^{n(k,l)} = Z^s \mathfrak{H}^{n(k-s,l-s)}$ can be computed from the formula*

$$P_s = \frac{(n+k+l-2s-1)!}{4^s(n+k+l-s-1)!S!} \sum_{r=s}^{q} (-1)^{r-s} \frac{(n+k+l-2-2s-r+s)!}{4^{r-s}(n+k+l-2-2s)!(r-s)!} Z^r \Delta^r. \tag{5}$$

COROLLARY.

$$\mathfrak{H}^{n(k,l)} = \sum_{k_1=0}^{k} \sum_{l_1=0}^{l} P_0 [z_n^{k-k_1} \bar{z}_n^{l-l_1} \mathfrak{H}^{(n-1)(k_1,l_1)}], \tag{6}$$

where $\mathfrak{H}^{(n-1)(k_1,l_1)}$ is a bihomogeneous harmonic polynomial in $\mathbf{z}'$ and $\overline{\mathbf{z}}'$; here $\mathbf{z}' = (z_1, \dots, z_{n-1})$.

The decomposition (6) and the results of [2] allow us, generalizing the technique of [1], to apply induction on the dimension of the group for the proof of facts 1°–4° and the construction of a basis in $\mathfrak{H}^{n(k,l)}$.

4. *Irreducible class* I *representations of* $SU(n)$. A quasi-regular representation of $SU(n)$ is realized in $\mathfrak{C}^2(S)$ by the operators $[L^n(u)f](\mathbf{z}, \bar{\mathbf{z}}) = f(u^{-1}\mathbf{z}, \bar{u}^{-1}\bar{\mathbf{z}})$, which are unitary with respect to the scalar product

$$(f_1, f_2) = \int f_1(z, \bar{z}) \cdot \overline{f_2(\mathbf{z}, \mathbf{z})} \cdot \delta(\sqrt{z} - 1) \frac{\Gamma(n)}{2\pi^n} \frac{i^{n^2}}{2^n} d\mathbf{z}\, d\bar{\mathbf{z}} = \int f_1(\varphi_{j_p}) \overline{f_2(\varphi_{j_p})}\, d\omega^{(n)}.$$

The bihomogeneity of the polynomials in $\mathfrak{R}^{n(k,l)}$ allows us to imbed it in $\mathfrak{C}^2(S)$. We have the decomposition

$$\mathfrak{C}^2(S) = \sum_{k,l=0}^{\infty} \mathfrak{F}^{n(k,l)} \text{ and } L^n(u) = \sum_{k,l=0}^{\infty} H^{n(k,l)}(u),$$

where $H^{n(k,l)}(u)$ is the representation induced in $\mathfrak{F}^{n(k,l)}$ by the operators $L^n(u)$. The representations $H^{n(k,l)}(u)$ are unitary with respect to the scalar product in $\mathfrak{C}^2(S)$.

1°. The $H^{n(k,l)}(u)$ are irreducible.

2°. The only (up to a constant factor) vector in $\mathfrak{F}^{n(k,l)}$ that is invariant with respect to $SU(n-1)$ is $P_0[z_n^k \bar{z}_n^l]$.

From 1°, 2° and the fact that the $H^{n(k,l)}(u)$ are unitary it follows that the $H^{n(k,l)}(u)$ are class I representations of $SU(n)$ relative to $SU(n-1)$.

3°. For $n \geqslant 3$ the representations $H^{n(k,l)}(u)$ are pairwise nonequivalent.

4°. Every class I representation is equivalent to some $H^{n(k,l)}(u)$.

5°. The functions

$$\Xi^{(k,l)n}_{(k_j,l_j,m)} = C^{n(k,l)}_{(k_j,l_j,m)} \prod_{j=0}^{n-3} e^{i[(k_j - l_j) - (k_{j+1} - l_{j+1})]\varphi_{1_{j+1}}}$$

$$\times e^{i(k_{n2} - l_{n-2} - m)\varphi_{1n-1}} e^{im\varphi_{1n}} \prod_{j=0}^{n-3} (\sin \varphi_{2_{j+1}})^{k_{j+1} + l_{j+1}} (\sin \varphi_{2_{n-1}})^m$$

$$\times \prod_{j=0}^{n-3} (\cos \varphi_{2_{j+1}})^{|(k_j - k_{j+1}) - (l_j - l_{j+1})|} P_{q_j}^{(k_{j+1} + l_{j+1} + n - j - 2, |(k_j - k_{j+1}) - (l_j - l_{j+1})|)}$$

$$\times (\cos 2\varphi_{2_{j+1}})(\cos \varphi_{2_{n-1}})^{|k_{n-2} - l_{n-2} - m|} P_{q_{n-2}}^{(|m|, |k_{n-2} - l_{n-2} - m|)}(\cos 2\varphi_{2_{n-1}}),$$

form an orthonormal basis in $\mathfrak{C}^2(S)$, and for fixed $k_0 = k$ and $l_0 = l$ they form an orthonormal basis in $\mathfrak{F}^{n(k,l)}$. Here $0 \leqslant k_j \leqslant k_{j-1}$, $0 \leqslant l_j \leqslant l_{j-1}$, $q_j = \min(k_j - k_{j+1}, l_j - l_{j+1})$, $C_{(k_j, l_j, m)}$ is a normalizing constant, m is any integer such that q_{n-2} is a nonnegative integer [2], and $P_m^{(\alpha,\beta)}(x)$ are the Jacobi polynomials.

BIBLIOGRAPHY

1. N. Ja. Vilenkin, *Special functions and the theory of group representations*, "Nauka", Moscow, 1965; English transl., Amer. Math. Soc., Providence, R. I., 1968.

2. N. Ja. Vilenkin, G. I. Kuznecov and Ja. A. Smorodinskiĭ, *Eigenfunctions of the Laplace operator providing representations of the groups* $U(2)$, $SU(2)$, $SO(3)$, $U(3)$ *and* $SU(3)$, *and the symbolic method,* Jadernaja Fiz. 2 (1965), 906–917; English transl. in Soviet J. Nuclear Phys. 2 (1965/66).

3. Francis D. Murnaghan, *The unitary and rotation groups,* Spartan Books, Washington, D. C., 1962.

Translated by R. W. SIZER

Amer. Math. Soc. Transl.
(2) Vol. 113, 1979

IRREDUCIBLE REPRESENTATIONS OF THE GROUP $SU(n)$ OF CLASS I RELATIVE TO $SU(n-1)$*

N. Ja. VILENKIN AND R. L. ŠAPIRO

A description is given to within equivalence of all irreducible unitary representations of the unimodular unitary group $SU(n)$, of class I relative to $SU(n-1)$. The proposed realization of the representations is a generalization of the class I representations of $SO(n)$ proposed by N. Ja. Vilenkin [1].

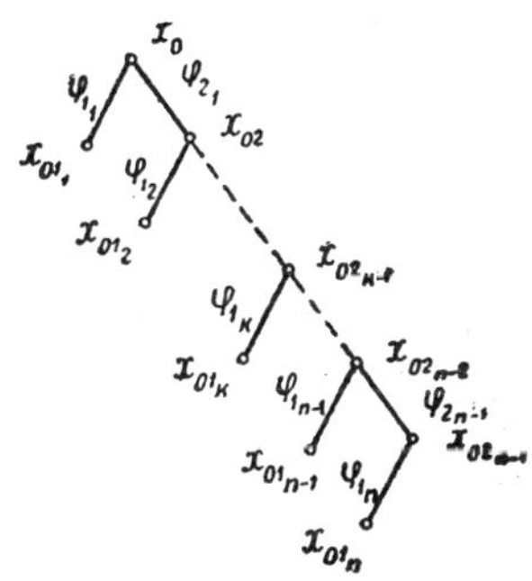

FIGURE 1

§1. Description of the group $SU(n)$

1. *Coordinate system in* C_n. We define a coordinate system in n-dimensional complex space C_n by the tree [2] depicted in Figure 1. Here

$$x_{0j_p} = x_{\underbrace{0\,2\ldots 2j}_{p}}, \quad \varphi_{j_p} = \varphi_{\underbrace{2\ldots 2j}_{p}}, \quad j = 1,\ 2, \tag{1.1}$$

where $0 \leqslant \varphi_{1_p} < 2\pi$ and $0 \leqslant \varphi_{2_p} \leqslant \pi/2$. If $z = (z_1, \ldots, z_n) \in C_n$ and $Z = z_1\bar{z}_1 + \cdots + z_n\bar{z}_n = r^2$, then, setting

$$z_p = x_{0\,2_{n-p}} + ix_{0\,1_{n-p+1}} \quad \text{for } p = 1,\ 2, \ldots, n-1; \ z_n = x_0 + ix_{0\,1_1}, \tag{1.2}$$

we obtain

AMS (MOS) subject classifications (1970). Primary 22E45.

*Translation of Izv. Vysš. Učebn. Zaved. Matematika 1967, no. 7 (62), 9–20. MR 36 #2741.

$$z_1 = r \sin\varphi_{2_1} \sin\varphi_{2_2} \dots \sin\varphi_{2_{n-1}} e^{i\varphi_{1_n}},$$

$$z_2 = r \sin\varphi_{2_1} \sin\varphi_{2_2} \dots \sin\varphi_{2_{n-2}} \cos\varphi_{2_{n-1}} e^{i\varphi_{1_{n-1}}},$$

$$\dots\dots\dots\dots\dots\dots\dots\dots\dots$$

$$z_n = r \cos\varphi_{2_1} e^{i\varphi_{1_1}}, \text{ where } \varphi_{1_k} = \arg z_{n-k+1}, \tag{1.3}$$

$$\sin\varphi_{2_k} = \frac{\sqrt{z_1\bar z_1 + \dots + z_{n-k}\bar z_{n-k}}}{\sqrt{z_1\bar z_1 + \dots + z_{n-k+1}\bar z_{n-k+1}}}, \quad \cos\varphi_{2_k} = \frac{\sqrt{z_{n-k+1}\bar z_{n-k+1}}}{\sqrt{z_1\bar z_1 + \dots + z_{n-k}\bar z_{n-k}}}.$$

The numbers φ_{j_k} are called the Euler angles of a vector $z \in C_n$; they are uniquely defined within the limits (1.1) for almost all z. Normalized measure on the sphere S

$$z_1\bar z_1 + \dots + z_n\bar z_n = 1 \tag{1.4}$$

in the coordinates (1.3) is computed by the formula

$$d\omega^{(n)} = \frac{\Gamma(n)}{2\pi^n} \prod_{m=1}^{n-1} (\sin\varphi_{2_m})^{2(n-m)-1} \cos\varphi_{2_m} d\varphi_{2_m} \prod_{m=1}^{n} d\varphi_{1_m}, \tag{1.5}$$

so that

$$\int d\omega^{(n)} = 1. \tag{1.5'}$$

2. *Euler angles on the group $SU(n)$*. Linear transformations of C_n which preserve the length of a vector z form a unitary group $U(n)$. We shall examine the unitary unimodular transformations $z' = uz$, in C_n (the group $SU(n)$), where $u \in SU(n)$ and $z \in C_n$, or

$$z'_p = \sum_{k=1}^{n} u_{pk} z_k, \quad u = \| u_{pk} \|_{p,\, k=1}^{n}. \tag{1.6}$$

The group $SU(n)$ operates transitively on S so that any vector z can be obtained from the vector $r\vec\omega_n = (0, \dots, 0, r)$, $r = |z|$, by some unitary transformation $z = ur\vec\omega_n$. Hence it is obvious that the space of right cosets $SU(n)/SU(n-1)$ can be identified with S, the transformations $z = ur\vec\omega_n \longrightarrow u^0 ur\vec\omega_n = u^0 z$ in C_n corresponding to the transformations $\{uh\} \longrightarrow \{u^0 uh\}$, $h \in SU(n-1)$, in the quotient space. This implies the geometric meaning introduced below of the parametrization of $SU(n)$. We denote by $u_{p,p+1}(\varphi, \theta, \psi)$, $0 \leqslant \varphi, \psi < 2\pi \pmod{2\pi}$, $0 \leqslant \theta \leqslant \pi/2$, the unitary transformation in the plane (z_p, z_{p+1}):

$$\begin{aligned} z'_p &= z_p \cos\theta e^{-i\varphi} + z_{p+1} \sin\theta e^{i\psi}, \\ z'_{p+1} &= -z_p \sin\theta e^{-i\psi} + z_{p+1} \cos\theta e^{i\varphi}. \end{aligned} \tag{1.7}$$

Utilizing the technique of introducing parameters on $SU(n)$ set forth in [5], we can easily prove

THEOREM 1. *Any unitary unimodular matrix $u \in SU(n)$ can be represented in the form $u = u^{(n)}u^{(n-1)} \cdots u^{(2)}$, where*

$$u^{(k)} = u_{12}(\varphi^{(k)}_{1_{n-1}}, \varphi^{(k)}_{2_{n-1}}, \varphi^{(k)}_{1_n}) u_{23}(\varphi^{(k)}_{1_{n-2}}, \varphi^{(k)}_{2_{n-2}}, 0) \cdots$$
$$\times u_{k-1k}(\varphi^{(k)}_{1_{n-k+1}}, \varphi^{(k)}_{2_{n-k+1}}, 0); \ 0 \leqslant \varphi^{(k)}_{1_p} < 2\pi, \ 0 \leqslant \varphi^{(k)}_{2_p} \leqslant \frac{\pi}{2}, \tag{1.8}$$

the parameters $\varphi^{(k)}_{j_p}$ being defined uniquely within the indicated limits for almost all $u \in SU(n)$.

We observe that if $z = ur\vec{\omega}_n$, then the Euler angles φ_{j_p} of z coincide with $\varphi^{(n)}_{j_p}$, while the angles $\varphi^{(k)}_{j_p}$ are the Euler angles of the vector $[u^{(n)} \cdots u^{(k+1)}]^{-1} ur\vec{\omega}_n$. In accordance with this we shall call the angles $\varphi^{(k)}_{j_p}$ the *Euler angles* on the group $SU(n)$.

3. *Invariant integration on the sphere S and the group $SU(n)$.* We define the invariant (relative to transformations (1.6)) integral on the sphere S of the function $f(z, \bar{z})$ by the formula

$$\frac{(i)^{n^2}}{2^n} A^{(n)} \int f(z, \bar{z}) \delta(\sqrt{Z} - 1)\, dz\, d\bar{z}, \tag{1.9}$$

where $A^{(n)}$ is a normalizing constant and $dz = dz_1 \cdots dz_n$. In coordinates (1.3) the invariant integral is written as

$$\int f(\varphi_{j_p})\, d\omega^{(n)}. \tag{1.9$'$}$$

We conclude from the geometric meaning of the Euler angles $\varphi^{(k)}_{j_p}$ that the expression for the invariant measure on $SU(n)$ is of the form

$$du^{(n)} = d\omega^{(n)} du^{(n-1)} = d\omega^{(n)} d\omega^{(n-1)} \cdots d\omega^{(2)}. \tag{1.10}$$

It is easy to see that $\int du^{(n)} = 1$ and

$$du^{(n)} = \prod_{k=2}^{n} \frac{\Gamma(k)}{2\pi^k} \prod_{m=n-k+1}^{n-1} (\sin \varphi^{(k)}_{2_m})^{2(n-m)-1} \cos \varphi^{(k)}_{2_m}\, d\varphi^{(k)}_{2_m} \prod_{m=n-k+1}^{n} d\varphi^{(k)}_{1_m}.$$

§2. Class I representations of $SU(n)$ relative to $SU(n-1)$

1. *Quasi-regular representation of $SU(n)$.* We define the quasi-regular representation of $SU(n)$ in the space of functions $f(z, \bar{z}) \in \mathfrak{L}^2(S)$, i.e. functions satisfying the condition

$$\int |f(z, \bar{z})|^2 \frac{(i)^{n^2}}{2^n} A^{(n)} \delta(\sqrt{Z} - 1)\, dz\, d\bar{z} < \infty, \tag{2.1}$$

by the formula

$$[L^n(u) f](z, \bar{z}) = f(u^{-1}z, \overline{u^{-1}z}). \tag{2.2}$$

It is obvious that $L^n(u)$ is a representation of $SU(n)$ which is unitary relative to the scalar product

$$(f_1, f_2) = \int f_1(z, \bar{z}) \overline{f_2(z, \bar{z})} \frac{(i)^{n^2}}{2^n} A^{(n)} \delta(\sqrt{Z} - 1)\, dz d\bar{z}. \tag{2.1'}$$

This representation is reducible. Spaces will be constructed below in which irreducible class I representations of $SU(n)$ relative to $SU(n-1)$ are realized, and the representation $L^n(u)$ will be factored into irreducible ones.

2. *Representations in spaces of bihomogeneous polynomials.* We denote by $\mathfrak{B}^{n(k,l)}$ the space of bihomogeneous degree (k, l) polynomials in z and $\bar{z}$, i.e. polynomials satisfying the condition

$$f(\lambda z, \bar{\lambda} \bar{z}) = \lambda^k \bar{\lambda}^l f(z, \bar{z}), \tag{2.3}$$

where k and l are nonnegative integers. Since each bihomogeneous polynomial is uniquely determined by its values on S (we denote vectors z of length 1 by $\vec{\omega}$) as follows:

$$f(z, \bar{z}) = f(r\vec{\omega}, r\vec{\bar{\omega}}) = r^{k+l} f(\vec{\omega}, \vec{\bar{\omega}}), \tag{2.3'}$$

we see that $\mathfrak{B}^{n(k,l)}$ can be considered as subspaces in $\mathfrak{L}^2(S)$. Transformations (2.2) carry $\mathfrak{B}^{n(k,l)}$ into itself in such a way that the representation $L^n(u)$ gives the representation $L^{n(k,l)}(u)$ in $\mathfrak{B}^{n(k,l)}$:

$$[L^{n(k,l)}(u) f](z, \bar{z}) = f(u^{-1}z, \overline{u^{-1}z}),\ f(z, \bar{z}) \in \mathfrak{B}^{n(k,l)}. \tag{2.4}$$

However, the representations $L^{n(k,l)}(u)$ are reducible. To decompose them into irreducible ones we establish certain properties of bihomogeneous polynomials.

3. *Bihomogeneous harmonic polynomials.* We denote by $\mathfrak{B}^n$ the space of all polynomials in z and $\bar{z}$:

$$\mathfrak{B}^n = \sum_{k=0}^{\infty} \sum_{l=0}^{\infty} \mathfrak{B}^{n(k,l)}. \tag{2.5}$$

We introduce two operators on $\mathfrak{B}^n$:

$$\Delta = 4 \sum_{i=1}^{n} \frac{\partial^2}{\partial z_i \partial \bar{z}_i} \tag{2.6a}$$

the Laplace operator, and

$$Z = z_1\bar{z}_1 + \dots + z_n\bar{z}_n \tag{2.6b}$$

the operator of multiplication by the function Z. It is obvious that

$$\Delta\colon \mathfrak{B}^{n(k,l)} \to \mathfrak{B}^{n(k-1,l-1)}, \quad \Delta\colon \mathfrak{B}^{n(k,0)} \to 0, \quad \Delta\colon \mathfrak{B}^{n(0,l)} \to 0,$$
$$Z\colon \mathfrak{B}^{n(k,l)} \to \mathfrak{B}^{n(k+1,l+1)}.$$

DEFINITION. A polynomial $f(z, \bar{z})$ is called *harmonic* if $\Delta f(z, \bar{z}) = 0$. Since $\mathfrak{B}^{n(k,0)}$ and $\mathfrak{B}^{n(0,l)}$ consist of harmonic polynomials and the representations $L^{n(k,0)}(u)$ and $L^{n(0,l)}(u)$ are irreducible [3], we shall assume in the sequel that k and l are simultaneously different from zero. The following relations can easily be established for the operators Δ and Z:

1°. For any $f(z, \bar{z}) \in \mathfrak{B}^{n(k,l)}$

$$[\Delta, Z] f(z, \bar{z}) = 4(n + k + l) f(z, \bar{z}). \tag{2.7}$$

2°. For any $f(z, \bar{z}) \in \mathfrak{B}^{n(k,l)}$ and any integer $r \geqslant 1$

$$[\Delta, Z^r] f(z, \bar{z}) = 4(n + k + l + r - 1) r Z^{r-1} f(z, \bar{z}). \tag{2.8}$$

3°. For any harmonic polynomial $f(z, \bar{z}) \in \mathfrak{B}^{n(k,l)}$ and any integer $r \geqslant 0$

$$\Delta^r Z^r f(z, \bar{z}) = 4^r \frac{(n + k + l + r - 1)!\, r!}{(n + k + l - 1)!} f(z, \bar{z}). \tag{2.9}$$

THEOREM 2. *The decomposition*

$$f(z, \bar{z}) = \sum_{s=0}^{n(k,l)} Z^s \hat{f}_s(z, \bar{z}), \tag{2.10}$$

holds for every polynomial $f(z, \bar{z}) \in \mathfrak{B}^{n(k,l)}$, *where*

a) $\Delta^{q+1} f(z, \bar{z}) = 0,\ 0 \leqslant q \leqslant \min(k, l)$;

b) $\Delta \hat{f}_s(z, \bar{z}) = 0,\ \hat{f}_s(z, \bar{z}) \in \mathfrak{B}^{n(k-s,\, l-s)}$.

The decomposition (2.10) *is unique.*

PROOF. Assuming that the decomposition (2.10) holds for any $f(z, \bar{z}) \in \mathfrak{B}^{n(k,l)}$, we shall prove its uniqueness; i.e., we shall prove that $f(z, \bar{z}) = 0$ implies $\hat{f}_s(z, \bar{z}) = 0$. We assume that in the sum $0 = \Sigma_0^q Z^s \hat{f}_s(z, \bar{z})$ not all the $\hat{f}_s(z, \bar{z})$ equal zero, and we denote $p = \max\{s \,|\, \hat{f}_s(z, \bar{z}) \neq 0\}$. Let us apply the operator Δ^p to both sides of the equality $0 = \Sigma_0^p Z^s \hat{f}_s(z, \bar{z})$. For all $s < p$ we have $\Delta^p Z^s \hat{f}_s(z, \bar{z}) = 0$ since either $\hat{f}_s(z, \bar{z}) = 0$ or, if $\hat{f}_s(z, \bar{z}) \neq 0$,

$$\Delta^p Z^s \hat{f}(z, \bar{z}) = \Delta^{p-s} 4^s \frac{(n + k + l - s - 1)!\, s!}{(n + k + l - 2s - 1)!} \hat{f}_s(z, \bar{z}) = 0.$$

for a harmonic bihomogeneous polynomial of degree $(k - s, l - s)$ as follows from (2.9). Consequently,

$$0 = \Delta^p Z^p \hat{f}_p(z, \bar{z}) = 4^p \frac{(n + k + l - p - 1)!\, p!}{(n + k + l - 2p - 1)!} \hat{f}_p(z, \bar{z}),$$

i.e. $\hat{f}_p(z, \bar{z}) = 0$, which contradicts the assumption. Thus, if the decomposition (2.10) holds, it is unique.

Now we prove the existence of the decomposition (2.10). Since $\Delta^q f_{n(k,l)}(z, \bar{z}) \in \mathfrak{B}^{n(k-q,l-q)}$ we obtain a harmonic polynomial for $q = \min(k, l)$. Thus for each $f_{n(k,l)}(z, \bar{z})$ an integer q, $0 \leqslant q \leqslant \min(k, l)$, exists such that $\Delta^{q+1} f_{n(k,l)}(z, \bar{z}) \neq 0$. We carry out the proof of the existence of (2.10) by induction on q. The decomposition (2.10) holds for all $f_{n(k,l)}(z, \bar{z}) \neq 0$ with the condition $\Delta f_{n(k,l)}(z, \bar{z}) = 0$: it is sufficient to set $\hat{f}_0(z, \bar{z}) = f_{n(k,l)}(z, \bar{z})$ (by virtue of uniqueness).

We now assume that the decomposition (2.10)

$$f(\boldsymbol{z}, \bar{\boldsymbol{z}}) = \sum_{s=0}^{q-1} Z^s \hat{f}_s(\boldsymbol{z}, \bar{\boldsymbol{z}})$$

holds for all $f_{n(k,l)}(z, \bar{z})$ satisfying $\Delta^q f_{n(k,l)}(z, \bar{z}) = 0$. Let $f_{n(k,l)}(z, \bar{z}) \in \mathfrak{B}^{n(k,l)}$, $\Delta^q f_{n(k,l)}(z, \bar{z}) \neq 0$ and $\Delta^{q+1} f_{n(k,l)}(z, \bar{z}) = 0$. This condition means that $\Delta^q f_{n(k,l)}(z, \bar{z})$ is a harmonic polynomial of degree $(k-q,\ l-q)$. Utilizing (2.9), we apply the operator $\Delta^q Z^q$ to it:

$$\Delta^q Z^q [\Delta^q f_{n(k,l)}(\boldsymbol{z}, \bar{\boldsymbol{z}})] = 4^q \frac{(n+k+l-q-1)!\, q!}{(n+k+l-2q-1)!} \Delta^q f_{n(k,l)}(\boldsymbol{z}, \bar{\boldsymbol{z}}),$$

so that

$$\Delta^q \left[1 - \frac{(n+k+l-2q-1)!}{4^q (n+k+l-q-1)!\, q!} Z^q \Delta^q \right] f_{n(k,l)}(\boldsymbol{z}, \bar{\boldsymbol{z}}) = 0.$$

This means that the polynomial

$$f^{(1)}_{n(k,l)}(\boldsymbol{z}, \bar{\boldsymbol{z}}) = \left[1 - \frac{(n+k+l-2q-1)!}{L^q (n+k+l-q-1)!\, q!} Z^q \Delta^q \right] f_{n(k,l)}(\boldsymbol{z}, \bar{\boldsymbol{z}}) \tag{2.11}$$

satisfies the induction hypothesis, and therefore

$$f^{(1)}_{n(k,l)}(\boldsymbol{z}, \bar{\boldsymbol{z}}) = \sum_{s=0}^{q-1} Z^s \hat{f}^{(1)}_s(\boldsymbol{z}, \bar{\boldsymbol{z}}).$$

Hence

$$f_{n(k,l)}(\boldsymbol{z}, \bar{\boldsymbol{z}}) = Z^q \frac{(n+k+l-2q-1)!}{4^q (n+k+l-q-1)!\, q!} \Delta^q f_{n(k,l)}(\boldsymbol{z}, \bar{\boldsymbol{z}}) + \sum_{s=0}^{q-1} Z^s \hat{f}_s(\boldsymbol{z}, \bar{\boldsymbol{z}}),$$

so that (by virtue of uniqueness) we can set

$$\begin{aligned} \hat{f}_q(\boldsymbol{z}, \bar{\boldsymbol{z}}) &= \frac{(n+k+l-2q-1)!}{4^q (n+k+l-q-1)!\, q!} \Delta^q f_{n(k,l)}(\boldsymbol{z}, \bar{\boldsymbol{z}}), \\ \hat{f}_s(\boldsymbol{z}, \bar{\boldsymbol{z}}) &= \hat{f}^{(1)}_s(\boldsymbol{z}, \bar{\boldsymbol{z}}), \quad s = 0, 1, \ldots, q-1, \end{aligned} \tag{2.12}$$

where $\Delta \hat{f}_q(z, \bar{z}) = \Delta \hat{f}^{(1)}_s(z, \bar{z}) = 0$ (by the induction hypothesis), which proves the theorem.

COROLLARY. *Denote the space of bihomogeneous harmonic polynomials of* *degree (i, j) in z and $\bar{z}$ by $\mathfrak{H}^{n(i,j)}$. The existence and uniqueness of* (2.10) *signifies a decomposition of $\mathfrak{B}^{n(k,l)}$ into a direct sum of subspaces of the form $Z^s\mathfrak{H}^{n(k-s,l-s)}$:*

$$\mathfrak{B}^{n(k,l)} = \sum_{s=0}^{q} Z^s \mathfrak{H}^{n(k-s,l-s)}, \quad q = \min(k, l). \tag{2.13}$$

From the theorem's proof it is easy to establish formulas for projection operators onto the subspaces $\mathfrak{H}^{n(k-s,l-s)}$. We shall show that an operator projecting onto $\mathfrak{H}^{n(k-s,l-s)}$ (we denote it by $H_{q,s}(\Delta, Z)$) is a polynomial in Δ and Z.

It is obvious that if $\Delta f_{n(k,l)}(z, \bar{z}) = 0$ then $\hat{f}_0(z, \bar{z}) = f_{n(k,l)}(z, \bar{z})$; that is, $H_{0,0}(\Delta, Z) = 1$. For $f_{n(k,l)}(z, \bar{z})$ satisfying $\Delta^q f_{n(k,l)}(z, \bar{z}) = 0$, $0 < q \leqslant \min(k, l)$, let projections onto $\mathfrak{H}^{n(k-s,l-s)}$ be defined by

$$\hat{f}_s(z, \bar{z}) = H_{q-1,s}(\Delta, Z) f_{n(k,l)}(z, \bar{z}), \quad s = 0, 1, \ldots, q-1.$$

Then, for $f_{n(k,l)}(z, \bar{z})$ satisfying $\Delta^q f_{n(k,l)}(z, \bar{z}) \neq 0$ and $\Delta^{q+1} f_{n(k,l)}(z, \bar{z}) = 0$, projections onto $\mathfrak{H}^{n(k-s,l-s)}$ have the form (cf. 2.12))

$$\hat{f}_q(z, \bar{z}) = \frac{(n+k+l-2q-1)!}{4^q (n+k+l-q-1)!\, q!} \Delta^q f_{n(k,l)}(z, \bar{z}),$$

$$f_s(z, \bar{z}) = H_{q-1,s}(\Delta, Z)\left[1 - \frac{(n+k+l-2q-1)!}{4^q (n+k+l-q-1)!\, q!} Z^q \Delta^q\right] f_{n(k,l)}(z, \bar{z}).$$

Thus

$$\begin{aligned}
H_{q,q}(\Delta, Z) &= \frac{(n+k+l-2q-1)!}{4^q (n+k+l-q-1)!\, q!} \Delta^q;\\
H_{q,s}(\Delta, Z) &= \frac{(n+k+l-2s-1)!}{4^s (n+k+l-s-1)!\, s!} \Delta^s \Bigg[1\\
&\quad - \frac{(n+k+l-2s-3)!}{4^{s+1} (n+k+l-s-2)!\, (s+1)!} Z^{s+1}\Delta^{s+1}\Bigg] \cdots \Bigg[1\\
&\quad - \frac{(n+k+l-2q-2)!}{4^q (n+k+l-q-1)!\, q!} Z^q \Delta^q\Bigg],\\
H_{q,0}(\Delta, Z) &= \left[1 - \frac{(n+k+l-2-1)!}{4 (n+k+l-1-1)!\, 1!} Z\Delta\right] \cdots \Bigg[1\\
&\quad - \frac{(n+k+l-2q-1)!}{4^q (n+k+l-q-1)!\, q!} Z^q \Delta^q\Bigg].
\end{aligned} \tag{2.14}$$

We note that if $\Delta^q f_{n(k,l)}(z, \bar{z}) = 0$ for $0 \leqslant q < \min(k, l)$, then

$$\prod_{j=q+1}^{\min(k,l)} \left[1 - \frac{(n+k+l-2j-1)!}{4^j (n+k+l-j-1)!\, j!} Z^j \Delta^j\right] f_{n(k,l)}(z, \bar{z}) = f_{n(k,l)}(z, \bar{z}),$$

so that it is possible to set $q = \min(k, l)$ in (2.14). Let us denote by P_s the projection operators onto the subspace $Z^s\mathfrak{H}^{n(k-s,l-s)}$. It is obvious that $P_s = Z^s H_{q,s}(\Delta, Z)$. Utilizing commutation relations and the condition $\Delta^{s+1}P_s = 0$, we can write the operators P_s in the form of polynomials with respect to the degrees $Z^r\Delta^r$. We omit the tedious proof of the following fact:

$$\prod_{j=s+1}^{q}\left[1-\frac{(n+k+l-2j-1)!}{4^j(n+k+l-j-1)!\,j!}Z^j\Delta^j\right]=1+\sum_{r=s+1}^{q}\alpha_r^s Z^r\Delta^r,$$

where the factors in the left side are arranged from left to right in order of increasing j. Since

$$P_s=\prod_{j=s+1}^{q}\left[1-\frac{(n+k+l-2j-1)!}{4^j(n+k+l-j-1)!\,j!}Z^j\Delta^j\right]$$
$$-\prod_{j=s}^{q}\left[1-\frac{(n+k+l-2j-1)!}{4^j(n+k+l-j-1)!\,j!}Z^j\Delta^j\right],$$

it follows that

$$P_s=\sum_{r=s}^{q}\beta_r^s Z^r\Delta^r,\quad \text{where } \beta_s^s=\frac{(n+k+l-2s-1)}{4^s(n+k+l-s-1)!\,s!},$$

and, in particular, for P_0 we have $\beta_0^0=1$. Now we utilize the fact that

$$\Delta^{s+1}P_s f_{n(k,l)}(z,\bar z)=0,$$

for any $f_{n(k,l)}(z,\bar z)$, so that

$$0=\Delta^{s+1}P_s f_{n(k,l)}(z,\bar z)=4^s\frac{(n+k+l-s-1)!\,s!}{(n+k+l-2s-1)!}\{\beta_s^s\Delta^{s+1}f_{n(k,l)}(z,\bar z)$$
$$+\sum_{r=s+1}^{q-1}\beta_r^s[4(n+k+l)-r-s-1)(r-s)Z^{r-s-1}$$
$$+Z^{r-s}\Delta]\Delta^r f_{n(k,l)}(z,\bar z)+$$
$$+\beta_q^s 4(n+k+l-q-s-1)(q-s)Z^{q-s-1}\Delta^q f_{n(k,l)}(z,\bar z)\}$$
$$=4^s\frac{(n+k+l-s-1)!\,s!}{(n+k+l-2s-1)!}\sum_{r=s+1}^{q}[\beta_{r-1}^s+\beta_r^s 4(n+k+l-r$$
$$-s-1)(r-s)]Z^{r-s-1}\Delta^r f_{n(k,l)}(z,\bar z).$$

For this equality to hold for any $f(z,\bar z)\in\mathfrak{B}^{n(k,l)}$ it is necessary and sufficient that β_r^s satisfy the following recursion relation:

$$\beta_{r-1}^s+\beta_r^s 4(n+k+l-r-s-1)(r-s)=0,\ r=s+1,\dots,q.$$

Taking into account that

$$\beta_s^s=\frac{(n+k+l-2s-1)!}{4^s(n+k+l-s-1)!\,s!},$$

we obtain

$$\beta_r^s=(-1)^{r-s}\frac{(n+k+l-2-2s-r+s)!}{4^{r-s}(n+k+l-2-2s)!\,(r-s)!}\beta_s^s. \tag{2.14}$$

In particular,

$$P_0=\sum_{r=0}^{q}(-1)^r\frac{(n+k+l-2-r)!}{4^r(n+k+l-2)!\,r!}Z^r\Delta^r. \tag{2.15}$$

We shall call the polynomial $\hat{f}_0(z,\bar{z})=P_0 f_{n(k,l)}(z,\bar{z})$ the *harmonic projection* of the polynomial $f_n(z,\bar{z})$ onto the subspace $\mathfrak{H}^{n(k,l)}$.

4. *Representations of SU(n) in the spaces of bihomogeneous harmonic polynomials.* In §2.2 the representations $L^{n(k,l)}(u)$ in $\mathfrak{B}^{n(k,l)}$ were defined (see (2.4)). These reprsentations are reducible: indeed, the subspace $Z\mathfrak{B}^{n(k-1,l-1)}$ is invariant relative to $L^{n(k,l)}(u)$. Consequently, $L^{n(k,l)}(u)$ gives a representation in the quotient space $\mathfrak{A}^{n(k,l)}=\mathfrak{B}^{n(k,l)}/Z\mathfrak{B}^{n(k-1,l-1)}$, which we shall denote by $T^{n(k,l)}(u)$. The decomposition (2.13) implies

$$\mathfrak{B}^{n(k,l)}=\mathfrak{H}^{n(k,l)}+Z\mathfrak{B}^{n(k-1,l-1)}, \tag{2.13'}$$

i.e., each bihomogeneous harmonic polynomial $h(z,\bar{z})\in\mathfrak{H}^{n(k,l)}$ of degree (k,l) uniquely defines a coset in $\mathfrak{A}^{n(k,l)}$. It is easy to see that the decompositions (2.13) and (2.13′) are invariant relative to $L^{n(k,l)}(u)$. Let us denote the restriction of $L^{n(k,l)}(u)$ to $\mathfrak{H}^{n(k,l)}$ by $H^{n(k,l)}(u)$:

$$[H^{1(k,l)}(u)f](z,\bar{z})=f(u^{-1}z,\overline{u}^{-1}\bar{z}),\quad f(z,\bar{z})\in\mathfrak{H}^{n(k,l)}. \tag{2.16}$$

The isomorphism of $\mathfrak{A}^{n(k,l)}$ and $\mathfrak{H}^{n(k,l)}$ implies the equivalence of $T^{n(k,l)}(u)$ and $H^{n(k,l)}(u)$. (2.13) implies

$$L^{n(k,l)}(u)=\sum_{s=0}^{\min(k,l)}H^{n(k-s,l-s)}(u). \tag{2.17}$$

It is possible to show that the representations $H^{n(k,l)}(u)$ are irreducible, and for $n\geqslant 3$ nonequivalent, so that (2.17) can be considered as a decomposition of $L^{n(k,l)}(u)$ into irreducible representations. In addition, since bihomogeneous harmonic polynomials are completely determined by their values on the sphere S, $\mathfrak{H}^{n(k,l)}$ can be considered a subspace in $\mathfrak{L}^2(S)$. Then

$$\mathfrak{B}^{n(k,l)}=\sum_{s=0}^{\min(k,l)}\mathfrak{H}^{n(k-s,l-s)}.$$

Since any function of $\mathfrak{L}^2(S)$ can be approximated with any degree of accuracy by polynomials, we have

$$\mathfrak{L}^2(S)=\sum_{k=0}^{\infty}\sum_{l=0}^{\infty}\mathfrak{H}^{n(k,l)} \tag{2.18}$$

and

$$L^n(S)=\sum_{k=0}^{\infty}\sum_{l=0}^{\infty} H^{n(k,l)}(u).$$

Thus a quasi-regular representation of $SU(n)$ can be split into irreducible (and for $n \geqslant 3$ mutually nonequivalent) representations. It will be shown that to within equivalence the class I representations of $SU(n)$ relative to $SU(n-1)$ are exhausted by the representations $H^{n(k,l)}(u)$.

5. *Decomposition of the restriction of the representation $H^{n(k,l)}(u)$ to the subgroup $SU(n-1)$.* Let us consider the representation $T^{n(k,l)}(u)$ realized in $\mathfrak{A}^{n(k,l)} \equiv \mathfrak{B}^{n(k,l)}/Z\mathfrak{B}^{n(k-1,l-1)}$. We denote by $\mathfrak{A}^{n(k,l)(k_1,l_1)}$ the subspace in $\mathfrak{A}^{n(k,l)}$ obtained from the subspace $z_n^{(k-k_1)}\bar{z}_n^{(l-l_1)}\mathfrak{H}^{(n-1)(k_1,l_1)} \subset \mathfrak{B}^{n(k,l)}$ on factoring $\mathfrak{B}^{n(k,l)}$ by $Z\mathfrak{B}^{n(k-1,l-1)}$. (Here $\mathfrak{H}^{(n-1)(k_1,l_1)}$ is the space of bihomogeneous harmonic polynomials of degree (k_1, l_1) in z' and $\bar{z}'$ where $z'=(z_1,\ldots,z_{n-1})$.) It is obvious that $\mathfrak{A}^{n(k,l)(k_1,l_1)}$ $(0 \leqslant k_1 \leqslant k; 0 \leqslant l_1 \leqslant l)$ is invariant relative to $T^{n(k,l)}(h)$, $h \in SU(n-1)$, while the representation induced in it by $T^{n(k,l)}(h)$ is equivalent to $H^{(n-1)(k_1,l_1)}(h)$, and consequently also to $T^{(n-1)(k_1,l_1)}(h)$. It is possible to prove that

$$T^{n(k,l)}(h)=\sum_{k_1=0}^{k}\sum_{l_1=0}^{l} T^{(n-1)(k_1,l_1)}(h), \tag{2.19}$$

or that

$$H^{n(k,l)}(h)=\sum_{k_1=0}^{k}\sum_{l_1=0}^{l} H^{(n-1)(k_1,l_1)}(h). \tag{2.20}$$

It is sufficient for this to show that the decomposition into the direct sum

$$\mathfrak{A}^{n(k,l)}=\sum_{k_1=0}^{k}\sum_{l_1=0}^{l} \mathfrak{A}^{n(k,l)(k_1,l_1)} \tag{2.21}$$

or

$$\mathfrak{B}^{n(k,l)}=Z\mathfrak{B}^{n(k-1,l-1)}+\sum_{k_1=0}^{k}\sum_{l_1=0}^{l} z_n^{k-k_1}\bar{z}_n^{l-l_1}\mathfrak{H}^{(n-1)(k_1,l_1)} \tag{2.21$'$}$$

holds. (2.21$'$) follows from the coincidence of the dimensions of the left and right sides and the possibility of representing any polynomial $f(z,\bar{z}) \in \mathfrak{B}^{n(k,l)}$ uniquely in the form

$$f(z,\bar{z})=Zf_1(z,\bar{z})+\sum_{k_1=0}^{k}\sum_{l_1=0}^{l} z_n^{k-k_1}\bar{z}_n^{l-l_1} h_{(n-1)(k_1,l_1)}(z',\bar{z}'). \tag{2.22}$$

Here we take into account the fact [1] that

$$\dim \mathfrak{B}^{n(k,l)}=\dim \mathfrak{B}^{n(k,0)} \dim \mathfrak{B}^{n(0,l)}=\frac{(n+k-1)!\,(n+l-1)!}{(n-1)!\,k!\,(n-1)!\,l!},$$

$$\dim \mathfrak{H}^{n(k,l)}=\dim \mathfrak{B}^{n(k,l)}-\dim \mathfrak{B}^{n(k-1,l-1)}$$

Decomposition (2.22) follows from the fact that any bihomogeneous polynomial of degree (k, l) can be represented uniquely in the form

$$f(z, \bar{z}) = ZF_1(z, \bar{z}) + \sum_{i=0}^{k-1} z_n^{k-i} \varphi_{(n-1)(i,l)}(z', \bar{z}') + \sum_{j=0}^{l-1} \bar{z}_n^{l-j} \varphi_{(n-1)(k,j)}(z', \bar{z}') + \varphi_{(n-1)(k,l)}(z', \bar{z}') \qquad (2.22')$$

by applying Theorem 2 to the polynomials $\varphi_{(n-1)(p,q)}(z', \bar{z}')$ analogously to the way this is done for $SO(n)$ in [1]. We observe further that (2.21) is equivalent to the decomposition

$$\mathfrak{H}^{n(k,l)} = \sum_{k_1=0}^{k} \sum_{l_1=0}^{l} P_0 [z_n^{k-k_1} \bar{z}_n^{l-l_1} \mathfrak{H}^{(n-1)(k_1,l_1)}]. \qquad (2.23)$$

The decompositions constructed here permit us to carry out further constructions by induction on the dimension of the group.

6. *Irreducibility of the representations* $H^{n(k,l)}(u)$. The realization of irreducible representations for $SU(2)$ and $SU(3)$ in $\mathfrak{H}^{n(k,l)}$ is described in [3]. The equivalence of $T^{n(k,l)}(u)$ and $H^{n(k,l)}(u)$ implies that $T^{n(k,l)}(u)$ is irreducible for $n = 2, 3$. Having assumed that $T^{(n-1)(k_1,l_1)}(h)$, $h \in SU(n-1)$, are irreducible and utilizing decomposition (2.21′), we can prove the irreducibility of $T^{n(k,l)}(u)$ (consequently also of $H^{n(k,l)}(u)$). We omit the tedious but simple proof of this fact since it is analogous to the proof of the irreducibility of the representations $H^{n(k,l)}(g)$ for $SO(n)$ [1].

7. *Equivalence relations in the set of representations* $H^{n(k,l)}(u)$. We realize $\mathfrak{H}^{n(k,l)}$ in the space $\mathfrak{L}^2(S)$. Any polynomial $f(\vec{\omega}, \bar{\vec{\omega}}) \in \mathfrak{H}^{n(k,l)}$ satisfies the equation

$$\Delta_0 f(\vec{\omega}, \bar{\vec{\omega}}) = -(k+l)(k+l+2n-2) f(\vec{\omega}, \bar{\vec{\omega}}),$$

where Δ_0 is the spherical part of the Laplace equation [2]. The invariance of the Laplace operator relative to unitary transformations implies [4] that the representations $H^{n(k,l)}(u)$ and $H^{n(p,q)}(u)$ are nonequivalent for $k + l \neq p + q$. Let us now suppose that $k + l = p + q$, and compare the dimensionality of $H^{n(k,l)}(u)$ and $H^{n(p,q)}(u)$. For $n = 2$ we have $\dim \mathfrak{H}^{2(k,l)} = \dim \mathfrak{H}^{2(p,q)}$; as is well known [3], these representations are equivalent. For $n = 3$ we have $\dim \mathfrak{H}^{3(k,l)} = \dim \mathfrak{H}^{3(p,q)}$ if only $kl = pq$. It is possible to prove by induction on n that equality of dimensions hold only when $kl = pq$. The conditions $k + l = p + q$ and $kl = pq$ are fulfilled if either $(p,q) = (k,l)$ (the representations coincide in this case) or $(p,q) = (l,k)$. However, $H^{n(k,l)}(u)$ and $H^{n(l,k)}(u)$ are nonequivalent. This can be proved by utilizing the orthogonality of the zonal spherical functions corresponding to the representations. We omit this proof, since the evaluation of zonal spherical functions [1] is connected with subsection 8, below, and is easily performed, and the invariant measure on $SU(n)$ is defined.

Thus a system of irreducible, and for $n \geqslant 3$ mutually nonequivalent, representations of $SU(n)$ is constructed:

$$[H^{n(k,l)}(u)f](\vec{\omega}, \overline{\vec{\omega}}) = f(u^{-1}\vec{\omega}, \bar{u}^{-1}\overline{\vec{\omega}}), \quad f(\vec{\omega}, \overline{\vec{\omega}}) \in \mathfrak{H}^{n(k,l)}.$$

These representations are unitary relative to the scalar product (2.1)

8. *Completeness of the system of representations $H^{n(k,l)}(u)$ in the set of class I representations of $SU(n)$ relative to $SU(n-1)$.* We shall show that the representations $H^{n(k,l)}(u)$ are class I representations relative to $SU(n-1)$ $(n \geqslant 3)$ (see [1]). Passing to the equivalent realization $T^{n(k,l)}(u)$ in $\mathfrak{A}^{n(k,l)}$, we conclude from (2.21) that $T^{n(k,l)}(h)$, $h \in SU(n-1)$, contains the only one-dimensional representation of $SU(n-1)$ which is realized in the coset mod $Z\mathfrak{B}^{n(k-1,l-1)}$, containing a function of the form $z_n^k \bar{z}_n^l$. In case $k = l$, there is another polynomial $(z_1\bar{z}_1 + \cdots + z_{n-1}\bar{z}_{n-1})^k$ invariant relative to $SU(n-1)$; however, it is easy to show that the coset defined by it coincides with $z_n^k \bar{z}_n^l + Z\mathfrak{B}^{n(k-1,l-1)}$ to within a proportionality coefficient. Thus in $\mathfrak{A}^{n(k,l)}$ there is a unique (to within a proportionality coefficient) coset invariant relative to $SU(n-1)$ and defined by the polynomial $z_n^k \bar{z}_n^l$. We conclude from the isomorphism between $\mathfrak{A}^{n(k,l)}$ and $\mathfrak{H}^{n(k,l)}$ that the unique (to within constant) polynomial in $\mathfrak{H}^{n(k,l)}$ invariant relative to $SU(n-1)$ has the form

$$\hat{f}_0(z, \bar{z}) = P_0[z_n^k \bar{z}_n^l]$$

$$= r_n^{k+l}\left(\frac{z_n}{r_n}\right)^k \left(\frac{\bar{z}_n}{r}\right)^l F\left(-k, -l; -n-k-l+2; \frac{r_n^2}{z_n\bar{z}_n}\right), \quad r_n = \sqrt{Z}.$$

In addition, the $H^{n(k,l)}(u)$ (consequently also the $H^{n(k,l)}(h)$) are unitary relative to the scalar product (2.1′), which signifies their membership in the set of class I representations of $SU(n)$. We conclude from (2.18) and (2.19) that the quasi-regular representation of $SU(n)$ which is realized in $\mathfrak{L}^2(S)$ decomposes into irreducible class I representations which are mutually nonequivalent for $n \geqslant 3$. (For $n = 2$ the representations $H^{n(k,l)}(u)$ and $H^{n(p,q)}(u)$ are equivalent if and only if $k + l = p + q$.) The completeness of the system of representations $H^{n(k,l)}(u)$ in the set of class I representations relative to $SU(n-1)$ is proved if we take into account that the manifolds $SU(n)/SU(n-1)$ and S coincide, exactly as was done for $SO(n)$ in [1].

9. *Orthonormal basis in the spaces $\mathfrak{H}^{n(k,l)}$.* In conclusion we describe an orthonormal basis in $\mathfrak{H}^{n(k,l)}$. Decomposition (2.23) implies that any bihomogeneous harmonic polynomial $h_{(k,l)}(z, \bar{z})$ can be represented uniquely as a linear combination of polynomials

$$P_0[z_n^{k-k_1} \bar{z}_n^{l-l_1} h_{(k_1,l_1)}(\mathbf{z}', \bar{\mathbf{z}}')], \quad 0 \leqslant k_1 \leqslant k, \ 0 \leqslant l_1 \leqslant l;$$

P_0 is an operator projecting $\mathfrak{B}^{n(k,l)}$ onto the harmonic subspace $\mathfrak{H}^{n(k,l)}$. Hence, applying (2.23) to $\mathfrak{H}^{(n-1)(k_1,l_1)}$, etc., we obtain a linearly independent system of polynomials in $\mathfrak{H}^{n(k,l)}$ such that any polynomial of $\mathfrak{H}^{n(k,l)}$ can be obtained as a linear combination of polynomials of that system.

Taking into account that the basis in $H^{2(k_{n-2},l_{n-2})}$ is described in [3], and denoting it by $h^m_{(k_{n-2},l_{n-2})}(z_1, z_2, \bar{z}_1, \bar{z}_2)$, we obtain the following result: each bihomogeneous harmonic polynomial in z and $\bar{z}$ of degree (k, l) can be uniquely represented as a linear combination of polynomials

$$\Xi_M(z,\bar{z}) = C_M^{n(k,l)} \prod_{j=0}^{n-3} r_{n-j}^{(k_j+l_j)-(k_{j+1}+l_{j+1})} \left(\frac{z_{n-j}}{r_{n-j}}\right)^{k_j-k_{j+1}} \left(\frac{\bar{z}_{n-j}}{r_{n-j}}\right)^{l_j-l_{j+1}}$$

$$F\left(-k_j+k_{j+1},\ -l_j+l_{j+1};\ -(n-j)-k_j-l_j+2;\ \frac{r_{n-j}^2}{z_{n-j}\bar{z}_{n-j}}\right) \tag{2.24}$$

$$\times\, r_2^{k_{n-2}+l_{n-2}} h^m_{(k_{n-2},\,l_{n-2})}\left(\frac{z_1}{r_2},\ \frac{z_2}{r_2};\ \frac{\bar{z}_1}{r_2},\ \frac{\bar{z}_2}{r_2}\right),$$

where M is a scheme of the form

$$M = \begin{pmatrix} k=k_0 & k_1 & \dots & k_{n-2} \\ & & & m \\ l=l_0 & l_1 & \dots & l_{n-2} \end{pmatrix},$$

the numbers k_j and l_j satisfy the conditions $0 \leqslant k_j \leqslant k_{j-1}$ and $0 \leqslant l_j \leqslant l_{j-1}$, and m is an integer such that the number

$$(k_{n-2}+l_{n-2}) - |m| - |k_{n-2}-l_{n-2}-m| = 2q_{n-2}$$

is even and nonnegative [3]; $C_M^{n(k,l)}$ is the normalizing constant.

When realizing $\mathfrak{H}^{n(k,l)}$ in $\mathfrak{L}^2(S)$ in the coordinate system (1.3), the functions Ξ_M take the following form (if account is taken of the relation between the hypergeometric polynomials $F(a, b; c; x)$ and the Jacobi polynomials $P_m^{(\alpha,\beta)}(y)$; see [5]):

$$\Xi_M = C_M^{n(k,l)} \prod_{j=0}^{n-3} \frac{\Gamma(k_j+l_j-q_j+n-j-1)\,\Gamma(q_j+1)}{\Gamma(k_j+l_j+n-j-1)}$$

$$\times \frac{\Gamma(q_{n-2}+1)\,\Gamma(|m|+1)}{\Gamma(q_{n-2}+|m|+1)} \prod_{j=0}^{n-3} \exp\left(i\left[(k_j-l_j)-(k_{j+1}-l_{j+1})\right]\varphi_{1_{j+1}}\right)$$

$$\times \exp\left(i(k_{n-2}-l_{n-2}-m)\varphi_{1_{n-1}}\right) \exp im\varphi_{1_n}$$

$$\times \prod_{j=0}^{n-3} (\sin\varphi_{2_{j+1}})^{k_{j+1}+l_{j+1}} (\cos\varphi_{2_{j+1}})^{|(k_j+k_{j+1})-(l_j-l_{j+1})|} (\sin\varphi_{2_{n-1}})^{|m|} \tag{2.25}$$

$$\times (\cos\varphi_{2_{n-1}})^{|k_{n-2}-l_{n-2}-m|}$$

$$\times \prod_{j=0}^{n-3} P_{q_j}^{(k_{j+1}+l_{j+1}+n-j-2),\,|(k_j-k_{j+1})-(l_j-l_{j+1})|)} (\cos 2\varphi_{2_{j+1}})$$

$$\times P_{q_{n-2}}^{(|m|,\,|k_{n-2}-l_{n-2}-m|)} (\cos 2\varphi_{2_{n-1}}),$$

where

$$q_j = \min(k_j-k_{j+1},\ l_j-l_{j+1}).$$

Using the orthogonality relations for Jacobi polynomials [4] and the condition that $\int_0^{2\pi} e^{ik\varphi}\,d\varphi = 0$ if and only if $k \neq 0$, it is not difficult to prove that for $M_1 \neq M_2$ the functions Ξ_{M_1} and Ξ_{M_2} are orthogonal relative to the scalar product (2.1′), and to evaluate

$$C_M^{n(k,l)} \ (p_j = \max(k_j - k_{j+1}, l_j - l_{j+1})):$$

$$C_M^{n(k,l)} = \prod_{j=0}^{n-3} \frac{\Gamma(k_j+l_j+n-j-1)}{\Gamma(k_j+l_j-q_j+n-j-1)\,\Gamma(q_j+1)} \, \frac{\Gamma(q_{n-2}+|m|+1)}{\Gamma(q_{n-2}+1)\,\Gamma(|m|+1)}$$

$$\times \left[\prod_{j=0}^{n-3} \frac{(k_j+l_j+n-j-1)\,\Gamma(q_j+1)\,\Gamma(p_j+k_{j+1}+l_{j+1}+n-j-1)}{\Gamma(k_j+l_j-p_j+n-j-1)\,\Gamma(p_j+1)}\right.$$

$$\left.\times \frac{(k_{n-2}+l_{n-2}+1)\,\Gamma(q_{n-2}+1)\,\Gamma(q_{n-2}+|m|+|k_{n-2}-l_{n-2}-m|+1)}{\Gamma(n)\,\Gamma(q_{n-2}+|m|+1)\,\Gamma(q_{n-2}+|k_{n-2}-l_{n-2}-m|+1)}\right]^{1/2}.$$

Thus for a fixed pair (k, l) the functions Ξ_M of (2.25) form an orthogonal basis in $\mathfrak{H}^{n(k,l)}$ considered as a subspace in $\mathfrak{L}^2(S)$, while for $k, l = 0, 1, \ldots,$ as follows from (2.18), they form an orthonormal basis in $\mathfrak{L}^2(S)$ relative to the scalar product (2.1′) in the coordinate system (1.3).

BIBLIOGRAPHY

1. N. Ja. Vilenkin, *Special functions and the theory of group representations*, "Nauka," Moscow, 1965; English transl., Amer. Math. Soc., Providence, R. I., 1968.

2. ———, *Polyspherical and orispherical functions*, Mat. Sb. 68 (110)(1965), 432–443. (Russian)

3. N. Ja. Vilenkin, G. I. Kuznecov and Ja. A. Smorodinskiĭ, *Eigenfunctions of the Laplace operator providing representations of the groups* U(2), SU(2), SO(3), U(3) *and* SU(3), *and the symbolic method*, Jadernaja Fiz. 2 (1965), 906–917; English transl. in Soviet J. Nuclear Phys. 2 (1965/66).

4. Gabor Szegö, *Orthogonal polynomials*, rev. ed., Amer. Math. Soc. Providence, R. I., 1959.

5. Francis D. Murnaghan, *The unitary and rotation groups*, Spartan Books, Washington, D. C., 1962.

Translated by F. M. GOLDWARE

Amer. Math. Soc. Transl.
(2) Vol. 113, 1979

SPECIAL FUNCTIONS CONNECTED WITH REPRESENTATIONS OF THE GROUP $SU(n)$ OF CLASS I RELATIVE TO $SU(n-1)$ $(n \geqslant 3)$*

R. L. ŠAPIRO

One of the important problems of representation theory is the study of special functions related to the representations of a *concrete* group (see the Preface to [1]).

Although this problem has already been solved for many concrete groups, results for new groups are important both from the point of view of representation theory and from the point of view of the theory of special functions.

Some new relations for the Jacobi polynomials related to the zonal and associated spherical functions in the aforementioned representations(1) are established in the present paper by means of group-theoretic methods (§1). The results are later utilized to study orthogonal polynomials on the circle (§2).

We manage to obtain an orthogonal polynomial system with weight $(1-x^2-y^2)^\beta$, $\beta > -1$, and not described in the literature, which, however, is easily explained by the fact that the weight function for the case of two or more arguments does not determine an orthogonal system uniquely ([4], §12.1).

§1. Zonal and associated spherical functions. Jacobi polynomials and some of their properties

1. *Matrix elements of the null column.* We realize the irreducible unitary class I representation $T^{n(k,l)}(u)$ of the group $SU(n)$ relative to $SU(n-1)$ in the space $\mathfrak{H}^{n(k,l)}$ of bihomogeneous harmonic polynomials of degree $(k-l)$ and in the basis $\Xi^{n(k,l)}_{M_1}(\vec{\omega}, \bar{\vec{\omega}})$ (see [3]), and indexing the matrix rows and columns by the schemes K_1M_1(2), we compute the matrix elements $t^{n(k,l)}_{K_1M_1}(u)$ of these representations. By definition ([1], Chapter I, §1.2)

$$T^{n(k,l)}(u)\,\Xi^{n(k,l)}_{M_1}(\vec{\omega}, \bar{\vec{\omega}}) = \sum_{K_1} t^{n(k,l)}_{K_1M_1}(u)\,\Xi^{n(k,l)}_{K_1}(\vec{\omega}, \bar{\vec{\omega}}). \tag{1.1}$$

AMS (MOS) subject classifications (1970). Primary 22E45, 33A75.

*Translation of Izv. Vysš. Učebn. Zaved. Matematika **1968**, no. 4 (71), 97–107. MR **37** #6512.

(1) The technique employed is a generalization to the case of n-dimensional complex space of the corresponding technique for the group $SO(n)$ ([1], Chapter IX, §§3–5). All terms are understood in the sense of [1] (specific references are indicated in the text of the article).

(2) Here

$$M_1 = (k_j,\ l_j,\ m) = \begin{pmatrix} k_1 \dots k_{n-2} & m \\ l_1 \dots l_{n-2} & \end{pmatrix}$$

(see [3]). K_1 is defined analogously.

In (1.1) we set $\vec{\omega} = \vec{\omega}_n = (0, \ldots, 0, 1)$; that is, on the sphere Ω_n we choose a point with hemispherical coordinates $\varphi_{j_p} = 0$ (see [3]). It is obvious that $\Xi^{n(k,l)}_{K_1}(\vec{\omega}_n, \bar{\vec{\omega}}_n) \neq 0$ only for $K_1 = O(\begin{smallmatrix}0 \cdots 0\\ 0 \cdots 0\end{smallmatrix}0)$, where

$$\Xi^{n\,(k,\,l)}_O(\vec{\omega}_n, \bar{\vec{\omega}}_n) = \left[\frac{(k+l+n-1)\,(n-1)\,\Gamma\,(k+n-1)\,\Gamma\,(l+n-1)}{\Gamma\,(n)\,\Gamma\,(k+1)\,\Gamma\,(n)\,\Gamma\,(l+1)}\right]^{\frac{1}{2}} = [\dim \mathfrak{H}^{n\,(k,\,l)}]^{\frac{1}{2}},$$

so that, utilizing the unitary property of the operators $T^{n(k,l)}(u)$, the geometric meaning of the Euler parameters on $SU(n)$ [3], and the formula defining the representation [3]

$$T^{n\,(k,\,l)}(u)\,f(\vec{\omega}, \bar{\vec{\omega}}) = f(u^{-1}\vec{\omega}, \bar{u}^{-1}\bar{\vec{\omega}}), \tag{1.2}$$

we obtain the formula

$$t^{n\,(k,\,l)}_{M,\,O}(u) = \frac{1}{[\dim \mathfrak{H}^{n\,(k,\,l)}]^{1/2}}\,\overline{\Xi^{n\,(k,\,l)}_{M_1}(u\vec{\omega}_n, \overline{u\vec{\omega}_n})} \tag{1.3}$$

for the matrix elements of the null column, the zonal and associated spherical functions ([1], Chapter I, §2.5, formula (6)).

We observe that the vector $u\vec{\omega}_n$ has spherical coordinates $\varphi^{(n)}_{j_p}$ if $\varphi^{(k)}_{j_p}$ are the Euler angles of the matrix u (see [3]). Consequently, (1.3) permits us to express the null column matrix elements by means of the Euler angles of the matrix u (from this obviously follows the group-theoretic meaning of the basis $\Xi^{n(k,l)}_{M_1}(\vec{\omega}, \bar{\vec{\omega}})$). In particular, if $M_1 = O$, the formula

$$\begin{aligned} t^{n\,(k,\,l)}_{OO}(u) = \mathfrak{P}^n_{k,\,l}(u) = \frac{\Gamma\,(n-1)\,\Gamma\,(q+1)}{\Gamma\,(q+n-1)}\,e^{-i\,(k-l)\,\varphi^{(n)}_{1_1}} \\ \times (\cos\varphi^{(n)}_{2_1})^{|k-l|}\,P^{(n-2,\,|k-l|)}_q(\cos 2\varphi^{(n)}_{2_1}) \end{aligned} \tag{1.4}$$

defines the zonal spherical function of the representation $T^{n(k,l)}(u)$.

REMARK 1. (1.3) and the results of [3] imply a relation between Jacobi polynomials and irreducible unitary class I representations of $SU(n)$, which permits one to obtain a number of properties for the former by group-theoretic methods.

REMARK 2. The collection of zonal spherical functions of the representations $T^{n(k,l)}(u)$ which is described by (1.4) will be used in §2 to construct an orthogonal polynomial basis in the disc $x^2 + y^2 \leqslant 1$, which makes it possible to establish from group-theoretic considerations a number of properties for the collection.

2. *Various expressions for the zonal spherical functions of the representations* $T^{n(k,l)}(u)$. Having introduced the notation

$$\varphi^{(n)}_{1_1} = \varphi, \quad \varphi^{(n)}_{2_1} = \theta, \quad 0 \leqslant \varphi < 2\pi\,(\text{mod}\,2\pi), \quad 0 \leqslant \theta \leqslant \frac{\pi}{2},$$

$$\vec{\omega} = u_{n-1\,n}(\varphi, \theta, 0)\,\vec{\omega}_n, \qquad \vec{z} = r\,\vec{\omega}, \text{ where } \vec{\omega}_n = (0, \ldots, 0, 1),$$

$$\omega = \frac{z_n}{r} = \cos\theta \cdot e^{i\varphi}, \quad r = |\boldsymbol{z}|, \ \boldsymbol{z} = (z_1, \ldots, z_n),$$

we indicate various expressions for the zonal spherical functions. Utilizing the relations between the hypergeometric polynomials

$$F(a, b; c; 1-z) = \frac{\Gamma(c)\Gamma(c-a-b)}{\Gamma(c-a)\Gamma(c-b)} F(a, b; -c+a+b+1; z); \text{(}^{3}\text{)}$$

$$F(a, b; c; z) = (1-z)^{-a} F\left(a, c-b; c; \frac{z}{z-1}\right) \quad \text{([6], 9.131)},$$

the relation between hypergeometric polynomials and Jacobi polynomials ([6], 8.962, formula 1)

$$P_m^{(\alpha, \beta)}(x) = \frac{\Gamma(m+1+\alpha)}{\Gamma(1+\alpha)\cdot m!}\left(\frac{x+1}{2}\right)^m F\left(-m, -m-\beta; \alpha+1; \frac{x-1}{x+1}\right)$$

and the formula [3]

$$P_0[z_n^k \bar{z}_n^l] = r^{k+l}\left(\frac{z_n}{r}\right)^k \left(\frac{\bar{z}_n}{r}\right)^l F\left(-k, -l; -n-k-l+2; \frac{r^2}{z_n\bar{z}_n}\right),$$

it is easy to establish the equivalence of the following formulas:

$$\mathfrak{P}_{k,l}^n(u) = \frac{\Gamma(q+1)\Gamma(n-1)}{\Gamma(q+n-1)} e^{-i(k-l)\varphi}(\cos\theta)^{|k-l|} P_q^{(n-2, |k-l|)}(\cos 2\theta), \tag{1.5}$$

$$\mathfrak{P}_{k,l}^n(u) = e^{-i(k-l)\varphi}(\cos\theta)^{k+l} F(-k, -l; n-1; -\operatorname{tg}^2\theta), \tag{1.6}$$

$$\mathfrak{P}_{k,l}^n(u) = e^{-i(k-l)\varphi}(\cos\theta)^{|k-l|} F(-q, p+n-1; n-1; \sin^2\theta), \tag{1.7}$$

$$r^{k+l}\mathfrak{P}_{k,l}^n(u) = \frac{\Gamma(n-1)\Gamma(k+l+n-1)}{\Gamma(k+n-1)\Gamma(l+n-1)} \overline{P_0[z_n^k \bar{z}_n^l]}. \tag{1.8}$$

We agree to designate $\mathfrak{P}_{k,l}^n(u) = \mathfrak{P}_{k,l}^n(\varphi, \theta)$ if $u = h_1 u_{n-1n}(\varphi, \theta, 0) h_2$, $h_1, h_2 \in SU(n-1)$, while $\mathfrak{P}_{k,l}^n(u)$ is defined by (1.5)–(1.7), and $\mathfrak{P}_{k,l}^n(u) = \mathfrak{P}_{k,l}^n(\omega, \bar{\omega})$ if $\mathfrak{P}_{k,l}^n(u)$ is defined by (1.8).

3. *Orthogonality relations and expansions in terms of zonal spherical functions.* Orthogonality relations for the functions $\Xi_{M_1}^{n(k,l)}(\bar{\omega}, \bar{\bar{\omega}})$, formula (1.3) and the expression for the measure $du(n)$ on $SU(n)$ (see [3]) imply that

$$\frac{n-1}{\pi}\int_0^{\frac{\pi}{2}}\int_0^{2\pi} \mathfrak{P}_{k,l}^n(\varphi, \theta)\overline{\mathfrak{P}_{r,s}^n(\varphi, \theta)}(\sin\theta)^{2(n-2)+1}\cos\theta\, d\varphi\, d\theta \tag{1.9}$$

$$= \begin{cases} 0, & \text{if } (k, l) \neq (r, s), \\ \dfrac{1}{\dim \mathfrak{H}^{n(k,l)}}, & \text{if } (k, l) = (r, s). \end{cases}$$

We observe that the zonal spherical functions are fixed on the double cosets of $SU(n)$ relative to $SU(n-1)$, the completeness of the system of representations $T^{n(k,l)}(u)$ in

(3) This relation for $c + \xi \neq 0, -1, \ldots$ and $c - a - b - \xi > 0$, where ξ is nonnegative and not greater than the degree of the hypergeometric polynomial, is easily proved with the help of formula (14) of [5] (§2.1.3) by comparing coefficients for identical degrees z.

the set of class I representations of $SU(n)$ relative to $SU(n-1)$ ([1], Chapter I, §2.5) implying that for any function $f(u)$ square integrable with respect to $du(n)$ and satisfying the conditions

$$f(h_1 uh_2) = f(u), \quad h_1, h_2 \in SU(n-1), \tag{1.10}$$

the expansion

$$f(u) = \sum_{k, l=0}^{\infty} a_{k, l} \mathfrak{P}^n_{k, l}(u) \tag{1.11}$$

holds, where

$$a_{k, l} = \dim \mathfrak{H}^{n(k, l)} \int_{SU(n)} f(u) \overline{\mathfrak{P}^n_{k, l}(u)}\, du(n). \tag{1.12}$$

Condition (1.10), the parametrization of $SU(n)$ and the formula for the invariant measure on $SU(n)$ (see [3]) imply that the class of functions under consideration is defined on the northern hemisphere S of the unit sphere S^2 ($0 \leqslant \varphi < 2\pi$ (mod 2π), $0 \leqslant \theta \leqslant \pi/2$) and satisfies the condition

$$\frac{n-1}{\pi} \int_0^{\frac{\pi}{2}} \int_0^{2\pi} |f(\varphi, \theta)|^2 (\sin\theta)^{2(n-2)+1} \cos\theta \, d\varphi\, d\theta < \infty, \tag{1.13}$$

so that (1.12) takes the form

$$a_{k, l} = \dim \mathfrak{H}^{n(k, l)} \frac{n-1}{\pi} \int_0^{\frac{\pi}{2}} \int_0^{2\pi} f(\varphi, \theta) \mathfrak{P}^n_{k, l}(\varphi, \theta) (\sin\theta)^{2(n-2)+1} \cos\theta \, d\varphi\, d\theta. \tag{1.14}$$

We conclude from this and the orthogonality relations (1.9) that the functions $\mathfrak{P}^n_{k,l}(\varphi, \theta)$ form an orthogonal basis in the space of functions square integrable on S with respect to the measure

$$ds = \frac{n-1}{\pi} (\sin\theta)^{2(n-2)+1} \cos\theta \, d\varphi\, d\theta. \tag{1.15}$$

4. *Differential equations for zonal spherical functions.* It follows from (1.8) that $r^{k+l} \mathfrak{P}^n_{k+l}(\varphi, \theta) \in \mathfrak{H}^{n(l,k)}$; i.e., it is a bihomogeneous harmonic polynomial of degree (l, k), so that

$$\Delta [r^{k+i} \mathfrak{P}^n_{k, l}(\varphi, \theta)] = 0 \quad (\Delta \text{ is the Laplacian operator}). \tag{1.16}$$

In addition, we conclude from (1.5)–(1.7) that

$$\frac{1}{i} \frac{\partial}{\partial\varphi} \mathfrak{P}^n_{k, l}(\varphi, \theta) = -(k-l) \mathfrak{P}^n_{k, l}(\varphi, \theta). \tag{1.17}$$

The system of equations (1.16), (1.17) defines $\mathfrak{P}^n_{k,l}(\varphi, \theta)$ uniquely to within a proportionality coefficient. In hemispherical coordinates (see [3]), equation (1.16) reduces to the form ([2], p. 435)

$$\left[\frac{1}{\cos^2\theta}\frac{\partial^2}{\partial\varphi^2}+\frac{1}{(\sin\theta)^{2(n-2)+1}\cos\theta}\frac{\partial}{\partial\theta}(\sin\theta)^{2(n-2)+1}\cos\theta\frac{\partial}{\partial\theta}\right]\mathfrak{P}^n_{k,l}(\varphi,\theta)$$
$$=-(k+l)(k+l+2n-2)\mathfrak{P}^n_{k,l}(\varphi,\theta). \tag{1.18}$$

5. *Addition theorem for zonal spherical functions.* The definition of the matrix elements implies that

$$\Xi_O^{n(k,l)}(u^{-1}\vec{\omega},\ \overline{u}^{-1}\vec{\overline{\omega}})=\sum_{M_1} t^{n(k,l)}_{M_1O}(u)\,\Xi^{n(k,l)}_{M_1}(\vec{\omega},\vec{\overline{\omega}}) \tag{1.19}$$

for any $u\in SU(n)$. Let $u=u_{n-1,n}(\varphi,\theta,0)$, let the vector $\vec{\omega}$ have hemispherical coordinates $\varphi_{j_p}=0$ for $p\neq 1,2$, i.e. $\omega_1=\cdots=\omega_{n-3}=0$, and let

$$\omega_{n-2}=\sin\varphi_{2_1}\sin\varphi_{2_2},\quad \omega_{n-1}=\sin\varphi_{2_1}\cos\varphi_{2_2}\cdot e^{i\varphi_{1_2}},\quad \omega_n=\cos\varphi_{2_1}\cdot e^{i\varphi_{1_1}}.$$

Then the nth coordinate of the vector $\hat{\vec{\omega}}=u^{-1}\vec{\omega}$ is computed from the formula

$$\begin{aligned}\hat{\omega}_n&=\omega_{n-1}\sin\theta+\omega_n\cos\theta\cdot e^{-i\varphi}\\&=\sin\varphi_{2_1}\cos\varphi_{2_2}\cdot e^{i\varphi_{1_2}}\sin\theta+\cos\varphi_{2_1}\cdot e^{i\varphi_{1_1}}\cos\theta\cdot e^{-i\varphi}.\end{aligned}$$

Let us set $\hat{\omega}_n=\cos\psi\cdot e^{i\alpha}$. Formula (1.3) implies that the matrix elements $t^{n(k,l)}_{M_1O}[u_{n-1,n}(\varphi,\theta,0)]$ differ from zero only for $M_1=M_1^0=\binom{k_1}{l_1}0$. As a result, we obtain from (1.19) the addition theorem for Jacobi polynomials ($n\geqslant 3$):[4]

$$e^{i(k-l)\alpha}(\cos\psi)^{|k-l|}P_q^{(n-2,\,|k-l|)}(\cos 2\psi)$$
$$=\sum_{k_1=0}^{k}\sum_{l_1=0}^{l}\frac{\Gamma(p+1)\Gamma(q_0+1)\Gamma(k+l-q_0+n-1)\Gamma(p_1+n-2)(k_1+l_1+n-2)}{\Gamma(p+n-1)\Gamma(p_0+1)\Gamma(k+l-p_0+n-1)\Gamma(p_1+1)}$$
$$\times e^{-i[(k-l)-(k_1-l_1)]\varphi}(\sin\theta)^{k_1+l_1}(\cos\theta)^{|(k-k_1)-(l-l_1)|}$$
$$\times P_{q_0}^{(k_1+l_1+n-2,\,|(k-k_1)-(l-l_1)|)}(\cos 2\theta)\cdot e^{i[(k-l)-(k_1-l_1)]\varphi_{1_1}}(\sin\varphi_{2_1})^{k_1+l_1}$$
$$\times(\cos\varphi_{2_1})^{|(k-k_1)-(l-l_1)|}P_{q_0}^{(k_1+l_1+n-2,\,|(k-k_1)-(l-l_1)|)}(\cos 2\varphi_{2_1})$$
$$\times e^{i(k_1-l_1)\varphi_{1_2}}(\cos\varphi_{2_2})^{|k_1-l_1|}P_{q_1}^{(n-3,\,|k_1-l_1|)}(\cos 2\varphi_{2_2}). \tag{1.20}$$

In terms of zonal spherical functions this takes the form

$$\mathfrak{P}^n_{k,l}(\alpha,\psi)=\sum_{k_1=0}^{k}\sum_{l_1=0}^{l}\frac{\Gamma(q+1)\Gamma(p+1)\Gamma(n-1)\Gamma(p_0+k_1+l_1+n-1)}{\Gamma(p+n-1)\Gamma(q+n-1)\Gamma(p_0+1)\Gamma(q_0+1)}$$
$$\times\frac{\Gamma(q_0+k_1+l_1+n-1)\Gamma(p_1+n-2)\Gamma(q_1+n-2)}{\Gamma(k_1+l_1+n-2)\Gamma(k_1+l_1+n-1)\Gamma(p_1+1)\Gamma(q_1+1)\Gamma(n-2)} \tag{1.21}$$
$$\times\mathfrak{P}^{k_1+l_1+n}_{k-k_1,\,l-l_1}(\varphi,\theta)(\sin\theta)^{k_1+l_1}\mathfrak{P}^{k_1+l_1+n}_{k-k_1,\,l-l_1}(\varphi_{1_1},\varphi_{2_1})(\sin\varphi_{2_1})^{k_1+l_1}\mathfrak{P}^{n-1}_{k_1,\,l_1}(\varphi_{1_2},\varphi_{2_2}),$$

which can be considered as the addition theorem for zonal spherical functions.

---

(4) For $n=3$ it must be taken into account that for $m=0$ we have $q_1+|k_1-l_1|=k_1+l_1-q_1=p_1=\max(k_1,l_1)$; i.e., (1.20) holds in this case.

Special cases of (1.20) yield a number of relations for Jacobi polynomials. For $\varphi = 0$, $\varphi_{1_1} = \varphi_{1_2} = \beta$ and $\varphi_{2_2} = 0$, it takes the form

$$[\cos(\varphi_{2_1} - \theta)]^{|k-l|} P_q^{(n-2,\,|k-l|)}[\cos 2(\varphi_{2_1} - \theta)]$$
$$= \sum_{k_1=0}^{k}\sum_{l_1=0}^{l} \frac{\Gamma(p+1)}{\Gamma(p+n-1)\Gamma(n-2)}$$
$$\times \frac{\Gamma(q_0+1)\Gamma(k+l-q_0+n-1)\Gamma(p_1+n-2)\Gamma(q_1+n-2)(k_1+l_1+n-2)}{\Gamma(p_0+1)\Gamma(k+l-p_0+n-1)\Gamma(p_1+1)\Gamma(q_1+1)}$$
$$\times P_{q_0}^{(k_1+l_1+n-2,\,|(k-k_1)-(l-l_1)|)}(\cos 2\theta)(\sin\theta)^{k_1+l_1}(\cos\theta)^{|(k-k_1)-(l-l_1)|} \tag{1.22}$$
$$\times P_{q_0}^{(k_1+l_1+n-2,\,|(k-k_1)-(l-l_1)|)}(\cos 2\varphi_{2_1})(\sin\varphi_{2_1})^{k_1+l_1}(\cos\varphi_{2_1})^{|(k-k_1)-(l-l_1)|}.$$

Setting $\varphi_{2_2} = \pi/2$, we obtain

$$\hat{\omega}_n = \cos\varphi_{2_1}\cos\theta\cdot e^{i(\varphi_{1_1}-\varphi)},$$

i.e.

$$e^{i\hat{\alpha}} = e^{i(\varphi_{1_1}-\varphi)}, \quad \cos\psi = \cos\varphi_{2_1}\cos\theta \text{ and } \cos 2\psi = 2\cos^2\varphi_{2_1}\cos^2\theta - 1.$$

Taking into account that in the right side of (1.20) the only nonzero addends are those in which $|k_1 - l_1| = 0$, we easily obtain

$$P_q^{(n-2,\,|k-l|)}(2\cos^2\varphi_{2_1}\cos^2\theta - 1) = \sum_{r=0}^{q}(-1)^r\frac{\Gamma(p+1)\Gamma(q_0+1)}{\Gamma(p+n-1)\Gamma(p_0+1)}$$
$$\times \frac{\Gamma(p_0+2r+n-1)\Gamma(r+n-2)(2r+n-2)}{\Gamma(q_0+2r+n-1)\Gamma(r+1)} P_{q_0}^{(2r+n-2,\,|k-l|)}(\cos 2\theta) \tag{1.23}$$
$$\times (\sin\theta)^{2r} P_{q_0}^{(2r+n-2,\,|k-l|)}(\cos 2\varphi_{2_1})(\sin\varphi_{2_1})^{2r},$$

where $p = q + |k - l|$, $q_0 = q - r$ and $p_0 = p - r$.

Now let $\varphi_{2_1} = \theta = \pi/2$; then $\hat{\omega}_n = \cos\varphi_{2_2}\cdot e^{i\varphi_{1_2}}$ i.e. $e^{i\alpha} = e^{i\varphi_{1_2}}$ and $\psi = \varphi_{2_2}$. Under these conditions the only nonzero addends in the right side of (1.20) are those in which $|(k - k_1) - (l - l_1)| = 0$; taking this into account, we obtain

$$P_q^{(n-2,\,|k-l|)}(\cos 2\varphi_{2_2}) = \sum_{r=0}^{q}\frac{\Gamma(q+|k-l|+1)\Gamma(r+|k-l|+n-2)}{\Gamma(q+|k-l|+n-1)\Gamma(r+|k-l|+1)} \tag{1.24}$$
$$\times (2r+|k-l|+n-2)P_r^{(n-3,\,|k-l|)}(\cos 2\varphi_{2_2}).$$

We observe that by utilizing analytic continuation it is possible to replace the integer $n - 2 > 0$ in (1.20)–(1.24) by the number $\beta > -1$ ([4], § 10.8).

6. *Multiplication theorem for zonal spherical functions.* Let us multiply both sides of (1.21) by

$$\overline{\mathfrak{P}_{r,s}^{n-1}(\varphi_{1_2}, \varphi_{2_2})}\,\frac{n-2}{\pi}(\sin\varphi_{2_2})^{2(n-3)+1}\cos\varphi_{2_2}$$

and, taking the orthogonality relations (1.9) into account, integrate with respect to φ_{1_2} from 0 to 2π and with respect to φ_{2_2} from 0 to $\pi/2$:

$$\frac{n-2}{\pi}\int_0^{\frac{\pi}{2}}\int_0^{2\pi} \mathfrak{P}^n_{k,l}(\alpha, \psi)\overline{\mathfrak{P}^{n-1}_{r,s}(\varphi_{1_2}, \varphi_{2_2})}(\sin\varphi_{2_2})^{2(n-3)+1}\cos\varphi_{2_2}\,d\varphi_{1_2}\,d\varphi_{2_2}$$
$$=[\Gamma(n-1)]^2\frac{\Gamma(p+1)\Gamma(q+1)\Gamma(p_0+r+s+n-1)\Gamma(q_0+r+s+n-1)}{\Gamma(p+n-1)\Gamma(q+n-1)\Gamma(p_0+1)\Gamma(q_0+1)[\Gamma(r+s+n-1)]^2}$$
$$\times\mathfrak{P}^{n+r+s}_{k-r,\,l-s}(\varphi_{1_1}, \varphi_{2_1})(\sin\varphi_{2_1})^{r+s}\overline{\mathfrak{P}^{n+r+s}_{k-r,\,l-s}(\varphi, \theta)}(\sin\theta)^{r+s}. \qquad (1.25)$$

We call (1.25) the multiplication theorem for zonal spherical functions. (In a sense it can be considered the multiplication theorem for Jacobi polynomials.) Setting $r = s = 0$, we obtain the special case of this formula:

$$\frac{n-2}{\pi}\int_0^{\frac{\pi}{2}}\int_0^{2\pi}\mathfrak{P}^n_{k,l}(\alpha, \psi)(\sin\varphi_{2_2})^{2(n-3)+1}\cos\varphi_{2_2}\,d\varphi_{1_2}\,d\varphi_{2_2}$$
$$=\mathfrak{P}^n_{k,l}(\varphi_{1_1}, \varphi_{2_1})\overline{\mathfrak{P}^n_{k,l}(\varphi, \theta)}. \qquad (1.26)$$

7. *Integral representation.* The equality

$$\mathfrak{P}^n_{k,l}(\varphi, \theta)=(i)^{-(k+l)}\frac{n-2}{\pi}\int_0^{\frac{\pi}{2}}\int_0^{2\pi}(\cos\varphi_{2_2}\cdot e^{i\varphi_{1_2}}\sin\theta+i\cos\theta\cdot e^{-i\varphi})^k$$
$$\times(\cos\varphi_{2_2}\cdot e^{-i\varphi_{1_2}}\sin\theta+i\cos\theta\cdot e^{i\varphi})^l(\sin\varphi_{2_2})^{2(n-3)+1}\cos\varphi_{2_2}\,d\varphi_{1_2}\,d\varphi_{2_2}, \qquad (1.27)$$

which can be considered an integral representation of the zonal spherical functions, is proved by direct computations. It can be obtained from group-theoretic considerations if we pass from the realization of the representations considered in $\mathfrak{H}^{n(k,l)}$ to an equivalent realization in the space of functions on the n-dimensional complex cone.

8. *Some recursion relations.*(5) We use (1.8), which implies that

$$\Delta[r^{k+l}\mathfrak{P}^n_{k,l}(\omega, \overline{\omega})]=0, \text{ where } \Delta=4\sum_{i=1}^n\frac{\partial^2}{\partial z_i\,\partial\overline{z}_i}, \qquad (1.28)$$

to derive certain recursion relations. In addition, we deduce the following formula from (1.8) and results of [3]:

$$\mathfrak{P}^n_{k,l}(\omega, \overline{\omega})=\frac{\Gamma(n-1)\Gamma(k+l+n-1)}{\Gamma(k+n-1)\Gamma(l+n-1)}\Big[\overline{\omega}^k\omega^l$$
$$-\frac{kl}{(n+k+l-2)\,1}\overline{\omega}^{k-1}\omega^{l-1}+\dots\Big], \qquad (1.29)$$

so that

(5) For $\varphi = 0$ we obtain recursion formulas automatically for the Jacobi polynomials.

$$\frac{\partial}{\partial \omega} \mathfrak{P}^n_{k,l}(\omega, \overline{\omega}) = \frac{l(k+n-1)}{n-1} \mathfrak{P}^{n+1}_{k,l-1}(\omega, \overline{\omega}), \tag{1.30}$$

$$\frac{\partial}{\partial \overline{\omega}} \mathfrak{P}^n_{k,l}(\omega, \overline{\omega}) = \frac{k(l+n-1)}{n-1} \mathfrak{P}^{n+1}_{k-1,l}(\omega, \overline{\omega}). \tag{1.30'}$$

Transforming (1.28) with the help of (1.30) and (1.30′), it is not difficult to obtain the recursion formula

$$\begin{aligned} &-\omega^2 l(l-1)(k+n-1)(k+n)\mathfrak{P}^{n+2}_{k,l-2}(\omega, \overline{\omega}) - \overline{\omega}^2 k(k-1)(l+n-1) \\ &\quad\times(l+n)\mathfrak{P}^{n+2}_{k-2,l}(\omega, \overline{\omega}) + 2(2-\omega\cdot\overline{\omega})kl(l+n-1)(k+n-1) \\ &\quad\times\mathfrak{P}^{n+2}_{k-1,l-1}(\omega, \overline{\omega}) + nl(k+n-1)(1-2n)\omega\mathfrak{P}^{n+1}_{k,l-1}(\omega, \overline{\omega}) \\ &\quad+ nk(l+n-1)(1-2n)\overline{\omega}\mathfrak{P}^{n+1}_{k-1,l}(\omega, \overline{\omega}) \\ &\quad+ n(n-1)(k+l)(k+l+2n-2)\mathfrak{P}^n_{k,l}(\omega, \overline{\omega}) = 0 \end{aligned} \tag{1.31}$$

for the zonal spherical functions.

Starting from (1.8), one can obtain a whole series of recursion relations after first rewriting this formula (cf. Theorem 2 of [3]) in the form

$$z_n^k \overline{z}_n^l = \frac{\Gamma(k+n-1)\Gamma(l+n-1)}{\Gamma(n-1)\Gamma(k+l+n-1)} r^{k+l} \mathfrak{P}^n_{l,k}(\omega, \overline{\omega}) + r^2 f_1(\omega, \overline{\omega}). \tag{1.32}$$

If both sides of (1.32) are multiplied by $z_n^p \overline{z}_n^s$, then (1.8), formula (5) of [3] for the operator P_0, and the uniqueness of expansion (6) in [3], we obtain a series of recursion formulas. Thus, for example, multiplying by z_n, we obtain

$$\begin{aligned} &2(n-1)(k+n-1)\mathfrak{P}^n_{k+1,l}(\omega, \overline{\omega}) - (n-1)(k+l+2n-2)\overline{\omega}\mathfrak{P}^n_{k,l}(\omega, \overline{\omega}) \\ &= k(l+n-1)\overline{\omega}^2\mathfrak{P}^n_{k-1,l}(\omega, \overline{\omega}) - l(k+n-1)(2-\omega\overline{\omega})\mathfrak{P}^{n+1}_{k,l-1}(\omega, \overline{\omega}), \end{aligned} \tag{1.33}$$

while after multiplying by $\overline{z}_n$ we get

$$\begin{aligned} &2(n-1)(l+n-1)\mathfrak{P}^n_{k,l+1}(\omega, \overline{\omega}) - (n-1)(k+l+2n-2)\omega\mathfrak{P}^n_{k,l}(\omega, \overline{\omega}) \\ &= l(k+n-1)\omega^2\mathfrak{P}^{n+1}_{k,l-1}(\omega, \overline{\omega}) - k(l+n-1)(2-\omega\overline{\omega})\mathfrak{P}^{n+1}_{k-1,l}(\omega, \overline{\omega}). \end{aligned} \tag{1.34}$$

§2. Orthogonal basis, with weight $(1-x^2-y^2)^\beta$, $\beta > -1$, in the disc $x^2+y^2 \leqslant 1$

1. *Description of the functions* $\mathfrak{P}^n_{k,l}(x, y)$. Let us consider the mapping

$$x = \cos\theta\cos\varphi, \quad y = \cos\theta\sin\varphi \tag{2.1}$$

of the northern hemisphere S of the unit sphere S^2 ($0 \leqslant \varphi < 2\pi \pmod{2\pi}$, $0 \leqslant \theta \leqslant \pi/2$) into the disc $\mathfrak{M}(x^2+y^2 \leqslant 1)$. Under it $L^2(S)$ turns into $L^2(\mathfrak{M})$, the measure $ds = (n-1)\pi^{-1}(\sin\theta)^{2(n-2)+1}\cos\theta\, d\varphi d\theta$ into $d\mu = (n-1)\pi^{-1}(1-x^2-y^2)^{n-2}dxdy$, and the functions $\mathfrak{P}^{(k,l)}_n(\varphi, \theta)$ turn into the functions

$$\begin{aligned} \mathfrak{P}^n_{k,l}(x, y) = &[x + i\operatorname{sgn}(l-k)y]^{|k-l|} \\ &\times F(-q, p+n-1; n-1; 1-x^2-y^2), \end{aligned} \tag{2.2}$$

or

$$\mathfrak{P}^n_{k,l}(x, y) = \frac{\Gamma(q+1)\Gamma(n-1)}{\Gamma(q+n-1)}[x + i\,\mathrm{sgn}(l-k)y]^{|k-l|} \times P_q^{(n-2,\,|k-l|)}[2(x^2+y^2)-1] \tag{2.3}$$

(cf. (1.5) and (1.7)). We conclude from (1.11) and (1.14) that the expansion

$$f(x, y) = \sum_{k,l=0}^{\infty} a_{k,l}\mathfrak{P}^n_{k,l}(x, y), \tag{2.4}$$

where

$$a_{k,l} = \dim \mathfrak{H}^{n(k,l)}\frac{n-1}{\pi}\int\limits_{x^2+y^2\leqslant 1} f(x, y)\overline{\mathfrak{P}^n_{k,l}(x, y)}(1-x^2-y^2)^{n-2}\,dx\,dy \tag{2.5}$$

holds for any function $f(x, y) \in L^2(\mathfrak{M})$.

Orthogonality relations are established in the following subsection for the functions $\mathfrak{P}^n_{k,l}(x, y)$, so that the following result holds: the functions

$$\frac{1}{[\dim \mathfrak{H}^{n(k,l)}]^{\frac{1}{2}}}\mathfrak{P}^n_{k,l}(x, y)$$

form a complete orthonormal system in the space of functions square integrable on the disc $x^2 + y^2 \leqslant 1$ with respect to the measure

$$d\mu = \frac{n-1}{\pi}(1-x^2-y^2)^{n-2}\,dx\,dy. \tag{2.6}$$

2. *Orthogonality relations for the functions* $\mathfrak{P}^n_{k,l}(x, y)$. From (1.9) and the relation between $\mathfrak{P}^n_{k,l}(x, y)$ and $\mathfrak{P}^n_{k,l}(\varphi, \theta)$ we obtain the orthogonality relations

$$\frac{n-1}{\pi}\int\limits_{x^2+y^2\leqslant 1}\mathfrak{P}^n_{k,l}(x, y)\overline{\mathfrak{P}^n_{r,s}(x, y)}(1-x^2-y^2)\,dxdy = \begin{cases} 0, & \text{if } (k, l)\neq(r, s), \\ \dfrac{1}{\dim \mathfrak{H}^{n(k,l)}}, & \text{if } (k, l) = (r, s). \end{cases} \tag{2.7}$$

3. *Differential equations.* By the change of variables (2.1) the system of equations (1.17), (1.18) reduces to the form

$$\left(-y\frac{\partial}{\partial x} + x\frac{\partial}{\partial y}\right)\mathfrak{P}^n_{k,l}(x, y) = -(k-l)\mathfrak{P}^n_{k,l}(x, y), \tag{2.8}$$

$$\begin{aligned}\Big\{&\frac{y^2+x^2(1-x^2-y^2)}{x^2+y^2}\frac{\partial^2}{\partial x^2} + \frac{x^2+y^2(1-x^2-y^2)}{x^2+y^2}\frac{\partial^2}{\partial y^2} - (2n-1)x\frac{\partial}{\partial x} \\ &- (2n-1)y\frac{\partial}{\partial y} - 2xy\frac{\partial^2}{\partial x\,\partial y}\Big\}\mathfrak{P}^n_{k,l}(x, y) \\ &= -(k+l)(k+l+2n-2)\mathfrak{P}^n_{k,l}(x, y).\end{aligned} \tag{2.9}$$

The functions $\mathfrak{P}^n_{k,l}(x, y)$ are defined uniquely to within a normalizing constant by (2.8) and (2.9).

4. *Addition theorem.* Let us denote

$$x_1 = \cos\varphi_{2_2}\cos\varphi_{1_2}, \quad y_1 = \cos\varphi_{2_2}\sin\varphi_{1_2},$$

$$x_2 = \cos\varphi_{2_1}\cos\varphi_{1_1}, \quad y_2 = \cos\varphi_{2_1}\sin\varphi_{2_1},$$

$$x_3 = \cos\theta\cos\varphi, \qquad y_3 = \cos\theta\sin\varphi,$$
$$x = \cos\psi\cos\alpha, \qquad y = \cos\psi\sin\alpha.$$

Then from the inequality

$$\cos\psi\cdot e^{i\alpha} = \sin\varphi_{2_1}\cos\varphi_{2_2}\cdot e^{i\varphi_{1_2}}\sin\theta + \cos\varphi_{2_1}\cdot e^{i\varphi_{1_1}}\cos\theta\cdot e^{-i\varphi}$$

we obtain

$$x = x_1\sqrt{1-x_2^2-y_2^2}\sqrt{1-x_3^2-y_3^2} + x_2x_3 + y_2y_3,$$

$$y = y_1\sqrt{1-x_2^2-y_2^2}\sqrt{1-x_3^2-y_3^2} + y_2x_3 - x_2y_3.$$

As a result (1.21) takes the form

$$\begin{aligned}\mathfrak{P}^n_{k,l}\Big(x_1\sqrt{1-x_2^2-y_2^2}\sqrt{1-x_3^2-y_3^2} + x_2x_3 + y_2y_3,\ y_1\sqrt{1-x_2^2-y_2^2} \\ \times\sqrt{1-x_3^2-y_3^2} + y_2x_3 - x_2y_3\Big) = \sum_{k_1=0}^{k}\sum_{l_1=0}^{l}\frac{\Gamma(q+1)\Gamma(p+1)\Gamma(n-1)}{\Gamma(q+n-1)\Gamma(p+n-1)} \\ \times\frac{\Gamma(k_1+l_1+p_0+n-1)\Gamma(k_1+l_1+q_0+n-1)\Gamma(p_1+n-2)\Gamma(q_1+n-2)(k_1+l_1+n-2)}{\Gamma(p_0+1)\Gamma(q_0+1)\Gamma(p_1+1)\Gamma(q_1+1)[\Gamma(k_1+l_1+n-1)]^2\Gamma(n-2)} \\ \times\mathfrak{P}^{n-1}_{k_1,l_1}(x_1,y_1)\,\mathfrak{P}^{k_1+l_1+n}_{k-k_1,l-l_1}(x_2,y_2)(1-x_2^2-y_2^2)^{\frac{k_1+l_1}{2}} \\ \times\overline{\mathfrak{P}^{k_1+l_1+n}_{k-k_1,l-l_1}(x_3,y_3)}(1-x_3^2-y_3^2)^{\frac{k_1+l_1}{2}}.\end{aligned} \tag{2.10}$$

We call (2.10) the addition theorem for the functions $\mathfrak{P}^n_{k,l}(x,y)$.

5. *Multiplication theorem.* Rewriting (1.25) and (1.26) in the coordinates (2.1), we arrive at the multiplication theorem for the functions $\mathfrak{P}^n_{k,l}(x,y)$:

$$\begin{aligned}\frac{n-2}{\pi}\int\limits_{x_1^2+y_1^2\leqslant 1}\mathfrak{P}^n_{k,l}(x,y)\,\overline{\mathfrak{P}^{n-1}_{r,s}(x_1,y_1)}(1-x_1^2-y_1^2)^{n-3}\,dx_1\,dy_1 \\ = [\Gamma(n-1)]^2\frac{\Gamma(p+1)\Gamma(q+1)\Gamma(p_0+r+s+n-1)\Gamma(q_0+r+s+n-1)}{\Gamma(p+n-1)\Gamma(q+n-1)\Gamma(p_0+1)\Gamma(q_0+1)[\Gamma(r+s+n-1)]^2} \\ \times\mathfrak{P}^{n+r+s}_{k-r,l-s}(x_2,y_2)(1-x_2^2-y_2^2)^{\frac{r+s}{2}}\,\overline{\mathfrak{P}^{n+r+s}_{k-r,l-s}(x_3,y_3)}(1-x_3^2-y_3^2)^{\frac{r+s}{2}};\end{aligned} \tag{2.11}$$

and a special case of it:

$$\begin{aligned}\frac{n-2}{\pi}\int\limits_{x_1^2+y_1^2\leqslant 1}\mathfrak{P}^n_{k,l}(x,y)(1-x_1^2-y_1^2)^{n-3}\,dx_1\,dy_1 \\ = \mathfrak{P}^n_{k,l}(x_2,y_2)\,\overline{\mathfrak{P}^n_{k,l}(x_3,y_3)}.\end{aligned} \tag{2.12}$$

6. *Integral representation*:

$$\mathfrak{P}^n_{k,l}(x, y) = (i)^{-(k+l)}\frac{n-2}{\pi} \int\limits_{x_1^2+y_1^2\leqslant 1} [(x_1+iy_1)\sqrt{1-x^2-y^2}+i(x-iy)]^k$$
$$\times [(x_1-iy_1)\sqrt{1-x^2-y^2}+i(x+iy)]^l (1-x_1^2-y_1^2)^{n-3}\,dx_1\,dy_1 \quad (2.13)$$

(cf. (1.27)).

7. *Recursion relations.* From (1.31), (1.33) and (1.34) we obtain the following recursion relations for the functions $\mathfrak{P}^n_{k,l}(x, y)$:

$$\begin{aligned} &-l(l-1)(k+n-1)(k+n)(x+iy)^2\,\mathfrak{P}^{n+2}_{k,l-2}(x, y) - k(k-1)\\ &\quad\times(l+n-1)(l+n)(x-iy)^2\,\mathfrak{P}^{n+2}_{k-2,l}(x, y)\\ &\quad+2(2-x^2-y^2)\,kl(l+n-1)(k+n-1)\,\mathfrak{P}^{n+2}_{k-1,l-1}(x, y)\\ &\quad+nl(k+n-1)(1-2n)(x+iy)\,\mathfrak{P}^{n+1}_{k,l-1}(x, y)\\ &\quad+nk(l+n-1)(1-2n)(x-iy)\,\mathfrak{P}^{n+1}_{k-1,l}(x, y)\\ &\quad+n(n-1)(k+l)(k+l+2n-2)\,\mathfrak{P}^{n}_{k,l}(x, y)=0; \end{aligned} \quad (2.14)$$

$$\begin{aligned} &2(n-1)(k+n-1)\,\mathfrak{P}^n_{k+1,l}(x, y) - (n-1)(k+l+2n-2)(x-iy)\\ &\quad\times\mathfrak{P}^n_{k,l}(x, y) = k(l+n-1)(x-iy)^2\,\mathfrak{P}^{n+1}_{k-1,l}(x, y)\\ &\quad-l(k+n-1)(2-x^2-y^2)\,\mathfrak{P}^{n+1}_{k,l-1}(x, y); \end{aligned} \quad (2.15)$$

$$\begin{aligned} &2(n-1)(l+n-1)\,\mathfrak{P}^n_{k,l+1}(x, y) - (n-1)(k+l+2n-2)(x+iy)\\ &\quad\times\mathfrak{P}^n_{k,l}(x, y) = l(k+n-1)(x+iy)^2\,\mathfrak{P}^{n+1}_{k,l-1}(x, y)\\ &\quad k(l+n-1)(2-x^2-y^2)\,\mathfrak{P}^{n+1}_{k-1,l}(x, y). \end{aligned} \quad (2.16)$$

REMARK. The results of §2 are proved for an integer $\beta = n-2 > 0$. Utilizing analytic continuation, it is possible to replace this β by any number $\beta > -1$ (cf. the Remark at the end of §1.5).

In conclusion the author thanks N. Ja. Vilenkin for useful suggestions.

BIBLIOGRAPHY

1. N. Ja. Vilenkin, *Special functions and the theory of group representations*, "Nauka", Moscow, 1965; English transl., Amer. Math. Soc., Providence, R. I., 1968.

2. ———, *Polyspherical and orispherical functions*, Mat. Sb. **68 (110)** (1965), 432–443. (Russian)

3. R. L. Šapiro, *Irreducible representations of the group* SU(n) *of class* I *with respect to* SU($n-1$), Uspehi Mat. Nauk **22** (1967), no. 1 (133), 180–182; English transl. in the present volume.

4. A. Erdélyi et al., *Higher transcendental functions.* Vol. II, McGraw-Hill, New York, 1953.

5. ———, *Higher transcendental functions.* Vo.. I, McGraw-Hill, New York, 1953.

6. I. S. Gradšteĭn and I. M. Ryžik, *Tables of integral, series and products*, 4th ed., Fizmatgiz, Moscow, 1963; English transl., Academic Press, New York, 1965.

Translated by F. M. GOLDWARE

Amer. Math. Soc. Transl.
(2) Vol. 113, 1979

IRREDUCIBLE REPRESENTATIONS OF POLYCYCLIC GROUPS OVER AN ABSOLUTE ALGEBRAIC FIELD OF PRIME CHARACTERISTIC*

E. M. LEVIČ [ĻEVIČS]

It is an interesting and important problem in the theory of groups to study properties of groups depending on properties of their representations. In particular, one can ask the question, what properties must a group have so that its irreducible representations over any field are finite-dimensional? It is known [1]–[4] that for all the irreducible representations of a group Γ to be finite-dimensional it is sufficient that Γ has a finitely generated Abelian normal subgroup of finite index. P. Hall has proved in [4] that every irreducible representation of a finitely generated nilpotent group over an absolute algebraic field of prime characteristic p is finite-dimensional (see also [8]). On the other hand, in the same paper he has shown that if a polycyclic group Γ does not have a finitely generated Abelian normal subgroup of finite index and if a field P is not absolutely algebraic of prime characteristic p, then Γ has an infinite-dimensional representation over P.

The present paper is devoted to a proof of the theorem [13] asserting that every irreducible representation of a polycyclic group over an absolutely algebraic field of prime characteristic p is finite-dimensional.(1) This theorem answers a question raised by P. Hall in [4]. The underlying idea of the proof is the same as in [8], where it is proved that all irreducible representations of a finitely generated nilpotent group over an absolutely algebraic field of prime characteristic p are finite-dimensional. In the present situation it consists in the following: to show that every irreducible representation of a polycyclic group over any field has a subspace that is maximal subject to being invariant under the center of the radical of the group (Theorem 10); the presence of this subspace enables us to show that if an irreducible representation of the group over an absolute field of prime characteristic p is faithful, then the center of the radical is finite (Theorem 11), from which it follows easily that the group in question is finite; that is, the representation is finite-dimensional.

The paper consists of four sections. In the first and second sections we give the basic definitions and notation from group theory, from the theory of group algebras of polycyclic groups, and from the theory of modules over principal ideal rings. Here we also compile a number of auxiliary assertions that are needed in what follows. In

---

AMS (MOS) subject classifications (1970). Primary 20E35, 16A26.

*Translation of Certain Questions of Group Theory (Proc. Riga Algebra Sem. No. 2, 1969/70), Latvijas Valsts Univ. Zinātn. Raksti **151** (1971), 104–124.

(1) Another proof of this assertion was recently obtained by A. E. Zalesskiĭ [14].

addition, the first section contains a new proof, due to the author, of Hall's theorem that the group algebra of a polycyclic group over any field is Noetherian. In the third section we study the group algebra of a polycyclic group as a module over a polynomial ring in one or several variables. In the fourth and last section we prove the main result.

§1. Basic definitions, notation, and some auxiliary assertions

Let Γ be a polycyclic group; that is, a group having a finite normal series with cyclic factors:

$$\{e\} = \Gamma_0 \subset \Gamma_1 \subset \cdots \subset \Gamma_n = \Gamma. \tag{1}$$

We mention that Γ is polycyclic if and only if it is soluble and satisfies the maximal conditions for subgroups [6]. Hence it follows that the Hirsch-Plotkin radical $R(\Gamma)$ of a polycyclic group Γ is finitely generated and nilpotent, so that the center $Z(R)$ of the radical is nontrivial.

We denote by γ_i an inverse image in Γ of a generating element of the factor Γ_i/Γ_{i-1}. It is easy to see that every element is uniquely representable in the form

$$\sigma = \prod_{i=1}^{n} \gamma_i^{\alpha_i} \qquad (\alpha_i \text{ integers}). \tag{2}$$

We assume that the series (1) is chosen so that $\gamma_1, \ldots, \gamma_l$ are generators of $Z(R)$.

LEMMA 1. *If $Z(R)$ is a finite group, then so is the polycyclic group Γ.*

The assertion of the lemma follows from [10] and [11], if we observe that the Hirsch-Plotkin radical $R(\Gamma)$ bounds the group; that is, that Γ/H, where H is the centralizer of $R(\Gamma)$ in Γ, is isomorphic to a group of automorphisms of $R(\Gamma)$. But $R(\Gamma)$ is finite, by hypothesis; consequently, Γ/H is also finite. Since H is a subgroup of $R(\Gamma)$, we see that H and hence Γ is finite.

Let $P\Gamma$ be the group algebra of a polycyclic group Γ over an arbitrary field P. Every element $h \in P\Gamma$ can be represented in the form

$$h = \sum_{i=s}^{m} \gamma_n^i h_i, \tag{3}$$

where $h_m \neq 0, h_s \neq 0$ and $h_i \in P\Gamma_{n-1}$ $(s \leqslant i \leqslant m)$.

We call the representation of h in the form (3) the expansion of h by powers of γ_n; we also call m the *upper γ_n-power* of h (notation: $\bar{l}(h) = m$), s the *lower γ_n-power* of h (notation: $\underline{l}(h) = s$), $m - s$ the *γ_n-length* of h (notation: $l(h) = m - s$), h_i the *ith coordinate* of h in the expansion (3) (notation: $w_i(h) = h_i 1$), h_m the *upper γ_n-projection* of h onto $P\Gamma_{n-1}$ (notation: $\bar{w}(h) = h_m 1$), *and* h_s the *lower γ_n-projection* of h onto $P\Gamma_{n-1}$ (notation: $\underline{w}(h) = h_s 1$). Let K be a set of elements of $P\Gamma$. We call $K^m = \{\bar{w}(h)\colon\ h \in K, \bar{l}(h) = m\}$ the *m-upper γ_n-projection* of K into $P\Gamma_{N-1}$, and $K_s = \{\underline{w}(h)\colon\ h \in K, \underline{l}(h) = s\}$ the *s-lower γ_n-projection* of K. If K is a right (or left, or two-sided) ideal in $P\Gamma$, then K^m and K_s are also right (or left, or two-sided) ideals in $P\Gamma_{n-1}$ for any integers m and s.

The following assertion holds:

LEMMA 2. *If K is a right ideal in* $P\Gamma$, *then* $K^{m+1} = \gamma_n^{-1}K^m\gamma_n$ *and* $K_s = \gamma_n K_{s+1}\gamma_n^{-1}$.

PROOF. Since the proofs of the two equalities are alike, we only prove the first. Let $h_m \in K^m$, where h is an element of K for which $\overline{w}(h) = h_m$ and $h = \Sigma_{i=s}^m \gamma_n^i h_i$. Then the upper γ_n-power of $h\gamma_n$ is $m + 1$. For

$$h\gamma_n = \left(\sum_{i=s}^m \gamma_n^i h_i\right)\gamma_n = \sum_{i=s}^m \gamma_n^i h_i\gamma_n = \sum_{i=s}^m \gamma_n^{i+1}(\gamma_n^{-1}h_i\gamma_n). \qquad (4)$$

From (4) it follows that $\gamma_n^{-1} h_m\gamma_n \in K^{m+1}$; that is, $K^{m+1} \subset \gamma_n^{-1}K^m\gamma_n$. The reverse inclusion is established similarly, and the lemma is proved.

LEMMA 3. *Let K be a right ideal in* $P\Gamma$ *and suppose that the maximal condition holds for right ideals of* $P\Gamma_{n-1}$. *Then there is a natural number C such that for any* $v \in P\Gamma$ *with the property* $\overline{w}(v) \in K^m$ *we can find for* $l(v) = m$ *an element* v' *such that* $v - v' \in K$ *and* $l(v') \leqslant \max\{l(v) - 1, C\}$.

PROOF. We fix a number m' and consider $K^{m'}$. Since $K^{m'}$ is a right ideal in $P\Gamma_{n-1}$, there are a finite number of generators $g_1, \dots, g_s$ for $K^{m'}$. In the ideal K we choose s elements with the properties $\overline{l}(h_j) = m'$ and $\overline{w}(h_j) = g_j$ $(j = 1, \dots, s)$. Let C be the largest γ_n-length of the elements h_j $(1 \leqslant j \leqslant s)$, i.e. $C = \max_{1\leqslant j\leqslant s} l(h_j)$. From Lemma 2 it follows that the elements $\gamma_n^{-(m-m')}(\overline{w}(h_j))\gamma_n^{m-m'}$ $(j = 1, \dots, S)$ are generators of the ideal K^m in $P\Gamma_{n-1}$. Since the upper γ_n-projection of v onto $P\Gamma_{n-1}$ lies in K^m, there are elements $q_j \in P\Gamma_{n-1}$ such that

$$\overline{w}(v) = \sum_{j=1}^s \gamma_n^{-(m'-m)}g_j\gamma_n^{m'-m}q_j.$$

Then the upper γ_n power of the element

$$v' = v - \sum_{j=1}^s \gamma_n^{-(m'-m)}h_j\gamma_n^{m'-m}q_j$$

is less than m. Evidently, v' satisfies all the conditions, and the lemma is proved.

THEOREM 1 (P. HALL [4]). *The group algebra* $P\Gamma$ *of a polycyclic group* Γ *over any field P satisfies the maximal condition for right ideals.*

To prove the theorem it is sufficient to show that every right ideal $K \subset P\Gamma$ is finitely generated. We conduct the proof by induction on the length of the series (1). If $n = 1$, the theorem is obvious. Suppose that $P\Gamma_{n-1}$ satisfies the maximal condition for right ideals, so that every right ideal in $P\Gamma_{n-1}$ is finitely generated. We assume that some right ideal K in $P\Gamma$ is not finitely generated and that $v_1, v_2, \dots$ is an infinite system of generators of K.

We consider K^m (where m is a natural number). Since K^m is a right ideal in $P\Gamma_{n-1}$ and, by the inductive hypothesis, $P\Gamma_{n-1}$ satisfies the maximal condition for right ideals, K^m has a finite system of generators $\overline{h}_1, \dots, \overline{h}_{s_1}$. Let $h_1, \dots, h_{s_1} \in K$ be such that $\overline{w}(h_i) = \overline{h}_i$ and $\overline{l}(\overline{h}_i) = m$. By Lemma 3, there are a number C and elements

$v_{01}, v_{02}, \ldots$ such that $v_i - v_{0i} \in K$ and $l(v_{0i}) \leqslant \max(l(v_i) - 1, C)$ $(i = 1, 2, \ldots)$. Applying Lemma 3 repeatedly, if necessary, we arrive at elements $v'_1, \ldots, v'_n$ such that $v_i - v'_i \in K$ and $l(v') \leqslant C$. From the proof of Lemma 3 it follows that $h_1, \ldots, h_{s_1}, v'_1, \ldots, v'_n$ is a system of generators of K. Let

$$L_C = \{\overline{w}(h);\ h \in K, \overline{l}(h) \leqslant C\}.$$

Clearly, L_c is a right ideal in $P\Gamma_{n-1}$ and, consequently, has a finite number of generators $\overline{h}_{s_1+1}, \ldots, \overline{h}_{s_2}$. If $h_{s_1+1}, \ldots, h_{s_2} \in K$ are such that $\overline{w}(h_j) = \overline{h}_j$, $\overline{l}(h_j) = C$ and $l(h) \leqslant C$, then we can construct a sequence $v''_1, \ldots, v''_n$ with the properties that $v'_i - v''_i \in K$, $l(v''_i) \leqslant C - 1$, and $h_1, \ldots, h_{s_2}, v''_1, \ldots, v''_n$ is a system of generators of K. Continuing the process described above, we construct a system $h_1, h_2, \ldots,$ of generators of K, such that $l(h_j) = 0$ and $h_j \in P\Gamma_{n-1}$ for $j \geqslant s_k + 1$. If we denote by K' the right ideal in $P\Gamma_{n-1}$ generated by the elements h_j $(j \geqslant s_k + 1)$, then we can find a finite system $h_{s_k+1}, \ldots, h_{s_{k+1}}$ of generators of the ideal K'. Clearly, then $h_1, \ldots, h_{s_{k+1}}$ is a system of generators of K. We have come to a contradiction, and the theorem is proved.

§2. Modules over principal ideal domains

Let J be a commutative principal ideal domain, but not a field, and $\omega_0 = \{P_\lambda, \lambda \in \Lambda\}$ the set of prime elements in J. Every element $x \in J$ $(x \neq 0)$ is uniquely representable, to within divisors of unity, as a product of primes:

$$x = \epsilon \prod_{\lambda \in \Lambda} p_\lambda^{n_\lambda},$$

where ϵ is a divisor of unity and only finitely many n_λ are different from zero.

We denote by $\mathfrak{A}(J)$ the class of all J-modules and let $V \in \mathfrak{A}(J)$ and $v \in V$. The order ideal $\nu(v)$ of v is defined as the collection of all elements in J that annihilate v: $\nu(v) = \{x \in J,\ vx = 0\}$. Since J is a principal ideal domain, $\nu(x)$ is a principal ideal; that is, it can be represented in the form $\nu(v) = x_0 J$. In this case the element $x_0 \in J$ is called the order of v. It is clear that $vJ \cong J/\nu(v)$.

A J-module V is said to be *torsion-free* if $\nu(v) = 0$ for every $v \in V$, and *periodic* if $\nu(v) \neq 0$ for every $v \in V$. Then (and only then) the J-module can be represented as a direct sum:

$$V = \bigoplus \sum_{\mu \in M} v_\mu J,$$

where $v_\mu J \cong J$; that is, $\nu(v_\mu) = \{0\}$. This is equivalent to the fact that the J-module V is generated by the set $\{v_\mu, \mu \in M\}$ consisting of elements that are linearly independent over J. If V is an arbitrary J-module and V has a free J-submodule V_0 generated by a maximal set of linearly independent elements (over J), then V/V_0 is a periodic module.

Let ω' be a subset of the set ω_0 of all prime elements in J. We say that a J-module V belongs to the class of modules $\mathfrak{A}(J, \omega')$ if V has a free submodule V_0 such that V/V_0 is an ω'-periodic J-module; that is, V/V_0 is a periodic J-module and the order of every element $v + V_0$ has as prime divisors only elements in ω'. If

$\omega' = \omega_0$, then the class $\mathfrak{A}(J, \omega')$ is the same as $\mathfrak{A}(J)$; and if ω' is the empty set, then $\mathfrak{A}(J, \omega')$ is the class of all free J-modules. On the other hand, if $V_0 = \{0\}$, then the J-module V is said to be ω'-periodic.

The following propositions are taken from [4], and we omit the proofs here.

LEMMA 4′. *The field of rational fractions R of J belongs to the class $\mathfrak{A}(J, \omega')$ if and only if $\omega' = \omega_0$.*

LEMMA 5′. *If $V \in \mathfrak{A}(J, \omega')$, then every J-submodule $V_0 \subset V$ also belongs to the class $\mathfrak{A}(J, \omega')$.*

LEMMA 6′. *Suppose that a J-module V has an ascending series of submodules $\{H_\alpha\colon\ \alpha \in M\}$ and that $V = H_\beta$. If $H_{\alpha+1}/H_\alpha \in \mathfrak{A}(J, \omega')$ for all $\alpha < \beta$, then $V \in \mathfrak{A}(J, \omega')$.*

We now introduce another class of J-modules, which we denote by $\mathfrak{A}(J, \omega', k)$. A J-module V belongs to $\mathfrak{A}(J, \omega', k)$ if and only if V has a free submodule V_0 such that V/V_0 is ω'-periodic and the order x of any element $v + V_0 \in V/V_0$ can be represented in the form

$$x = \prod_{\lambda \in \Lambda} p_\lambda^{n_\lambda},$$

where only finitely many n_λ are different from zero and all the n_λ are less than or equal to k.

With due account of Lemmas 4′, 5′, and 6′ from [4] it is easy to obtain the following assertions:

LEMMA 4. *The field of rational fractions R of J does not belong to any class $\mathfrak{A}(J, \omega', k)$ for all finite k.*

LEMMA 5. *If $V \in \mathfrak{A}(J, \omega', k)$, then every submodule $V_0 \subset V$ also belongs to $\mathfrak{A}(J, \omega', k)$.*

LEMMA 6. *If a J-module V has a finite ascending series of submodules $\{0\} = H_0 \subset H_1 \subset \cdots \subset H_n = V$, with $H_{i+1}/H_i \in \mathfrak{A}(J, \omega', k_i)$, then $V \in \mathfrak{A}(J, \omega', k)$, where $k = \Sigma_0^{n-1} k_i$.*

LEMMA 7. *A J-module V belongs to $\mathfrak{A}(J, \omega', k)$ only when each of its proper submodules belongs to some class $\mathfrak{A}(J, \omega', k')$ with $k' < k$.*

Let G be a vector space over a field P and σ an endomorphism of it. Then we can regard G as a J-module, where $J = P[X]$ is the polynomial ring in a single variable X with coefficients in P, provided that we specify a representation of J relative to G by the mapping $X \longrightarrow \sigma$. In that case we say for convenience that G is a J_σ-module.

An element $y \in G$ is called σ-algebraic if the order ideal $\nu(y)$ is nonzero. It is easy to see that an element $y \in G$ is algebraic if and only if the minimal σ-invariant subspace containing y is finite-dimensional. When $\nu(y) = 0$, y is called σ-transcendental. This means that the minimal σ-invariant subspace containing $y \in G$ is infinite-dimensional.

If the J_σ-module G is periodic, σ is called a locally algebraic endomorphism; when it is torsion-free, σ is called transcendental.

A J_σ-module G is said to be *divisible* if an equation of the form $x\lambda = y$ is soluble in G for every $y \in G$ and an arbitrary $\lambda \in J$. It is easy to see that this happens if and only if $G\lambda = G$ for every $\lambda \in J$.

LEMMA 8. *An endomorphism σ has a proper maximal σ-invariant subspace in G if and only if G is not a divisible J_σ-module.*

PROOF. Suppose that G_0 is a proper maximal σ-invariant subspace of G. Then σ induces in G/G_0 an endomorphism $\bar{\sigma}$, and the minimal polynomial $\lambda \in J$ satisfied by $\bar{\sigma}$ is irreducible. Evidently, then $G\lambda \subset G_0 \neq G$. This means that G is not a divisible J_σ-module. Then, by definition, there is an irreducible element $\lambda \in J$ such that $G\lambda \neq G$. Hence it follows that $\bar{x}\lambda = 0$ for every $\bar{x} \in G/G\lambda$. Since λ is an irreducible element of J, it follows from Theorem 7.4.1 of [9] that in $G/G\lambda$ there exists a proper maximal $\bar{\sigma}$-invariant subspace. But then G also has a proper maximal σ-invariant subspace, and the lemma is proved.

LEMMA 9. *If $G \in \mathfrak{A}(J_\sigma, \omega', k)$ for some subset $\omega' \subset \omega_0$, then the endomorphism σ has a proper maximal σ-invariant subspace.*

According to Theorem 7.4.1 of [9], the J_σ-module G, by the conditions of the lemma, can be represented as a direct sum of finite-dimensional J_σ-submodules. The assertion of the lemma follows immediately from this.

Let G be a J-module (where J is a principal ideal domain). A J-submodule $H \subset G$ is called *pure* if an equation $x\lambda = g$ ($\lambda \in J$, $g \in H$) can be solved in H whenever it can be solved in G.

Here are a few examples of pure J-submodules.

1. The collection of all algebraic elements of a J-module G, i.e. the maximal periodic J-submodule, is pure in G.
2. Every divisible J-submodule of an arbitrary J-module is pure.

It is easy to verify the following simple assertions:

LEMMA 10. *Every pure J-submodule ω' of a divisible J-module ($\omega' \subset \omega_0$) is ω'-divisible.*

LEMMA 11. *A minimal pure J-submodule of H containing an algebraic element $x \in G$ is an infinite-dimensional vector space if and only if H is a divisible J-submodule of G (see* [9], *Theorem* 6.5).

LEMMA 12. *Let H be a pure J-submodule of G, and F a pure J-submodule of H. Then F is pure in G.*

Let G and G_1 be $P\Gamma$-modules, where $P\Gamma$ is the group algebra of a group Γ over a field P, and let φ be some automorphism of $P\Gamma$. Then G and G_1 are called *weakly isomorphic* if there is an isomorphism φ_1 between the additive groups of G and G_1 and an automorphism φ_2 of $P\Gamma$ such that $(x\lambda)^{\varphi_1} = x^{\varphi_1}\lambda^{\varphi_2}$ for any $x \in G_1$ and $\lambda \in P\Gamma$.

LEMMA 13. *Let φ be an automorphism of $P\Gamma$ and K a right ideal of $P\Gamma$. Then the $P\Gamma$ modules $P\Gamma/K$ and $P\Gamma/K^{\varphi}$ are weakly isomorphic.*

All the above assertions remain true when instead of $P[x]$ we consider the group algebra $J = P\{x\}$ of a cyclic torsion-free group over P.

If σ is an arbitrary element of $P\Gamma$, then every $P\Gamma$-module can be regarded naturally as a J_σ-module, provided that a representation of $P\{x\}$ is specified by means of the mapping $x \longrightarrow \sigma$.

The following assertion is easy to verify:

LEMMA 14. *Let $P\Gamma/K$ be a $P\Gamma$-module, φ an automorphism of $P\Gamma$, and σ an element of $P\Gamma$. If every periodic J_σ-submodule $P\Gamma$ of $P\Gamma/K$ belongs to $\mathfrak{A}(J_\sigma, \omega', k)$, then every periodic $J_{\sigma\varphi}$-submodule of the $P\Gamma$-module $P\Gamma/K^{\varphi}$ belongs to $\mathfrak{A}(J_{\sigma\varphi}, \omega', k)$.*

§3. On some submodules of the group algebra of a polycyclic group

Let $P\Gamma$ be the group algebra of a group Γ over an arbitrary field P, and let K be a right ideal in it. We say that a property Θ of elements of $P\Gamma$ is a *K-property* if from the fact that an element $h \in P\Gamma$ has the property Θ it follows that $h - g$ for every $g \in K$ also has this property. A K-property Θ of elements of $P\Gamma$ is said to be *additive* if from the fact that two elements h and g have the property Θ it follows that all elements of the form $\alpha h + \beta g$, where α and β are elements of P, also have this property. A subset of elements of $P\Gamma$ is said to be a *Θ-set* if every element of this subset has the property Θ. A Θ-set M is called *K-closed* if the following two conditions hold:

a) $\overline{w}(h) \in K^m$, where $m = \overline{l}(h)$, for every $h \in M$;

b) $\overline{w}(h-g) \in K^m$, where $m = \overline{l}(h-g)$, for any $h \in M$ and $g \in K$.

A Θ-set M is called *almost K-closed* if it is K-closed with the exception of finitely many elements.

Evidently, every K-closed subset of K for which every element has a K-property Θ coincides with K.

Let $P\Gamma$ be the group algebra of a polycyclic group Γ, and K a right ideal of $P\Gamma$. Then we can introduce in K a γ_n-filtration when we write

$$L_s(K) = \{h \in K \colon\ l(h) \leqslant s\}.$$

It is easy to see that K has the finite series of $P\Gamma_{n-1}$-modules; $K \supset L_s(K) \supset L_{s-1}(K) \supset \cdots \supset L_0(K)$.

We say that a K-closed Θ-set M, where Θ is some K-property, is *closely connected with the ideal K* if from $\overline{w}(h) \in K^m$ for every $h \in M$ it follows that $\overline{w}(h) \in L^m_{ms}(K)$, where $m = \overline{l}(h)$ and $s = l(h)$.

LEMMA 15. *Let $P\Gamma$ be the group algebra of a polycyclic group Γ, K a right ideal in $P\Gamma$, Θ a K-property, and M a K-closed Θ-set closely connected with K. Then for every element $h \in M$ there is an element $f \in K$ such that $h - f \in P\Gamma_{n-1}$.*

PROOF. By means of Lemma 3 and by using induction we can find for every $h \in M$ an element $f' \in K$ such that $l(h - f') \leqslant C$, where C is independent of the choice of $h \in M$. If $h - f' \in P\Gamma_{n-1}$ for every $h \in M$, then the assertion of the lemma is true. Suppose that $h - f' \notin P\Gamma_{n-1}$ for at least one $h \in M$.

We denote by L the $P\Gamma_{n-1}$-submodule of K generated by all the elements $g \in K$ with $\bar{l}(g) \leqslant C$ and $l(g) \leqslant C$. Now L is not empty, because M is closely connected with K. L has a finite basis as $P\Gamma_{n-1}$-module, which we denote by $\bar{g}_1, \ldots, \bar{g}_s$. Let $g_1, \ldots, g_s$ be elements in L for which $\bar{w}(g_i) = \bar{g}_i$. Then for every $h - f'$ there are elements $u_i \in P\Gamma_{n-1}$ such that $\bar{w}(h - f') = \Sigma \bar{g}_i u_i$. Therefore,

$$l(h - f' - \Sigma g_i u_i) \leqslant C - 1, \qquad f' + \Sigma g_i u_i \in K.$$

Applying induction, we obtain the assertion we need, and the lemma is proved.

COROLLARY. *Let $P\Gamma$ be the group algebra of a polycyclic group Γ, K a right ideal in $P\Gamma$, Θ a K-property, and M an almost K-closed Θ-set closely connected with K. Then for every element $h \in M$, with finitely many possible exceptions, there is an element $f \in K$ such that $h - f \in P\Gamma_{n-1}$.*

A representation (G, Γ) of a group Γ by automorphisms of a vector space G is called *cyclic* if G contains an element x such that the minimal Γ-admissible subspace containing x coincides with G; in other words, if for a suitable element x we have $xP\Gamma = G$. If $K = \{g \in P\Gamma\colon\ xg = 0\}$, then the two $P\Gamma$-modules G and $P\Gamma/K$ are isomorphic ([12], Chapter I, §3).

LEMMA 16. *Let (G, Γ) be a cyclic representation of a polycyclic group Γ, where G is a vector space over an arbitrary field P. Then for every $\sigma \in Z(R)$ there is a finite set $\omega_\sigma \subset \omega_0$ such that every irreducible J_σ-submodule $G_0 \subset G$ belongs to $\mathfrak{A}(J_\sigma, \omega_\sigma)$.*

PROOF. We apply induction on the length of the series (1). If this length is 1, the assertion is obvious. Suppose that the assertion of the lemma is true for Γ_{n-1} but not for Γ. This means that there is a cyclic representation (G, Γ) of Γ and an element $\sigma \in Z(R)$ such that G has a periodic J_σ-submodule G_0 with the property that there is no finite subset $\omega_\sigma \in \omega_0$ such that $G_0 \in \mathfrak{A}(J_\sigma, \omega_\sigma)$.

According to the definition of a cyclic representation, we may assume, without loss of generality, that $G = P\Gamma/K$, where K is a right ideal in $P\Gamma$. By assumption, $P\Gamma$ then has a subsequence M of elements x_i such that

1) $x_i \notin K$, $i = 1, 2, \ldots$, and

2) $x_i \lambda_i \in K$, $\lambda_i \in \omega_\sigma$, and $\lambda_i \neq \lambda_j$ if $i \neq j$.

We consider $P\Gamma_{n-1}/K$. Bearing the inductive hypothesis in mind, we can find a finite subset $\omega'_\sigma \subset \omega_0$ such that every periodic J_σ-submodule H of $P\Gamma_{n-1}/K$ belongs to $\mathfrak{A}(J_\sigma, \omega_\sigma)$ for $\sigma \in Z(R)$. Without loss of generality, we may assume that $\lambda_i \notin \omega'_\sigma$. Then $\bar{w}(x_i) \in K^{m_i}$, where $m_i = \bar{l}(x_i)$. For suppose that $\bar{w}(x_i) \notin K^{m_i}$. But $\bar{w}(x_i)\lambda_i \in K^{m_i}$; that is, $\lambda_i \in \omega'_\sigma$, and this contradiction shows that $\bar{w}(x_i) \in K^{m_i}$. It is easy to verify that M is an almost K-closed Θ-set, where Θ is the property 1) and 2) above. Applying Lemma 14, we can find a infinite set M' of elements of $P\Gamma_{n-1}$ having the property Θ. But then we arrive at a contradiction to the inductive hypothesis, and the lemma is proved.

LEMMA 17. *Let (G, Γ) be a cyclic representation of a polycyclic group Γ, where G is a vector space over an absolutely algebraic field of prime characteristic p. Then there is a finite set $\omega \subset \omega_0$ such that for every $\sigma \in Z(R)$ each periodic J_σ-submodule $G_0 \subset G$ belongs to $\mathfrak{A}(J_\sigma, \omega)$.*

The proof is completely analogous to that of Lemma 16.

THEOREM 2. *Let (G, Γ) be a cyclic representation of a polycyclic group Γ, where G is a vector space over an arbitrary field p. Then for every $\sigma \in Z(R)$ there is a finite set $\omega_\sigma \subset \omega_0$ such that every minimal pure J_σ-submodule $G_0 \subset G$ containing $x_0 \in G$ belongs to $\mathfrak{A}(J_\sigma, \omega_\sigma)$.*

PROOF. We apply induction on the length of the series (1). If this length is 1, the assertion is obvious. Suppose that the theorem holds for Γ_{n-1}, but not for Γ. This means that there exists a cyclic representation (G, Γ) in which for some $\sigma \in Z(R)$ there is an element x_0 in G such that the minimal pure J_σ-submodule $G_0 \subset G$ containing x_0 does not belong to $\mathfrak{A}(J_\sigma, \omega')$ for any finite subset $\omega' \subset \omega_0$. Taking Lemma 16 into account, we may consider the case when the J_σ-module G_0 is torsion-free.

Without loss of generality we may assume that $G = P\Gamma/K$, where K is a right ideal in $P\Gamma$. By hypothesis, $P\Gamma$ then has a sequence of elements $\{x_i\}$ with the following properties:

1) $x_i \notin K$, $i = 1, 2, \ldots$;

2) $x_i\lambda_i - x_i \in K$, $\lambda_i \in \omega_0$, and $\lambda_i \neq \lambda_j$ if $i \neq j$.

Considering $P\Gamma_{n-1}/K^m$ and bearing the inductive hypothesis in mind, we can find a finite subset $\omega'_\sigma \subset \omega_0$ such that every minimal pure module containing an arbitrary element of $P\Gamma_{n-1}/K^m$ belongs to $\mathfrak{A}(J_\sigma, \omega'_\sigma)$. Without loss of generality we may assume that $\lambda_i \notin \omega'_\sigma$. Hence and from Lemma 16 it follows at once that $\bar{l}(x_i) \leqslant \bar{l}(x_0)$ for almost all x_i (that is, with finitely many exceptions). If $\bar{l}(x_i) = \bar{l}(x_0)$ for infinitely many x_i, then by considering $P\Gamma_{n-1}/K^m$ ($m = \bar{l}(x_0)$) and applying the inductive hypothesis, we come at once to a contradiction. This means that $\bar{l}(x_0) > \bar{l}(x_i)$ for almost all x_i. But then $\bar{w}(x_0) \in K^m$, and consequently, taking Lemma 3 into account, we can find an element $x' \in K$ such that $\bar{l}(x_0 - x') < m$. Applying induction on m and repeating the preceding arguments, we easily arrive at a contradiction to the inductive hypothesis, and so the theorem is proved.

The proof of the next theorem is quite similar to that of Theorem 2.

THEOREM 3. *Let (G, Γ) be a cyclic representation of a polycyclic group Γ, where G is a vector space over an absolutely algebraic field of prime characteristic p. Then there is a finite subset $\omega_1 \subset \omega_0$ such that for every $\sigma \in Z(R)$ each minimal pure*

J_σ-submodule $G_0 \subset G$ containing $x_0 \in G$ belongs to $\mathfrak{A}(J_\sigma, \omega_1)$.

THEOREM 4. *Let (G, Γ) be a cyclic representation of a polycyclic group Γ over an arbitrary field P. Then for every $\sigma \in Z(R)$ each periodic J_σ-submodule G_0 belongs to $\mathfrak{A}(J_\sigma, \omega, k)$, where $\omega \subset \omega_0$ and k is a natural number.*

PROOF. We apply induction on the length of (1). For $n = 1$ the assertion of the theorem is obvious. Suppose that the theorem holds for Γ_{n-1}, but not for Γ. This means that there is a cyclic representation (G, Γ) in which for some $\sigma \in Z(R)$ and $\omega \subset \omega_0$ there is a periodic J_σ-submodule $G_0 \subset G$ such that $G_0 \notin \mathfrak{A}(J_\sigma, \omega, k)$ for any natural number k. Without loss of generality we may assume that $G = P\Gamma/K$, where K is a right ideal in $P\Gamma$. By hypothesis, $P\Gamma$ then has a subsequence M of elements x_i with the following property Θ:

$$x_i\lambda^i \in K, \qquad x_i\lambda^{i-1} \notin K, \qquad \lambda_i \in \omega.$$

It is easy to verify that M is an almost K-closed Θ-set closely connected with K. According to Lemma 14, then there are elements $y_i \in K$ such that for almost all $x_i \in M$

$$x_i - y_i \in P\Gamma_{n-1}, \qquad (x_i - y_i)\lambda^i \in K, \qquad (x_i - y_i)\lambda^{i-1} \notin K, \qquad \lambda_i \in \omega.$$

So we have come to a contradiction to the inductive hypothesis, and the theorem is proved.

THEOREM 5. *Let (G, Γ) be a cyclic representation of a polycyclic group Γ over an arbitrary field P. Then for every $\sigma \in Z(R)$ each minimal pure J_σ-submodule G_0 containing x_0 belongs to $\mathfrak{A}(J_\sigma, \omega, k)$.*

The proof is completely analogous to that of Theorems 2 and 4.

All the subsequent theorems of this section are proved on the same lines as applied above; therefore we omit the proofs.

THEOREM 6. *Let (G, Γ) be a cyclic representation of a polycyclic group Γ over an arbitrary field P. Then there is a natural number k such that for every $\sigma \in Z(R)$ each J_σ-submodule $G_0 \subset G$ belongs to $\mathfrak{A}(J_\sigma, \omega, k)$.*

THEOREM 7. *Let (G, Γ) be a cyclic representation of a polycyclic group Γ over an arbitrary field P. Then for every element $\sigma \in Z(R)$ there is a finite set $\omega_\sigma \subset \omega_0$ and a natural number k such that a J_σ-submodule $G_0 \subset G$ belongs to $\mathfrak{A}(J_\sigma, \omega_\sigma, k)$.*

The next theorem follows from Theorems 3 and 5.

THEOREM 8. *Let (G, Γ) be a cyclic representation of a polycyclic group Γ over an absolutely algebraic field P of prime characteristic p. Then there exists a finite set $\omega_1 \subset \omega_0$ and a natural number k such that for every $\sigma \in Z(R)$ each J_σ-submodule $G_0 \subset G$ belongs to $\mathfrak{A}(J_\sigma, \omega_1, k)$.*

Let (G, Γ) be a representation of a polycyclic group Γ. We write

$$G(\sigma_0, \lambda_0) = \{x\colon\ x \in G, x = y(\sigma_0 - \lambda_0 e), y \in G\} \qquad (\sigma_0 \in Z(R), \lambda_0 \in P)$$

and we denote by $G(\sigma_1, \ldots, \sigma_m, \lambda_1, \ldots, \lambda_m)$ the linear hull of the spaces $G(\sigma_1, \lambda_1)$, $\ldots$, $G(\sigma_m, \lambda_m)$. Clearly, every $G(\sigma_i, \lambda_i)$ is a $PZ(R)$-submodule of G.

THEOREM 9. *Let (G, Γ) be a cyclic representation of a polycyclic group Γ, and let $\sigma_1, \ldots, \sigma_m \in Z(R)$ and $\lambda_i \in P$. Then every minimal pure J_σ-submodule of the $PZ(R)$-module $G/G(\sigma_1, \ldots, \sigma_m, \lambda_1, \ldots, \lambda_m)$ containing x_0 belongs to $\mathfrak{A}(J_\sigma, \omega_1, k)$, where $\omega_1 \subset \omega_0$ and k does not depend on the choice of $\sigma_1, \ldots, \sigma_m$, but only on m and $\sigma \in Z(R)$.*

THEOREM 10. *Let (G, Γ) be a cyclic representation of a polycyclic group Γ. Then G has a proper maximal $PZ(R)$-submodule.*

PROOF. Let $\sigma_1, \ldots, \sigma_m$ be generators of $Z(R)$. According to Theorem 7 there is an element $\lambda_1 \in P$ for which $G(\sigma_1 - \lambda_1) \neq G$. Suppose that we have already found numbers $\lambda_1, \ldots, \lambda_{m-1}$ such that

$$G \neq G(\sigma_1, \ldots, \sigma_{m-1}, \lambda_1, \ldots, \lambda_{m-1}).$$

By Theorem 9 we can then find a number λ_m such that

$$[G/G(\sigma_1, \ldots, \sigma_{m-1}, \lambda_1, \ldots, \lambda_{m\ 1}] (\sigma_m - \lambda_m \bar{e}) \neq G/G(\sigma_1, \ldots, \sigma_{m-1}, \lambda_1, \ldots, \lambda_{m-1}).$$

But then $G \neq G(\sigma_1, \ldots, \sigma_m, \lambda_1, \ldots, \lambda_m)$. Since in the quotient

$$G/G(\sigma_1, \ldots, \sigma_m, \lambda_1, \ldots, \lambda_m)$$

every $\sigma \in Z(R)$ induces an algebraic automorphism, G has a proper maximal $PZ(R)$-submodule (Lemma 9). This proves the theorem.

§4. Proof of the main theorem (positive solution of a problem of P. Hall [4])

In this section we consider only representations of groups over an absolutely algebraic field P of prime characteristic p.

THEOREM 11. *Every irreducible representation of a polycyclic group Γ by automorphisms of a vector space G over an absolutely algebraic field P of prime characteristic p is finite-dimensional.*

PROOF. Let (G, Γ) be a faithful irreducible representation of Γ. Then $xP\Gamma = G$ for every $x \in G$. Let $Q = \{g \in P\Gamma\colon\ xg = 0\}$. Then Q is a right ideal in $P\Gamma$, and the $P\Gamma$-modules G and $P\Gamma/Q$ are isomorphic. According to Theorem 10, G has a maximal $Z(R)$-invariant subspace H. Since $Z(R)$ is a finitely generated Abelian group, the quotient space G/H is finite-dimensional, and this means that every element $\sigma \in Z(R)$ induces in G/H an algebraic endomorphism. From the fact that P is an absolutely algebraic field of prime characteristic p it follows that for every σ there is a natural number n such that σ^n induces the identity automorphism in G/H. Since $Z(R)$ is a finitely generated Abelian group, there is a natural number k such that σ^k lies in the kernel of the representation $(G/H, Z(R))$ for every $\sigma \in Z(R)$. We write $Z^k(R)$ for the subgroup of $Z(R)$ generated by the elements of the form σ^k, where $\sigma \in Z(R)$. It is easy to see that $Z^k(R)$ is a characteristic subgroup of $Z(R)$, and therefore a normal subgroup of Γ. Bearing in mind that (G, Γ) is an irreducible representation of Γ, we see that G, by Remak's theorem, is a subdirect sum of subspaces H_γ, where γ ranges over a transversal to $Z(R)$ in Γ. Since $Z^k(R)$ lies in the kernel of $(G/H_\gamma, Z(R))$ for every member γ of the transversal, it lies in the kernel of (G, Γ). Therefore, $Z(R)$ is a finite group; but then Γ itself is also finite. Therefore, the vector space G is finite-dimensional, and the theorem is proved.

BIBLIOGRAPHY

1. Irving Kaplansky, *Groups with representations of bounded degree,* Canad. J. Math. **1** (1949), 105–112.

2. A. I. Mal'cev, *On the faithful representation of infinite groups by matrices,* Mat. Sb. **8 (50)** (1940), 405–422; English transl. in Amer. Math. Soc. Transl. (2) **45** (1965).

3. S. A. Amitsur, *Groups with representations of bounded degree.* II, Illinois J. Math. **5** (1961), 198–205.

4. Philip Hall, *On the finiteness of certain soluble groups,* Proc. London Math. Soc. (3) **9** (1959), 595–622.

5. ———, *Finiteness conditions for soluble groups,* Proc. London Math. Soc. (3) **4** (1954), 419–436.

6. Kurt A. Hirsch, *On infinite soluble groups.* III, Proc. London Math. Soc. (2) **49** (1946), 184–194.

7. B. I. Plotkin, *Groups of automorphisms of algebraic systems,* "Nauka", Moscow, 1966; English transl., Wolters-Noordhoff, Groningen, 1972.

8. E. M. Levič [Ļevics], *On a theorem of P. Hall,* Inter-University Sci. Sympos. General Algebra, Tartu. Gos. Univ., Tartu, 1966, pp. 63–65. (Russian) MR **34** #2721.

9. ———, *Modules over the ring of polynomials in one endomorphism,* Latvian Math. Yearbook, no. 2, "Zinatne", Riga, 1966, pp. 127–174. (Russian)

10. B. I. Plotkin, *Radical groups,* Mat. Sb. **37 (79)** (1955), 507–526; English transl. in Amer. Math. Soc. Transl. (2) **17** (1961).

11. ———, *Radical and semisimple groups,* Trudy Moskov. Mat. Obšč. **6** (1957), 299–336. (Russian)

12. Nathan Jacobson, *Structure of rings,* Amer. Math. Soc., Providence, R. I., 1956.

13. E. M. Levič [Ļevičs], *On a problem of P. Hall,* Dokl. Akad. Nauk SSSR **188** (1969), 1241–1243; English transl. in Soviet Math. Dokl. **10** (1969).

14. A. E. Zalesskiĭ, *The construction of a finitely generated module over the group algebra of a polycyclic group,* Dokl. Akad. Nauk BSSR **14** (1970), 977–980. (Russian)

Translated by K. A. HIRSCH

Amer. Math. Soc. Transl.
(2) Vol. 113, 1979

ON TORSION-FREE GROUPS WHOSE IRREDUCIBLE REPRESENTATIONS OVER SOME FIELD ARE ALL FINITE-DIMENSIONAL*

E. M. LEVIČ [ĻEVIČS]

One of the interesting problems in the theory of group representations is the problem of studying properties of a group as they depend on properties of its representations. In particular, one can ask: what properties must a group have so that all its irreducible representations over some field are finite-dimensional? It is known [1]–[4] that for all irreducible representations of a group Γ over any field to be finite-dimensional it is sufficient that Γ has a finitely generated Abelian normal subgroup of finite index. It is natural to make the following conjecture: for a torsion-free group Γ to have only finite-dimensional irreducible representations over any field it is necessary and sufficient that Γ contains a finitely generated Abelian normal subgroup of finite index.

When the group is not torsion-free, this conjecture does not hold, in general, because an infinite Abelian group of exponent p has only finite-dimensional irreducible representations [5]. In support of the conjecture we can point to the theorem of P. Hall in [4], which asserts that a polycyclic group Γ having no finitely generated abelian normal subgroups of finite index has an infinite-dimensional irreducible representation over any field other than an absolute algebraic field of prime characteristic p.

The aim of the present note is a proof of the conjecture for the case of locally nilpotent torsion-free groups. In addition, we investigate problems concerning the structure of groups for which all irreducible representations over some field are finite-dimensional.

We say that a group Γ has the property $(A)_P$ if all its irreducible representations over a field P are finite-dimensional; that it has property (A) if all its irreducible representations over any field are finite-dimensional; and that it has property (B) if it is a finite extension of a finitely-generated Abelian group.

PROPOSITION 1. *Let Γ be a group, and H a subgroup of Γ having an infinite-dimensional irreducible representation over a field P. Then Γ itself also has an infinite-dimensional irreducible representation over P.*

AMS (MOS) subject classifications (1970). Primary 20E35; Secondary 20E25, 20K05, 20F25.

*Translation of Certain Questions of Group Theory (Proc. Riga Algebra Sem., No. 2, 1969/70), Latvijas Valsts Univ. Zinātn. Raksti 151 (1971), 125–139.

PROOF. Let (G, H) be an infinite-dimensional irreducible representation of H over P. Then the group algebra PH of H over P contains a maximal right ideal J_H such that the representations (G, H) and $(PH/J_H, H)$ are equivalent. We denote by $\bar{J}_H$ the right ideal in the group algebra $P\Gamma$ of Γ over P generated by J_H, and we claim that $J_H \neq P\Gamma$. We can represent $P\Gamma$ as a direct sum of vector spaces over P:

$$P\Gamma = \sum_{J \in M} PHg_\nu,$$

where g_ν is a representative of the right coset Hg_ν of H in Γ, g_α and g_β for $\alpha \neq \beta$ do not lie in the same coset, and $\{g_\nu, J \in M\}$is a right transversal. Then

$$\bar{J}_H = \sum_{J \in M} J_H g_\nu.$$

But the unit element of Γ does not belong to $\bar{J}_H$, and consequently $\bar{J}_H \neq P\Gamma$. If J is a maximal right ideal of $P\Gamma$ containing $\bar{J}_H$, then $J \cap PH = J_H$, because $(PH/J_H, H)$ is irreducible. But then $(P\Gamma/J, \Gamma)$ is an infinite-dimensional representation over P, because $P\Gamma/J$ contains a subspace isomorphic to the infinite-dimensional space PH/J_H.

COROLLARY 1. *If a group Γ has the property $(\mathrm{A})_P$, then so has every subgroup and factor group.*

This follows immediately from Proposition 1 and the perfectly obvious remark that if a factor group of some group has an infinite-dimensional irreducible representation over some field, then the group itself has an irreducible infinite-dimensional representation over the same field.

We denote by $\mathfrak{A}_P$ the class of all groups having the property $(\mathrm{A})_P$, and by $\mathfrak{A}$ the class of all groups having the property (A).

The following proposition holds (compare with Theorem 5 of [6]).

THEOREM 1. *If a finitely generated group Γ lies in $\mathfrak{A}_P$, where P is an arbitrary field of characteristic other than 2, then it is residually finite.*

PROOF. We claim that for every element $g \in \Gamma$ there is a normal subgroup H_g such that $g \notin H_g$, and Γ/H_g is a finite group. Observe that in the group algebra of the cyclic group $\{g\}$ over P we can always find for any element $h \in \{g\}$ a maximal right ideal J_h that does not contain the element $h - e$. We denote by $\bar{J}_g$ a maximal right ideal of $P\Gamma$ containing J_g. Just as in the proof of Proposition 1, we can show that $\bar{J}_g \neq P\Gamma$. Since $(P\Gamma/\bar{J}_g, \Gamma)$ is an irreducible representation, $P\Gamma\,\bar{J}_g$ is an infnite-dimensional vector space over P, because $\Gamma \in \mathfrak{A}_P$. We write $\bar{H}_g$ for the kernel of this representation. Then $\Gamma/\bar{H}_g$ can be regarded as a finitely generated matrix group over P, and the image of g in this factor group is nontrivial. According to [7], $\Gamma/\bar{H}_g$ is residually finite. Consequently, we can find a normal subgroup H_g of Γ such that $g \notin H_g$ and Γ/H_g is finite.

This proves the theorem.

Applying Remak's theorem, we obtain immediately from Theorem 1 the following proposition:

COROLLARY 2. *If a finitely generated group* Γ *lies in* $\mathfrak{A}_P$, *where* P *is an arbitrary field of characteristic other than* 2, *then* Γ *is a subdirect product of finite groups.*

If P is a finite field of characteristic $p \neq 2$, then Theorem 1 can be strengthened by omitting the condition that Γ is finitely generated, because every matrix group over a finite field is finite. Thus, we have the following assertion:

COROLLARY 3. *If* $\Gamma \in \mathfrak{A}_P$, *where* P *is a finite field of characteristic* $p \neq 2$, *then* Γ *is residually finite.*

We denote by fRad$P\Gamma$ the intersection of all kernels of the finite-dimensional representations of the group algebra $P\Gamma$ of Γ over P.

The following proposition holds (compare with Theorem 3 of [6]):

PROPOSITION 2. *If* $\Gamma \in \mathfrak{A}_P$, *where* P *is any field of characteristic other than* 2, *then* fRad$P\Gamma = \{0\}$,

PROOF. Let $h \in$ fRad$P\Gamma$. Then $h = \Sigma_1^n \alpha_i g_i$, where $g_i \in \Gamma$. From Theorem 1 it follows that Γ is residually finite. We denote by H_g a normal subgroup of Γ of finite index that does not contain g_i, and we write $H = \bigcap_1^n H_{g_i}$. Clearly H is a normal subgroup of finite index, and $g_i \notin H$, $i = 1, \ldots, n$. If $\Delta(H)$ is the two-sided ideal in $P\Gamma$ generated by all the elements of the form $g - e$, where $g \in H$, then $P\Gamma/\Delta(H)$ is isomorphic to the group algebra $P(\Gamma/H)$ of Γ/H over P. Hence it follows immediately that $\bar{h} = \Sigma_1^n \alpha_i \bar{g}_i \neq 0$, where $\bar{g}_i$ is the image of g_i under the natural homomorphism of $P\Gamma$ onto $P\Gamma/\Delta(H)$, because $g_i \notin H$. But $(p\Gamma/\Delta(H), \Gamma)$ is a finite-dimensional representation, because Γ/H is a finite group. So we have come to a contradiction, and the proposition is proved.

PROPOSITION 3. *If all the irreducible representations of a group* Γ *are finite-dimensional over the field of rational numbers and their dimensions are bounded, then* Γ *is a periodic group of finite exponent.*

PROOF. First of all, we observe that over the field of rational numbers a torsion-free cyclic group has irreducible representations of every sufficiently high dimension. Hence, and from Corollary 1, it follows that Γ is periodic. From Lemma 1 below and Jordan's theorem (see [8]) it follows that Γ has an Abelian normal subgroup H such that Γ/H is of finite exponent. Applying Lemma 2 below, we find that H, and consequently also Γ, has finite exponent. This proves the proposition.

LEMMA 1. *If all the irreducible representations of a group* Γ *over the field* Q *of rational numbers are finite-dimensional and their dimensions are bounded, then* Γ *is a subdirect product of matrix groups of bounded dimensions.*

The proof of the lemma follows easily from that of Theorem 1. The latter asserts that Γ is residually an irreducible matrix group. But by the conditions of the lemma, all irreducible representations of Γ are of bounded dimension. Consequently, Γ is a subdirect product of matrix groups of bounded dimensions.

LEMMA 2. *If an Abelian group Γ lies in $\mathfrak{A}_Q$, then its periodic part $\Pi(\Gamma)$ is of finite exponent, and $\Gamma/\Pi(\Gamma)$ is a finitely generated torsion-free Abelian group.*

PROOF. Suppose that $\Pi(\Gamma)$ is not of finite exponent. Then it splits into a direct sum of primary groups relative to the various prime numbers p (see [9], §19). This means that $\Pi(\Gamma)$ contains a normal subgroup A of one of the following forms:

a) A is a group of type p^∞;

b) A is a direct product of infinitely many cyclic groups for various primes p;

c) A is a direct product of infinitely many cyclic groups of order p^n, where $n = 1, 2, \dots$.

We analyse these cases in turn.

a) A is a group of type p^∞. There is an isomorphic embedding μ of A into the multiplicative group of the field of complex numbers such that to every element of A there corresponds a root of unity whose degree is a power of p. Let G_C be a one-dimensional vector space over the field C of complex numbers. We define a representation of A relative to G_C: $g \cdot a = a^\mu g$ for all $g \in G_C$ and $a \in A$, where a^μ is the image of a under μ. We can regard G_C as an infinite-dimensional vector space G_Q over the field Q of rational numbers. The representation of A relative to G_C induces a representation of A relative to G_Q. In G_Q we choose an A-composition factor $F \supset H$; that is, F and H are A-invariant subspaces and $(F/H, A)$ is an irreducible representation. Suppose that $(F/H, A/\Sigma)$ is a one-dimensional representation, where Σ is the kernel of $(F/H, A)$. Then, by Schur's theorem (see [10], Chapter IX, §2.1), A/Σ is a finite group. But A is a group of type p^∞; therefore $A = \Sigma$. This means that F/H is a one-dimensional factor; that is, $F = H \oplus \{g_0\}$, where $g_0 \in F \setminus H$. Since $g_0 \cdot a - g_0 \in H$, we see that $g_0 \cdot (a-1) \in H$ for every $a \in A$, and if a is an element of order p^n, then

$$-\sum_{i=1}^{p^n} g_0 \cdot (a^i - 1) = p^n g_0 \in H;$$

that is, $g_0 \in H$ and $F = H$. So we have come to a contradiction, from which it follows that $(F/H, A)$ is an infinite-dimensional representation, and this means that $A \notin \mathfrak{A}_Q$. But then also $\Gamma \notin \mathfrak{A}_Q$ (Corollary 1).

b) Let A be a direct product of infinitely many cyclic groups for various primes p. By arguments similar to the above, we obtain a representation $(F/H, A/\Sigma)$, where A/Σ is a finite group. This means that Σ contains a cyclic group A_1 of order p_1. Now $g \cdot (a-1) \in H$ for every $a \in \Sigma$ and every $g \in F \setminus H$, and hence

$$-\sum_{i=1}^{p_1} g \cdot (a^i - 1) = p_1 g \in H;$$

that is, $g \in H$, provided that a is a generator of A_1. But for g we can take an arbitrary element of F. Consequently, $(F/H, A/\Sigma)$ is an infinite-dimensional irreducible representation, and this means that $A \notin \mathfrak{A}_Q$; that is, $\Gamma \notin \mathfrak{A}_Q$ (Corollary 1).

c) Let A be a direct product of infinitely many cyclic groups A_n of order p^n, where $n = 1, 2, \ldots$. We denote by μ_n an isomorphic embedding of A_n into the multiplicative group of the field of complex numbers C under which to every element $a \in A_n$ of order p^m there corresponds a nontrivial root of unity of degree p^m. Repeating the preceding arguments in their entirety, we easily verify that A has an infinite-dimensional irreducible representation over the field of rational numbers.

Thus, from the cases analyzed so far it follows that if $\Pi(\Gamma)$ contains a normal subgroup of one of the types a)–c), then $\Pi(\Gamma)$ has an infinite-dimensional irreducible representation over Q; that is, $\Gamma \notin \mathfrak{A}_Q$, which contradicts the condition of Lemma 1. Therefore, $\Pi(\Gamma)$ is of finite exponent.

The proof of the lemma will be complete if we can show that $\Gamma/\Pi(\Gamma)$ is a fintely-generated Abelian group. For if $\Gamma/\Pi(\Gamma)$ is a torsion-free Abelian group and not finitely generated, then it contains a normal subgroup A of one of the following forms:

d) A is a free Abelian group (with countably many generators) (this case easily reduces to a));

e) A contains a finitely generated free Abelian group H and A/H is a periodic group, but not of finite exponent.

Since in both cases $\Gamma/\Pi(\Gamma)$ has a normal subgroup or a factor group H such that $H \notin \mathfrak{A}_Q$, we also have $\Gamma \notin \mathfrak{A}_Q$.

So we have come to a contradiction, which proves the lemma.

COROLLARY 3. *An Abelian group Γ belongs to the class $\mathfrak{A}$ if and only if it is a direct product of a finitely generated torsion-free Abelian group and a periodic Abelian group of bounded exponent.*

PROPOSITION 4. *A periodic Abelian group Γ belongs to the class $\mathfrak{A}_P$, where the field P contains all roots of unity.*

PROOF. Let (G, Γ) be an irreducible representation of Γ. Then every element $\sigma \in \Gamma$ induces on G a locally algebraic automorphism. Applying Lemma 4 in [12], we obtain the required assertion.

PROPOSITION 5. *An Abelian group Γ containing a free Abelian group H does not belong to the class $\mathfrak{A}_P$, where the field P has the same cardinality as H.*

PROOF. We claim that $H \notin \mathfrak{A}_P$. We denote by R the field of rational functions of a single variable over P, and we regard R as a vector space R_P over P, clearly of infinite dimension. If Σ is the multiplicative group of R, then the representation (R_P, Σ) evidently is irreducible. Since H and Σ have the same cardinality, there is an isomorphism of H onto Σ. Thus, H has an infinite-dimensional irreducible representation over P; that is, $H \notin \mathfrak{A}_P$. But then also $\Gamma \notin \mathfrak{A}_P$, which proves the proposition.

PROPOSITION 6. *An Abelian group Γ containing a free Abelian group H for which Γ/H is periodic belongs to the class $\mathfrak{A}_P$, where P is any algebraically closed field whose cardinality exceeds that of H.*

PROOF. Let (G, Γ) be an irreducible representation of Γ. Then, by Schur's lemma ([13], Chapter II, §1), $P\Gamma$ has a maximal ideal J such that $P\Gamma/J$ is a field. Since the dimension of PH, as a vector space over P, is less than the cardinality of P, we see that $P\Gamma/J$ is an algebraic extension of P. Hence it follows that every $\sigma \in \Gamma$ induces on G an algebraic automorphism. Applying Lemma 4 of [12], we obtain the required assertion.

PROPOSITION 7. *A nilpotent torsion-free group Γ belongs to the class $\mathfrak{A}_P$, where P is any field other than an absolute algebraic field of prime characteristic p, if and only if it is Abelian and of cardinality less than that of P.*

PROOF. In one direction the assertion of the theorem is obvious. If Γ is not an Abelian group, then it has a subgroup that does not contain Abelian normal subgroups of finite index. But then, by a theorem of P. Hall in [4], the group has an infinite-dimensional irreducible representation over P. Consequently it is torsion-free Abelian. Applying Propositions 5 and 6, we obtain the final part of the proposi tion.

COROLLARY 4. *Suppose that a locally nilpotent torsion-free group Γ lies in $\mathfrak{A}_P$, where P is any field other than an absolutely algebraic field of prime characteristic p. Then Γ is Abelian and of cardinality less than that of P.*

PROOF. From Proposition 7 it follows that every finitely generated subgroup of Γ is Abelian, by Corollary 1. But then the whole group Γ is Abelian. Applying Propositions 5 and 6, we come to the required assertion.

COROLLARY 5. *A locally nilpotent torsion-free group Γ belongs to the class $\mathfrak{A}$ if and only if it is finitely generated and Abelian.*

PROOF. In one direction the assertion is obvious. The converse follows from Corollary 4 and Lemma 2.

PROPOSITION 8. *There exist periodic nilpotent groups of class 2 and of bounded exponent that do not belong to $\mathfrak{A}_P$ for any field P.*

PROOF. S. N. Černikov has constructed an example of an infinite periodic metanilpotent group with a cyclic center of order p. Every nontrivial irreducible representation of the center induces, as is easy to see, an infinite-dimensional representation of Černikov's group, and this proves the proposition.

COROLLARY 6. *If a periodic nilpotent group* Γ *belongs to* $\mathfrak{A}_Q$, *then it is of finite exponent.*

PROOF. Suppose that Γ has a finite central series $\Gamma = \Gamma_0 \supset \Gamma_1 \supset \cdots \supset \Gamma_n = \{e\}$. From Lemma 2 it follows that every factor of this series is of finite exponent; consequently, so is the whole group Γ.

COROLLARY 7. *If a periodic soluble group* Γ *belongs to* $\mathfrak{A}_Q$, *then it is of finite exponent.*

PROPOSITION 9. *Let* Γ *be a finitely generated torsion-free soluble group belonging to* $\mathfrak{A}_Q$. Then Γ *has the property* (B).

PROOF, We denote by $R(\Gamma)$ the Hirsch-Plotkin radical of Γ. By Corollary 5, $R(\Gamma)$ is a finitely generated torsion-free Abelian group. By a theorem of Plotkin ([10], Chapter V, §3.4) $R(\Gamma)$ bounds Γ; that is, the centralizer $\mathfrak{Z}(R)$ of the radical in Γ is contained in $R(\Gamma)$. But then the factor group $\Gamma/\mathfrak{Z}(R)$ can be regarded as a group of automorphisms of $R(\Gamma)$. By a theorem of Mal'cev ([14]; see also [10], Chapter IX, §2.3), $\Gamma/\mathfrak{Z}(R)$ is a polycyclic group. Since $\mathfrak{Z}(R)$ is finitely generated and Abelian, Γ is polycyclic. Applying once more the theorem of P. Hall [4], we see that Γ has the property (B).

PROPOSITION 10. *The discrete wreath product of an Abelian group* Γ *of exponent p and an infinite Abelian group H of exponent p over an algebraically closed field has an infinite-dimensional irreducible representation.*

PROOF. It is sufficient to prove that the discrete wreath product of a cyclic group $\{q\}$ of order p and an infinite Abelian group H of exponent p has an infinite-dimensional irreducible representation over an algebraically closed field P. Let $\mathfrak{G} = \{e\} \sim H$, $P\mathfrak{G}$ the group algebra of $\mathfrak{G}$ over P and $P\{\mathfrak{g}\}$ the group algebra of $\{\mathfrak{g}\}$ over P. We denote by J_0 the ideal in $P\{\mathfrak{g}\}$ generated by $g - \lambda$, where g is a generator of $\{\mathfrak{g}\}$ and λ a pth root of unity, and by J a maximal right ideal in $P\mathfrak{G}$. Then $(P\mathfrak{G}/J, \mathfrak{G})$ is an irreducible representation of $\mathfrak{G}$ over P. We claim that it is infinite-dimensional. Suppose, to the contrary, that $(PG/J, \mathfrak{G})$ is finite-dimensional. The kernel F of this representation evidently intersects H nontrivially. But then it follows from [15] that F contains the basis group $\{\mathfrak{g}\}^H$, and hence also $\{\mathfrak{g}\}$. We have come to a contradiction, which proves the proposition.

We denote by $\mathfrak{B}$ the class of all groups that are generated by cyclic normal subgroups.

LEMMA 3. *If a group* Γ *is torsion-free and not finitely generated, and if* $\Gamma \in \mathfrak{B}$, *then* Γ *has an Abelian normal subgroup that cannot be finitely generated.*

PROOF. Let $\{g_0\}$ be an arbitrary cyclic normal subgroup of Γ generated by g_0. Then every element $h \in \Gamma$ induces in $\{g_0\}$ an automorphism of order 2, because $\{g_0\}$ is torsion-free. Consequently, the subgroup H of Γ generated by the squares of all elements of Γ lies in the centralizer of $\{g_0\}$ in Γ. Since Γ is generated by cyclic normal subgroups and $\{g_0\}$ is an arbitrary cyclic normal subgroup, it follows that H lies in the center of Γ and is, therefore, an Abelian group. We denote by $R(\Gamma)$ the Hirsch-Plotkin radical of Γ. Clearly, $H \subset R(\Gamma)$. It is easy to see that Γ/H is Abelian, because it is generated by cyclic normal subgroups of order 2. Hence Γ is soluble. Therefore, $R(\Gamma)$ bounds Γ. If $R(\Gamma)$ is finitely generated Abelian, then Γ is polycyclic (see the proof of Proposition 9). But Γ is not finitely generated, nor is, therefore, $R(\Gamma)$. This proves the lemma.

LEMMA 4. *Suppose that Γ is a locally soluble torsion-free group and that every finitely generated subgroup of it has the property* (B). *If Γ is not finitely generated, then it has a torsion-free Abelian normal subgroup that is not finitely generated.*

PROOF. Let Γ_0 be a finitely generated subgroup of Γ. By the condition of the lemma, Γ_0 has a finitely generated Abelian normal subgroup $\overline{\Gamma}_0$ of finite index in Γ_0. We denote by F_0 the subgroup generated by $\overline{\Gamma}$ and $g_0 \in \Gamma$. Then F_0 also has a finitely generated Abelian normal subgroup $\overline{F}_0$ of finite index. There is an n such that $g_0^n \in \overline{F}_0$. We consider the subgroup $\widetilde{F}_0$ generated by g_0^n and $\overline{\Gamma}_0$. Let $h \in \overline{\Gamma}_0$. Since $g_0^n \in \overline{F}_0$, we see that $h^{-1}g_0^nh = g_0^\alpha f$, where $f \in \overline{\Gamma}_0 \cap \overline{F}_0$ and α is an integer. If $\alpha = 0$, then $h^{-1}g_0^nh \in \overline{\Gamma}_0$ and $h^{-k}g_0^nh^k \in \overline{\Gamma}_0$ for every k. Bearing in mind that the restriction of the inner automorphism induced by h to $\overline{F}_0$ is of finite order, we come to a contradiction. This means that $\alpha \neq 0$. Since $f \in \overline{\Gamma}_0$, we have $h^{-1}fh = f$. If $|\alpha| > 1$, then, using the facts that Γ is torsion-free and that the restriction of the inner automorphism induced by h to F_0 is of finite order, we obtain a contradiction. Therefore, $|\alpha| = 1$. Similarly the case $\alpha = 1$ and $f \neq e$ can be excluded. There remain two possibilities: either $\alpha = -1$, or $f = e$ and $\alpha = 1$. Hence it follows that in any case $h^{-2}g_0^nh^2 = g_0^n$. Consequently, the group H_0 generated by g_0^n and the square of $\overline{\Gamma}_0$ is Abelian. We denote by $\widetilde{F}_0^2$ the square of $\widetilde{F}_0$. Clearly $\widetilde{F}_0^2 \in H_0$.

But the element g_0 was chosen arbitrarily in Γ, and therefore for every element $g \in \Gamma$ the group constructed by the same principle as H_0 is Abelian.

We denote by F_1 the subgroup generated by H_0 and $g_1 \in \Gamma$. Then F_1 has a finitely generated Abelian normal subgroup $\overline{F}_1$ of finite index. There exists an n_1 such that $g_1^{n_1} \in \overline{F}_0$. We consider the subgroup $\widetilde{F}_1$ generated by H_0 and $g_1^{n_1}$. Bearing in mind the remark made above, we can claim that $\overline{\Gamma}_0^2$, the square of $\overline{\Gamma}_0$, commutes with the element $g_1^{n_1}$. By arguments similar to the ones above we can show that the element g_0^{2n} commutes with $g_1^{n_1}$. Therefore, $\widetilde{F}_0^2$ commutes with $g_1^{n_1}$. This means that the subgroup generated by $\widetilde{F}_0^2$ and $g_1^{n_1}$ is Abelian. So is the group $\widetilde{F}_1^2$ generated by $\widetilde{F}_0^2$ and $g_1^{2n_1}$.

Repeating this process, we obtain a sequence of Abelian groups, each embedded in the next:

$$\widetilde{F}_0^2 \subset \widetilde{F}_1^2 \subset \cdots \subset \widetilde{F}_n^2 \subset \cdots .$$

We denote by F the union of this sequence, and claim that F cannot be finitely generated. Suppose the contrary, namely that F has not more than n generators. Then we can find a subgroup $B \subset \Gamma$ having not more than m generators (where m is a sufficiently large number, which we determine later) such that it has a normal subgroup B_0 of finite index contained in F. Since F and B_0 are torsion-free Abelian groups, B_0 has not more than n generators. We denote by $R(B)$ the Hirsch-Plotkin radical of B. Now B_0 is contained in $R(B)$. Bearing in mind that B_0 is of finite index in B, we see that $R(B)$ is an Abelian group with not more than n generators. By the theorem of Plotkin quoted above, $R(B)$ bounds B; that is, the centralizer of $R(B)$ lies in $R(B)$, because Γ is a locally soluble torsion-free group. Since $B/R(B)$ is periodic, it can be regarded as a matrix group of order not exceeding n over the ring of integers. But it is standard knowledge that an integral periodic group of matrices has an order bounded above by a number that depends only on the order of the matrices. Choosing m larger than this number, we come to a contradiction. Consequently F cannot be finitely generated, and the lemma is proved.

LEMMA 5. *If a soluble torsion-free group Γ does not have an Abelian normal subgroup H such that Γ/H is periodic, then it has an infinite-dimensional irreducible representation over any field P other than an absolutely algebraic field of prime characteristic p.*

The proof is quite similar to that of Theorem 3.3 in P. Hall's paper [4].

THEOREM 2. *If a locally soluble torsion-free group Γ belongs to $\mathfrak{A}_P$, where P is a field other than an absolutely algebraic field of prime characteristic p, then it is a periodic extension of an Abelian group.*

COROLLARY 8. *If a locally soluble torsion-free group Γ belongs to $\mathfrak{A}_Q$, then it has the property* (B).

The theorem follows from Lemma 5 and the proof of Lemma 4, and Corollary 8 from Theorem 2 and Proposition 9.

BIBLIOGRAPHY

1. S. A. Amitsur, *Groups with representations of bounded degree.* II, Illinois J. Math. 5 (1961), 198–205.

2. A. I. Mal′cev, *On the faithful representation of infinite groups by matrices,* Mat. Sb. 8 (50) (1940), 405–422; English transl. in Amer. Math. Soc. Transl. (2) 45 (1965).

3. Irving Kaplansky, *Groups with representations of bounded degree,* Canad. J. Math. 1 (1949), 105–112.

4. Philip Hall, *On the finiteness of certain soluble groups,* Proc. London Math. Soc. (3) 9 (1959), 595–622.

5. S. D. Berman, *Group algebras of countable Abelian p-groups,* Publ. Math. Debrecen 14 (1967), 365–405.

6. D. S. Passman, *On groups with enough finite representations,* Proc. Amer. Math. Soc. 14 (1963), 782–787.

7. A. I. Tokarenko, *A remark on finitely generated linear groups,* Proc. Riga Sem. Algebra, Latv. Gos. Univ., Riga, 1969, pp. 280–281. (Russian)

8. Charles W. Curtis and Irving Reiner, *Representation theory of finite groups and associative algebras,* Interscience, New York, 1962.

9. A. G. Kuroš, *Theory of groups,* 3rd ed., "Nauka", Moscow, 1967; English transl. of 2nd ed., vols. I, II, Chelsea, New York, 1955, 1956, 1960.

10. B. I. Plotkin, *Groups of automorphisms of algebraic systems,* "Nauka", Moscow, 1966; English transl., Wolters-Noordhoff, Groningen, 1972.

11. ———, *Normal divisors which bound a group,* Mat. Sb. 53 (95) (1961), 343–352. (Russian)

12. E. M. Levič [Ļevičs], *Triangulability of groups,* Izv. Vysš. Učebn. Zaved. Matematika 1967, no. 7 (62), 70–77. (Russian)

13. Nathan Jacobson, *Structure of rings,* Amer. Math. Soc., Providence, R. I., 1956.

14. A. I. Mal'cev, *On certain classes of infinite solvable groups,* Mat. Sb. 28 (70) (1951), 567–588; English transl. in Amer. Math. Soc. Transl. (2) 2 (1956).

15. Peter M. Neumann, *On the structure of standard wreath products of groups,* Math. Z. 84 (1964), 343–373.

Translated by K. A. HIRSCH

Amer. Math. Soc. Transl.
(2) Vol. 113, 1979

SYMMETRIC GROUPOIDS*

A. I. MAL'CEV

A set of elements G is called a *groupoid* if every ordered pair of its elements a, b combine to form a uniquely determined element $c \in G$, called the product of a and b and denoted by ab. A groupoid is said to be *associative* if $x(yz) = (xy)z$ for any of its elements x, y, z. Associative groupoids are also called *associative systems*. The notions of subgroupoid, isomorphism and homomorphism are defined in the usual manner.

Single-valued mappings of an arbitrary set M into itself are called *transformations* or *substitutions* of the elements of that set. In addition, if A is some transformation of the set M and $m \in M$, then we denote by mA the image of the element m under the mapping A. The composition (or product) of the transformations A and B of the set M is defined, as usual, by the formula $m(AB) = (mA)B$. Moreover, the multiplication of transformations is associative, and so the set of all transformations of a set M is an associative groupoid, which will be called a *symmetric groupoid* and designated by C_M.

Besides the obvious significance of symmetric groupoids for the theory of transformations, their value in the general theory is determined by the fact that any associative groupoid is isomorphic to a subgroupoid of a suitably chosen symmetric groupoid.

In the present paper we consider the simplest characteristics of finite as well as infinite symmetric groupoids. In §1 we give an abstract property of symmetric groupoids, from which we deduce as a consequence the well-known fact that any automorphism of a symmetric groupoid is inner. In §2 we find all factor-groupoids of symmetric groupoids.

As is well known, factor-groupoids cannot be characterized by a system of a groupoid's elements; recall the instance of normal divisors in groups. But the simplest factor-groupoids are defined by means of ideals and normal subgroupoids, being the preimages of a zero and identity respectively (see §1). All normal subsystems of a finite symmetric groupoid C_M were found in Vorob'ev's note [2]. They were shown to coincide with the normal divisors of the symmetric *group* of transformations of the set M. However, up to now the general problem of finding all factor-groupoids of a symmetric groupoid has apparently remained unexamined. Since every factor-groupoid is uniquely determined by a corresponding equivalence relation, which is compatible with the operation of multiplication, the problem then consists in finding all such relations on a symmetric groupoid. The enumeration of these relations is given by the

AMS (MOS) subject classifications (1970). Primary 20L05.
*Translation of Mat. Sb. 31 (73)(1952), 136–151. MR 14, 349.

theorem in §2 of the present paper. It is shown that, except for the normal divisors of a symmetric group of transformations of the same degree, for the given groupoid it is necessary to take into account both the normal divisors of a symmetric group of lower degree and, for an infinite groupoid, moreover, to consider the sum of products of normal and ideal equivalence relations.

§1. Right zeros and automorphisms

The elements e and 0 are called the *identity* and *zero* of the groupoid G if $ex = xe = x$ and $x0 = 0x = 0$ for all x in G. An element n of a groupoid G is called a *right zero* if it satisfies the relation $xn = n$ for any $x \in G$. A groupoid containing a single right zero is obviously associative and so to within isomorphism is determined by the number of its elements.

THEOREM. *Let G be an arbitrary associative groupoid and N the set of all its right zeros. In order that G be isomorphic to a symmetric groupoid it is necessary and sufficient that*

1) *if $na = nb$ for all right zeros $n \in N$, then $a = b$;*

2) *for every single-valued mapping U of the set of right zeros N into itself there exists an element u in G with the property $nU = nu$ for all n in N.*

PROOF. *Necessity.* The right zeros of the symmetric groupoid C_M are the transformations mapping all elements of the set M to some single element of this set. In fact, let P_a denote the transformation sending all elements of M to $a \in M$, and let B be an arbitrary right zero in C_M. Then for arbitrary $m \in M$ and any fixed $c \in M$ we have

$$mB = mP_cB = cB = mP_{cB},$$

i.e., $B = P_{cB}$. Conversely, any transformation P_c is a right zero in C_M, since $m(BP_c) = (mB)P_c = c = mP_c$ $(B \in C_M)$.

A symmetric groupoid satisfies property 1), since if $P_aA = P_aB$ for all right zeros P_a, then for any $m \in M$ we have $mP_aA = mP_aB$ and so $aA = aB$, whence $A = B$.

Now let U represent any single-valued mapping of the set of all right zeros of C_M into itself. The right zeros of C_M are in one-to-one correspondence with the elements of the set M. Therefore the transformation U of the set of (right) zeros determines a certain transformation A of M, subject to the relation $P_mU = P_{mA}$ $(m \in M)$, whence

$$a \cdot P_mA = mA = aP_{mA} = a \cdot P_mU,$$

i.e., $P_mU = P_mA$. Consequently, symmetric groupoids possess property 2).

Sufficiency. Let G be an associative groupoid satisfying 1) and 2). If n is a right zero and a an arbitrary element of G, then na is again a right zero of G. Therefore the product of a right zero by an arbitrary fixed element a of G induces a well-defined transformation A_a $(nA_a = na)$ on the set of all right zeros. The mapping $a \longrightarrow A_a$ is a homomorphism of G into the symmetric groupoid C_N of all transformations of the set N of right zeros of G. Condition 2) shows that this mapping is a mapping of G *onto* C_N, and from 1) it follows that distinct elements of G induce distinct transformations on N. Consequently, G is isomorphic to C_N.

An isomorphic mapping of some groupoid onto itself is called an *automorphism.* A mapping of the form $x \longrightarrow a^{-1}xa$, where a is a fixed unit of the associative groupoid G with identity, is called an *inner automorphism* induced on G by the element a.

As a corollary we shall prove that *all automorphisms of a symmetric groupoid are inner* (see [4]).

Let N be the set of all right zeros of the symmetric groupoid G. Any automorphism A of G maps a right zero of G onto another and so induces a certain inverse transformation on the set N. By the theorem, there are elements a and b in G satisfying, for all right zeros n, the relations $nA = na$ and $nA^{-1} = nb$. Hence we have

$$n \cdot a\,(bA) = nA \cdot bA = (nb)\,A = n, \quad nA \cdot (bA \cdot a) = nA,$$

i.e., $a \cdot bA = bA \cdot a = e$ and $bA = a^{-1}$. Since, for an arbitrary $g \in G$, the product ng is a right zero, it follows that

$$nA \cdot gA = (ng)\,A = nga = na \cdot a^{-1}ga.$$

Hence, taking into consideration the arbitrary nature of nA, and property 1) stated in the conditions of the theorem, we obtain $gA = a^{-1}ga$, as required.

Let us agree to call the cardinality of the subset M_1 of M, onto which the mapping A maps M, the *rank of the mapping A.* Since the product of two transformations of rank not exceeding ξ is a transformation of rank not greater than ξ, the set of all transformations on the set M having rank less than ξ is a subgroupoid C_M^ξ of the symmetric groupoid C_M. The right zeros of C_M^ξ are exactly the right zeros of C_M. Therefore, if in the theorem we replace the arbitrary mapping U by a mapping of rank not greater than ξ, we obtain a characteristic of C_M^ξ.

Analogous remarks may be made in regard to the corollary of the theorem. In that corollary, it was actually established that every one-to-one mapping of the set of right zeros of C_M onto itself is uniquely extendable to an automorphism of C_M. Since the sets of right zeros of C_M and C_M^ξ are one and the same, it is possible to assert, by virtue of property 1), that *every one-to-one mapping of the set of all right zeros of C_M^ξ onto itself may be extended in one and only one way to an automorphism of that groupoid.* In this way, the groups of automorphisms of all subgroupoids C_M^ξ are exactly the same, i.e., isomorphic to the symmetric group on the elements of C_M^2.

§2. Homomorphisms of a symmetric groupoid

The determination of all factor-groupoids on a given groupoid G is equivalent to finding all equivalence relations between the elements of G compatible with the groupoid's principal operation, i.e., having the following properties: for all a, b, c, d in G, 1) $a \sim a$; 2) if $a \sim b$, then $b \sim a$; 3) if $a \sim b$ and $b \sim c$, then $a \sim c$; 4) if $a \sim b$, then $ad \sim bd$ and $da \sim db$.

A set of elements I of the groupoid G is called an *ideal* in G, if from $a \in I$ and $x \in G$ it follows that $ax \in I$ and $xa \in I$. The elements a and b of G are said to be *congruent with respect to the ideal I* if they are either equal or both belong to I. It is clear that congruences with respect to ideals are equivalence relations on G. The corresponding factor-groupoid G/I is obtained from G by discarding all elements of the ideal I and replacing them by a zero element. If the elements of G satisfy certain identities, the same identities are also satisfied by the elements of G/I. Therefore, if G is associative, so is G/I.

As a general rule, ideal-factor-groupoids represent very special cases of factor-groupoids. However, for the determination of all factor-groupoids of a symmetric groupoid ideal-factor-groupoids will have an essential significance.

LEMMA 1. *All ideals of the symmetric groupoid C_M are exhausted by the subgroupoids of the form C_M^ξ.*[1]

Since the product of any transformation of the set M by a transformation of rank less than ξ is a transformation of rank less than ξ, all subgroupoids C_M^ξ are ideals in C_M. It remains to show that if some ideal I of C_M contains a transformation A of rank ξ, then I contains all transformations whose rank does not exceed ξ.

Let B be an arbitrary transformation of the set M of rank ξ. Let us index the elements of the sets MA and MB by the very same subscripts $\alpha, \beta, \ldots$, and let M_λ denote the set of elements of M mapped under the substitution B to $b_\lambda \in MB$ ($\lambda = \alpha, \beta, \ldots$). For each λ, select an arbitrary element $a'_\lambda \in M$ which is mapped by the substitution A to $a_\lambda \in MA$. Now it is easily seen that by multiplying A from the right by a substitution sending a_λ to the element b_λ for each λ, and from the left by the substitution sending the set M_λ to the element a'_λ for each λ ($\lambda = \alpha, \beta, \ldots$), we obtain B. Consequently, since $A \in I$, also $B \in I$. If the rank of B is less than ξ, then, multiplying A by an appropriate substitution from the right, we obtain a substitution in I with the same rank as B. By repeating the preceding argument we obtain $B \in I$, which was required.

The union of all elements of M which are mapped by A to one and the same element $a \in M$, we shall call a *residue class with respect to A*. Obviously, distinct residue classes are disjoint, and their union forms a partition of the underlying set M.

Let n be a finite natural number, not exceeding the number of elements of M. Let us denote by $\mathfrak{N}_n$ one of the normal divisors of the symmetric group of permutations of the numbers $1, \ldots, n$; for $n > 4$, $\mathfrak{N}_n$ is either the symmetric group itself, or the set of all even permutations, or the identity permutation. Let us agree to call the transformations A and B of the set M *congruent with respect to* $\mathfrak{N}_n$, in symbols $A \equiv B \pmod{\mathfrak{N}_n}$, if one of the following conditions is realized:

a) A and B are equal;

b) A and B have rank less than n;

c) the rank of A and B is equal to n; $MA = MB$; and the residue classes with respect to A and with respect to B are the same, where we index these residue classes in an arbitrary manner with the numbers $1, \ldots, n$ and write the substitutions A and B in the form

$$A = \begin{pmatrix} M_1 & M_2 & \ldots & M_n \\ a_1 & a_2 & \ldots & a_n \end{pmatrix}, \quad B = \begin{pmatrix} M_1 & M_2 & \ldots & M_n \\ a_{i_1} & a_{i_2} & \ldots & a_{i_n} \end{pmatrix}, \tag{1}$$

[1] We have been informed that a similar proposition has been proved by N. N. Vorob'ev.

where the M_i are the corresponding residue classes and the permutation

$$P = \begin{pmatrix} 1 & 2 & \dots & n \\ i_1 & i_2 & \dots & i_n \end{pmatrix}$$

belongs to $\mathfrak{N}_n$.

By changing the numeration of the residue classes the permutation P is replaced by a conjugate. Therefore in condition c), instead of requiring that P should belong to $\mathfrak{N}_n$ for any numeration of the residue classes, it suffices that P belong to $\mathfrak{N}_n$ for only one numeration.

Congruences with respect to normal divisors $\mathfrak{N}_n$ are equivalence relations on the symmetric groupoid C_M, compatible with the operation of multiplication.

Reflexivity, symmetry, and transitivity of congruences with respect to $\mathfrak{N}_n$ follow directly from the fact that $\mathfrak{N}_n$ is a group. Therefore it suffices to only consider the possibility of multiplying congruences by an arbitrary substitution D from C_M. Let $A \equiv B \pmod{\mathfrak{N}_n}$. If A and B satisfy condition a) or b), then AD, BD and DA, DB obviously satisfy the same conditions; consequently, in this case $AD \equiv BD$ and $DA \equiv DB$. Let us now suppose that A and B have the form (1) and

$$D = \begin{pmatrix} a_1 & a_2 & \dots & a_n \dots \\ d_1 & d_2 & \dots & d_n \dots \end{pmatrix}.$$

Then

$$AD = \begin{pmatrix} M_1 & M_2 & \dots & M_n \\ d_1 & d_2 & \dots & d_n \end{pmatrix}, \quad BD = \begin{pmatrix} M_1 & M_2 & \dots & M_n \\ d_{i_1} & d_{i_2} & \dots & d_{i_n} \end{pmatrix}.$$

If some of the elements $d_1, \dots, d_n$ are equal, then the ranks of AD and BD are less than n and $AD \equiv BD$. Even if all d_i are distinct, then AD and BD satisfy the whole of condition c). In order to prove the congruence $DA \equiv DB$, let us denote by L_i the set of elements of M mapped by D into the set M_i ($i = 1, \dots, n$). If some of the sets L_i are empty, then the ranks of DA and DB are less than n, and consequently $DA \equiv DB$. If each of the sets L_i is nonempty, then DA and DB satisfy condition c), and so, again, $DA \equiv DB$.

It is easily seen that similarly-defined types of congruence with respect to normal divisors of an infinite symmetric group are not compatible with multiplication and are therefore unsuitable for constructing factor-groupoids. Instead of these we shall define congruences with respect to $\mathfrak{N}_\xi$ of a particular type, where ξ is some infinite cardinal number and where here $\mathfrak{N}_\xi$ will not designate a normal divisor, but will be the notation for the equivalence itself. Namely, let A and B be given transformations of the set M. Let us denote by M_0 the set of those elements of M for which $mA \neq mB$. Let us agree to say that the substitutions A and B differ at ξ places if ξ is the greater cardinality of the sets M_0A and M_0B. It is clear that the number of places at which the substitutions differ cannot exceed the maximum rank of those substitutions. If A is the identity, then the number of places at which A and B differ equals the number of elements of M moved by B. We shall say the transformations A and B of M are *congruent with respect to* $\mathfrak{N}_\xi$ if the number of places at which they differ is less than ξ.

Congruences with respect to $\mathfrak{N}_\xi$ *on the groupoid* C_M *are equivalences compatible with the operation of multiplication.*

We shall prove, for example, that it is possible to multiply this congruence from the right by any transformation D. In fact, the relation $AD \not\equiv BD$ means that the cardinal of one of the sets M_0AD or M_0BD is not less than ξ, where M_0 consists of those elements m for which $mAD \neq mBD$. But in this case the cardinal of the corresponding set M_0A or M_0B will also not be less than ξ, and from $m \in M_0$ it will follow that $mA \neq mB$, i.e., $A \not\equiv B \pmod{\mathfrak{N}_\xi}$. The remaining properties are shown in an analogous manner.*

As has already been noted, the set of substitutions whose rank is less than a given cardinal number ξ forms an ideal in the groupoid C_M. This ideal will be denoted by I_ξ. If the set M has cardinality μ and ν is the next larger cardinal, then the congruences with respect to the ideal I_ν and with respect to $\mathfrak{N}_\nu$ are obviously equivalent, since in both cases all substitutions of C_M are equivalent. The congruence with respect to the ideal I_n of finite index n is equivalent to the congruence with respect to the normal divisor $\mathfrak{N}_n$ consisting of only the identity permutation. In particular, the congruences with respect to I_2 and with respect to $\mathfrak{N}_2$ are equal in the ordinary sense. For $\aleph_0 \leqslant \xi \leqslant \mu$ the congruences with respect to I_ξ and with respect to $\mathfrak{N}_\xi$ are different.

The equivalence relation θ_1 is said to *contain the relation* θ_2, in symbols $\theta_1 \geqslant \theta_2$, if from $x \equiv y \pmod{\theta_1}$ for any x and y it follows that $x \equiv y \pmod{\theta_2}$. The equivalence θ is called *the sum*** *of the arbitrary system of equivalences* $\{\theta_i\}$ provided that $x \equiv y \pmod{\theta}$ if and only if $x \equiv y \pmod{\theta_i}$ for every θ_i from the given set. The equivalence θ' is called *the product*** *of the equivalences* $\{\theta_i\}$ provided that $x \equiv y \pmod{\theta'}$ if and only if there exists a finite sequence of elements $x_1, \ldots, x_{n-1}$ satisfying the congruences

$$x \equiv x_1 \pmod{\theta_{i_1}}, \quad x_1 \equiv x_2 \pmod{\theta_{i_2}}, \ldots, x_{n-1} \equiv y \pmod{\theta_{i_n}},$$

where $\theta_{i_1}, \ldots, \theta_{i_n}$ are certain equivalences from the given system.

Denoting by the symbols $\mathfrak{N}_\xi$ and I_ξ the congruences modulo $\mathfrak{N}_\xi$ and the ideal I_ξ, we will obviously have $\mathfrak{N}_\xi \geqslant \mathfrak{N}_{\xi'}$, $I_\xi \geqslant I_{\xi'}$ and $I_\xi > \mathfrak{N}_{\xi'}$ $(\xi \leqslant \xi')$. Let

$$\aleph_0 \leqslant \xi_1 < \xi_2 < \ldots < \xi_k \leqslant \eta_k < \eta_{k-1} < \ldots < \eta_1 < \eta_0 \leqslant \nu. \tag{2}$$

Then the equivalence θ, defined by the formula

$$\theta = I_{\eta_0} + I_{\eta_1}\mathfrak{N}_{\xi_1} + I_{\eta_2}\mathfrak{N}_{\xi_2} + \ldots + I_{\eta_k}\mathfrak{N}_{\xi_k},$$

has the following meaning: $A \equiv B \pmod{\theta}$ if and only if one of the following conditions is realized: a) the ranks of A and B are less than η_k; b) the ranks of A and B equal ρ, $\eta_{i+1} \leqslant \rho < \eta_i$, and the number of places at which the substitutions A and B differ is less than ξ_{i+1}; c) $A = B$.

---

*Translators' note. Transitivity is not entirely obvious: see [5], p. 229.

**Translators' note. The words "sum" and "product" would be better interchanged.

THEOREM. *Every equivalence* on the symmetric groupoid C_M is either a congruence with respect to a normal divisor $\mathfrak{N}_\xi$ of finite index ξ or an equivalence of the form* $\theta = I_{\eta_0} + I_{\eta_1}\mathfrak{N}_{\xi_1} + \cdots + I_{\eta_k}\mathfrak{N}_{\xi_k}$, *where the subscripts are subject to condition* (2). *In particular, all equivalences* on a finite symmetric groupoid C_k are exhausted by the congruences with respect to the normal divisors $\mathfrak{N}_n$* $(n = 1, 2, \ldots, k+1)$.

The proof of this theorem has a computational nature; it is convenient to split it up into a series of lemmas. In all that follows, by an equivalence is to be understood an equivalence on the groupoid C_M compatible with the operation of multiplication.

LEMMA 2. *Under any equivalence different from the identity, all transformations of rank* 1 *are mutually equivalent. If in some equivalence a transformation of rank ρ is equivalent to a transformation of smaller rank, then all transformations of rank not exceeding ρ are equivalent.*

Let A and B be two different but equivalent transformations of the groupoid C_M. Select some element m in M for which $mA \neq mB$, and denote by P_m the transformation of rank 1 sending all elements of M to m, and by D a transformation sending mA and mB to arbitrary predetermined elements a and b respectively. Then $P_mAD \sim P_mBD$, i.e., $P_a \sim P_b$, and the first assertion is proved. Hence, in particular, the set of substitutions which are equivalent to a substitution of rank 1 obviously forms an ideal in C_M, and so, by virtue of Lemma 1, *if some substitution of rank ρ is equivalent to a substitution of rank* 1, *then all substitutions of rank not more than ρ are equivalent.*

The proof of the lemma's second assertion in the case of infinite ρ is obvious. In fact, denote by D the transformation mapping all elements of the set MB to some single element m and leaving all the remaining elements fixed; then from $A \sim B$ we obtain $AD \sim BD$, i.e., $AD \sim P_m$. Since the rank of AD remains equal to ρ, and the rank of P_m equals 1, it follows, according to certain remarks above, that all transformations of rank not exceeding ρ are equivalent.

Now let ρ be finite. Rewrite the equivalence $A \sim B$ explicitly in the form

$$A = \begin{pmatrix} M_1 & M_2 & \ldots & M_k & M_{k+1} & \ldots & M_\rho \\ a_1 & a_2 & \ldots & a_k & c_1 & \ldots & c_{\rho-k} \end{pmatrix}$$
$$\sim \begin{pmatrix} N_1 & \ldots & N_m & N_{m+1} & \ldots & N_{\rho+m-k} \\ b_1 & \ldots & b_m & c_1 & \ldots & c_{\rho-k} \end{pmatrix} = B,$$

where $c_1, \ldots, c_{\rho-k}$ are the elements common to MA and MB, and multiply it from the right by the substitution sending $b_1, \ldots, b_m$ to c_1 and leaving the remaining elements fixed. We thus obtain

$$A \sim \begin{pmatrix} N'_{m+1} & N_{m+2} & \ldots & N_{\rho+m-k} \\ c_1 & c_2 & \ldots & c_{\rho-k} \end{pmatrix} = B_1. \tag{3}$$

**Translators' note.* For "equivalence" read "congruence"—see the next paragraph.

Further multiply (3) from the right by the substitutions D_i $(i = 1, \ldots, \rho - k)$, sending $c_1, \ldots, c_b$ to a_1 and leaving the remaining elements fixed; we obtain the sequence of equivalences

$$A \sim B_1, A_1 \sim B_2, \ldots, A_{\rho-k} \sim B_{\rho-k+1}.$$

Since the rank of $A_{\rho-k}$ is k and the rank of $B_{\rho-k+1}$ is 1, all transformations of rank k are equivalent. In particular, the transformation $B_{\rho-k}$ of rank 2 will be equivalent to $B_{\rho-k+1}$, and the transformation $A_{\rho-k-1}$ of rank $k+1$, equivalent to $B_{\rho-k}$, will be equivalent to $B_{\rho-k+1}$. If we continue this process, we see that A is equivalent to a transformation of rank 1.

LEMMA 3. *If a transformation A of finite rank is not equivalent to a transformation of rank* 1 *and $A \sim B$, then $MA = MB$, and A and B generate identical partitions of the set M.*

First suppose that $MA \neq MB$. Then, if we multiply A and B from the right by a transformation D, mapping some pair of elements $m_1 \in MB$, $m_2 \in MB$, $m_2 \notin MA$, to m_1 and leaving the rest fixed, we obtain the equivalence $A \sim BD$, where the rank of BD is less than the rank of A, so that, by virtue of Lemma 1,* we have as a consequence that A is equivalent to a transformation of rank 1. Now let $MA = MB$ and

$$A = \begin{pmatrix} M_1 \ldots M_\rho \\ a_1 \ldots a_\rho \end{pmatrix} \sim \begin{pmatrix} N_1 \ldots N_\rho \\ a_1 \ldots a_\rho \end{pmatrix} = B. \tag{4}$$

If the partitions $M_1, \ldots, M_\rho$ and $N_1, \ldots, N_\rho$ are different, then we may find elements n_i and n_j belonging to distinct sets N_i and N_j in the second partition and to one and the same set M_k of the first partition. If we choose from the remaining sets of the second partition elements $n_p \in N_p$, and multiply (4) from the right by a substitution sending $a_1, \ldots, a_\rho$ to $n_1, \ldots, n_\rho$ respectively, and square both sides of the new equivalence, we obtain a substitution of rank ρ equivalent to a substitution of smaller rank, which contradicts the condition of the lemma.

LEMMA 4. *If a transformation A of finite rank ρ is not equivalent to a transformation of rank* 1 *and $A \sim B$, $A \neq B$, then the equivalence under consideration is a congruence with respect to one of the normal divisors $\mathfrak{N}_\rho$.*

We shall prove first that all substitutions of rank less than ρ are equivalent. According to Lemma 3, the equivalence $A \sim B$ has the form

$$\begin{pmatrix} M_1 & M_2 & \ldots & M_\rho \\ a_1 & a_2 & \ldots & a_\rho \end{pmatrix} \sim \begin{pmatrix} M_{i_1} & M_{i_2} \ldots & M_{i_\rho} \\ a_1 & a_2 & \ldots & a_\rho \end{pmatrix}. \tag{5}$$

In the case when $\rho = 2$, the assertion is contained in Lemma 2. But if $\rho > 2$, we may find a pair of subscripts k, j such that $\{k, j\} \neq \{i_k, i_j\}$, whence $M_k + M_j \neq M_{i_k} + M_{i_j}$. If we multiply both sides of (5) from the right by a substitution sending a_k and a_j to

*Translators' note. For "Lemma 1" read "Lemma 2".

a_k and leaving the remaining elements fixed, we obtain an equivalence between substitutions of rank $\rho - 1$ generating different partitions of the set M. Hence, by virtue of Lemma 3, all substitutions of rank not more than $\rho - 1$ are equivalent.

Under the conditions of Lemma 4, *substitutions of rank greater than* ρ *are equivalent only if they are equal.* In fact, in the contrary case, as has been proved, all substitutions of rank less than $\rho + 1$, among them both A and B, will be equivalent to a substitution of rank 1.

Now let $M = M_1 + \cdots + M_\rho$ be some partition of the set M into ρ nonempty, pairwise disjoint classes. Consider some ordered pair of transformations A, B, appertaining to the stated partition. These transformations can be written in the form

$$A = \begin{pmatrix} M_1 & M_2 & \dots & M_\rho \\ a_1 & a_2 & \dots & a_\rho \end{pmatrix}, \quad B = \begin{pmatrix} M_1 & M_2 & \dots & M_\rho \\ a_{i_1} & a_{i_2} & \dots & a_{i_\rho} \end{pmatrix}.$$

The permutation sending $1, \dots, \rho$ to $i_1, \dots, i_\rho$ respectively will be said to correspond to the above pair A, B. Denote by $\mathfrak{N}_\rho$ the set of permutations corresponding to all possible equivalent pairs A, B. This set contains the identity substitution, since the latter arises from the equivalence $A \sim A$. Furthermore, let α and β be the permutations corresponding to the equivalences

$$\begin{pmatrix} M_1 & \dots & M_\rho \\ a_1 & \dots & a_\rho \end{pmatrix} \sim \begin{pmatrix} M_1 & \dots & M_\rho \\ a_{i_1} & \dots & a_{i_\rho} \end{pmatrix}, \quad \begin{pmatrix} M_1 & \dots & M_\rho \\ b_1 & \dots & b_\rho \end{pmatrix} \sim \begin{pmatrix} M_1 & \dots & M_\rho \\ b_{j_1} & \dots & b_{j_\rho} \end{pmatrix}. \tag{6}$$

If we multiply the second of these by the transformation sending the elements $b_1, \dots, b_\rho$ to $a_1, \dots, a_\rho$ respectively, and compare the result with the first equivalence, we obtain

$$\begin{pmatrix} M_1 & \dots & M_\rho \\ a_{i_1} & \dots & a_{i_\rho} \end{pmatrix} \sim \begin{pmatrix} M_1 & \dots & M_\rho \\ a_{j_1} & \dots & a_{j_\rho} \end{pmatrix},$$

which corresponds to the permutation $\beta\alpha^{-1}$. Finally, if we multiply the first equivalence in (6) on the left by the transformation sending $M_1, \dots, M_\rho$ to $m_{j_1}, \dots, m_{j_\rho}$ respectively ($m_k \in M_k$), we obtain an equivalent pair corresponding to the permutation $\beta\alpha\beta^{-1}$. Thus, $\mathfrak{N}_\rho$ is a normal divisor of the symmetric group of permutations on the set of numbers $1, \dots, \rho$.

It remains to prove that two transformations of rank ρ which map one and the same partition of the set M into itself and which consequently have the form

$$A = \begin{pmatrix} N_1 & \dots & N_\rho \\ a_1 & \dots & a_\rho \end{pmatrix}, \quad \begin{pmatrix} N_1 & \dots & N_\rho \\ a_{i_1} & \dots & a_{i_\rho} \end{pmatrix} = B,$$

are equivalent if and only if the permutation

$$\alpha = \begin{pmatrix} 1 & 2 & \dots & \rho \\ i_1 & i_2 & \dots & i_\rho \end{pmatrix}$$

belongs to $\mathfrak{N}_\rho$. If $A \sim B$, then, multiplying this equivalence from the left by the substitution sending $M_1, \ldots, M_\rho$ to the elements $n_1, \ldots, n_\rho$ respectively ($n_k \in N_k$), we obtain an equivalence arising from the partition $M_1, \ldots, M_\rho$ and corresponding to the permutation α, whence $\alpha \in \mathfrak{N}_\rho$. Conversely, if $\alpha \in \mathfrak{N}_\rho$, then, if we take an equivalence arising from the partition $M_1, \ldots, M_\rho$ and corresponding to the permutation α, and multiply it from the left by an appropriate substitution, we obtain the equivalence $A \sim B$. Thus we have proved both Lemma 4 itself and our theorem for finite groupoids. Therefore in the following it will be assumed that the underlying set M is infinite.

LEMMA 5. *If the equivalent substitutions A and B have infinite rank ρ and the set of points on which they differ has infinite cardinality ξ, then any two substitutions with exactly the same rank not exceeding ρ, and differing on a set of points of cardinality less than or equal to ξ, are equivalent.*

By the conditions of the lemma, A and B can be represented in the form

$$A = \begin{pmatrix} P_1 & P_2 & \ldots & R_1 & R_2 & \ldots \\ a_1 & a_2 & \ldots & c_1 & c_2 & \ldots \end{pmatrix}, \quad B = \begin{pmatrix} Q_1 & Q_2 & \ldots & R_1 & R_2 & \ldots \\ b_1 & b_2 & \ldots & c_1 & c_2 & \ldots \end{pmatrix},$$

where the elements in the lower line are distinct, and their inverse images are written above. Since ξ is infinite, it can be assumed that the sets of elements $\{a_i\}$ and $\{b_i\}$ have cardinal ξ, and the set of elements $\{c_i\}$ is either empty in the case $\xi = \rho$ or has cardinal ρ if $\xi < \rho$. Multiplying the equivalence $A \sim B$ from the right by the substitution sending the elements $c_1, c_2, \ldots$ to $r_1, r_2, \ldots$ respectively ($r_i \in R_i$), and leaving the remaining elements fixed, we obtain an equivalence of the form

$$\begin{pmatrix} P_1 & P_2 & \ldots & R_1 & R_2 & \ldots \\ a_1 & a_2 & \ldots & r_1 & r_2 & \ldots \end{pmatrix} \sim \begin{pmatrix} Q_1 & Q_2 & \ldots & R_1 & R_2 & \ldots \\ b_1 & b_2 & \ldots & r_1 & r_2 & \ldots \end{pmatrix}. \tag{7}$$

Referring to the case when ξ elements of the set $\{r_i\}$ are necessarily in the set $\{a_i\}$ and the set $\{b_i\}$, we can assume that, with the notation of (7), the elements a_i and b_i are not contained in the set $R = R_1 + R_2 + \cdots$. Let $M' = P_1 + P_2 + \cdots$. To any substitution A on the elements of the set M, such that $M'A \subset M'$ and $R_iA = r_i$ ($i = 1, 2, \ldots$), there corresponds a transformation A' of M' induced by the transformation A. This correspondence is one-to-one and preserves the operation of multiplication. Therefore in what follows, instead of A we shall consider the induced transformation A'. Thus we have two equivalent transformations of the set M':

$$\begin{pmatrix} P_1 & P_2 & \ldots \\ a_1 & a_2 & \ldots \end{pmatrix} \sim \begin{pmatrix} Q_1 & Q_2 & \ldots \\ b_1 & b_2 & \ldots \end{pmatrix} \tag{8}$$

of rank ξ and differing on a set of points of cardinal ξ. Present the set of lower elements in the form

$$\{a_i\} = \{c_i\} + \{d_i\}, \; \{b_i\} = \{c_i\} + \{e_i\}$$

where $\{c_i\} = \{a_i\} \cap \{b_i\}$. If the cardinal of the set $\{d_i\}$ equals ξ, then, multiplying the equivalence (8) from the right by the substitution sending all elements b_i to the element b_1 and leaving the remaining elements fixed, we obtain on the left a substitution of rank ξ, and on the right a substitution of rank 1. Hence, because of Lemma 2, it follows that all substitutions of the set M' of rank not more that ξ are equivalent. The same is correct if ξ is the cardinal of the set $\{e_i\}$. We shall prove that in the case when the cardinals of the sets $\{d_i\}$ and $\{e_i\}$ are less than ξ, all substitutions on the elements of the set M' are also equivalent. Multiplying (8) from the right by the substitution sending all elements of the sets $\{d_i\}$ and $\{e_i\}$ to c_1 and leaving the remaining elements fixed, we obtain an equivalence of the form

$$\begin{pmatrix} M_1 & M_2 & \dots \\ c_1 & c_2 & \dots \end{pmatrix} \sim \begin{pmatrix} N_1 & N_2 & \dots \\ c_1 & c_2 & \dots \end{pmatrix}. \tag{9}$$

By the conditions, the substitutions in this equivalence differ at ξ places. Consequently, representatives $m_i \in M_i$ can be chosen from the sets M_i in such a way that for ξ of them the images in the left-hand substitution differ from the images in the right. If we multiply (9) from the right by a substitution sending $c_1, c_2, \dots$ to $m_1, m_2, \dots$ respectively and multiply both sides of the equivalence obtained from the left by its left-hand substitution, we will have

$$\begin{pmatrix} M_1 & M_2 & \dots \\ m_1 & m_2 & \dots \end{pmatrix} \sim \begin{pmatrix} M_1 & M_2 & \dots \\ m_{\sigma_1} & m_{\sigma_2} & \dots \end{pmatrix}, \tag{10}$$

where the substitutions, by choice of the representatives m_i, still differ at ξ places. The sets M_i can contain other elements, besides the elements $m_1, m_2, \dots$, which later on will not be written out explicitly, so that the equivalence (10) will have the form

$$\begin{pmatrix} m_1 & m_2 & \dots \\ m_1 & m_2 & \dots \end{pmatrix} \sim \begin{pmatrix} m_{\alpha_1} & m_{\alpha_2} & \dots & X_1 & X_2 & \dots \\ m_{\beta_1} & m_{\beta_2} & \dots & m_{\gamma_1} & m_{\gamma_2} & \dots \end{pmatrix}, \tag{11}$$

where all m_β and m_γ are distinct and each set X_i contains more than one element: $X_i = \{m_{\mu_i}, m_{\nu_i}, \dots\}$. If the number of sets X_i equals ξ, then, multiplying the equivalence (11) from the left by substitutions sending $m_1, m_2, \dots$ to $m_{\mu_1}, m_{\mu_2}, \dots$ and to $m_{\nu_1}, m_{\nu_2}, \dots$ respectively, and observing that in both cases after multiplication the very same substitution will turn out to be on the right, we obtain

$$\begin{pmatrix} m_1 & m_2 & \dots \\ m_{\mu_1} & m_{\mu_2} & \dots \end{pmatrix} \sim \begin{pmatrix} m_1 & m_2 & \dots \\ m_{\nu_1} & m_{\nu_2} & \dots \end{pmatrix}.$$

Since here the sets of lower elements have no elements in common, it follows that each of the substitutions is equivalent to a substitution of rank 1, and so all substitutions on the elements $m_1, m_2, \dots$ are equivalent. It will be the same if the number of sets X_i is less than ξ but their sum contains ξ elements. In that case it suffices to multiply (11) from the left by a substitution sending $m_1, m_2, \dots$ to distinct elements of that sum. As a result we obtain that a substitution of rank ξ will be equivalent to a substitution of smaller rank, so that, by Lemma 2, we consequently have that all substitutions are equivalent to one another. Finally, consider the case when the sum of the sets X_i contains less than ξ elements. Denote by $\{m_{k_1}, m_{k_2}, \ldots\}$ the set of all those elements which finite positive powers of the right-hand substitution of (11) send to an element of $\{m_{\gamma_1}, m_{\gamma_2}, \ldots\}$, and by $\{m_{i_1}, m_{i_2}, \ldots\}$ the set of the remaining elements. Rewrite (11) in the form

$$\begin{pmatrix} m_1 & m_2 & \ldots \\ m_1 & m_2 & \ldots \end{pmatrix} \sim \begin{pmatrix} m_{i_1} & m_{i_2} & \ldots & m_{k_1} & m_{k_2} & \ldots \\ m_{j_1} & m_{j_2} & \ldots & m_{l_1} & m_{l_2} & \ldots \end{pmatrix}. \tag{12}$$

The set $\{m_{i_1}, m_{i_2}, \ldots\}$ contains $\{m_{j_1}, m_{j_2}, \ldots\}$, and its cardinal equals ξ since the cardinal of $\{m_{k_1}, m_{k_2}, \ldots\}$ equals the cardinal of the sum $X = X_1 + X_2 + \cdots$ and is consequently smaller than ξ. Multiplying (12) from the left by a substitution sending $m_1, m_2, \ldots$ to $m_{i_1}, m_{i_2}, \ldots$ respectively and simultaneously from the right by a substitution sending $m_{i_1}, m_{i_2}, \ldots$ to $m_1, m_2, \ldots$, we obtain

$$\begin{pmatrix} m_1 & m_2 & \ldots \\ m_1 & m_2 & \ldots \end{pmatrix} \sim \begin{pmatrix} m_1 & m_2 & \ldots \\ m_{p_1} & m_{p_2} & \ldots \end{pmatrix}, \tag{13}$$

where all m_p are distinct and also ξ of them are different from the element above them. We can assume that the number of elements of $\{m_1, m_2, \ldots\}$ which are not contained among $m_{p_1}, m_{p_2}, \ldots$ is smaller than ξ, since, otherwise, if we multiply both sides by the substitution sending $m_{p_1}, m_{p_2}, \ldots$ to m_{p_1} and leaving the remaining elements fixed, and apply Lemma 2, we again obtain all substitutions equivalent to one another. Decompose the right-hand substitution in (13) in the usual manner into a product of disjoint cycles and semicycles, i.e., substitutions of the form

$$[a_1, a_2, \ldots) = \begin{pmatrix} a_1 & a_2 & a_3 & \ldots \\ a_2 & a_3 & a_4 & \ldots \end{pmatrix}.$$

The semicycles start with elements absent from the lower line, and therefore the number of them is less than ξ. Rewriting (13) in the form

$$E \sim \prod_\lambda (\ldots, a^\lambda_{-1}, a^\lambda_0, a^\lambda_1, \ldots) \prod_\mu [b^\mu_1, b^\mu_2, \ldots), \tag{14}$$

whence

$$\prod_\mu [b^\mu_1, b^\mu_2, \ldots) \sim \prod_\lambda (\ldots, a^\lambda_1, a^\lambda_0, a^\lambda_{-1}, \ldots), \tag{15}$$

and multiplying (14) from the right by

$$\prod_\mu \begin{pmatrix} b^\mu_1 & b^\mu_2 & b^\mu_3 & \ldots \\ b^\mu_1 & b^\mu_1 & b^\mu_2 & \ldots \end{pmatrix},$$

we obtain

$$\prod_{\mu}\begin{pmatrix} b_1^{\mu} & b_2^{\mu} & b_3^{\mu} & \ldots \\ b_1^{\mu} & b_1^{\mu} & b_2^{\mu} & \ldots \end{pmatrix} \sim \prod_{\lambda}(\ldots, a_{-1}^{\lambda}, a_0^{\lambda}, a_1^{\lambda}, \ldots),^{*} \tag{16}$$

Since the number of semicycles is less than ξ, and the number of elements moved equals ξ, the number of elements encountered in the cycles equals ξ. If we partition them into two equivalent sets $\{c_i^{\lambda}\}$, $\{d_i^{\lambda}\}$** of cardinal ξ and transform (15) and (16) by means of suitable permutations, we obtain the equivalences

$$\prod_{\mu}[b_1^{\mu}, b_2^{\mu}, \ldots) \sim \prod_{\lambda}(\ldots, c_1^{\lambda}, c_0^{\lambda}, c_{-1}^{\lambda}, \ldots),$$

$$\prod_{\mu}\begin{pmatrix} b_1^{\mu} & b_2^{\mu} & b_3^{\mu} & \ldots \\ b_1^{\mu} & b_1^{\mu} & b_2^{\mu} & \ldots \end{pmatrix} \sim \prod_{\lambda}(\ldots, d_1^{\lambda}, d_0^{\lambda}, d_{-1}^{\lambda}, \ldots),$$

which when multiplied together will of necessity be an equivalence of the form $U \sim E$, where E is the identity permutation of the elements $m_1, m_2, \ldots$, and U is an invertible substitution on those same elements which moves ξ elements. The set of all permutations equivalent to the identity permutation is a normal divisor of the permutation group. Since this normal divisor contains a permutation U moving ξ elements, it must coincide with the whole group (see [3]), i.e., the equivalence $U \sim E$ holds for any permutation of $m_1, m_2, \ldots$.

We shall prove that the equivalence $U \sim E$ holds for any injective mapping of the set $m_1, m_2, \ldots$ into itself. For this we decompose U into the disjoint product of cycles and semicycles. By what has been proved, a product of cycles is equivalent to E. Consequently we can assume that U consists only of semicycles. We have

$$E \sim \prod_{\mu}(\ldots, a_3^{\mu}, a_1^{\mu}, a_2^{\mu}, a_4^{\mu}, \ldots),$$

whence

$$E \sim \prod_{\mu}\begin{pmatrix} a_1^{\mu} & a_2^{\mu} & a_3^{\mu} & \ldots \\ a_2^{\mu} & a_4^{\mu} & a_6^{\mu} & \ldots \end{pmatrix}(\ldots, a_3^{\mu}, a_1^{\mu}, a_2^{\mu}, a_4^{\mu}, \ldots)\begin{pmatrix} \ldots & a_3^{\mu} & a_1^{\mu} & a_2^{\alpha} & a_4^{\mu} & a_6^{\mu} & \ldots \\ \ldots & a_1^{\mu} & a_1^{\mu} & a_1^{\mu} & a_2^{\mu} & a_3^{\mu} & \ldots \end{pmatrix},$$

i.e.,

$$E \sim \prod_{\mu}[a_1^{\mu}, a_2^{\mu}, \ldots),$$

as required.

Now we shall separate the elements $m_1, m_2, \ldots$ into two equivalent classes $m_{\lambda_1}, m_{\lambda_2}, \ldots$ and $m_{\mu_1}, m_{\mu_2}, \ldots$. By what has been established,

$$E \sim \begin{pmatrix} m_1 & m_2 & \ldots \\ m_{\lambda_1} & m_{\lambda_2} & \ldots \end{pmatrix} \sim \begin{pmatrix} m_1 & m_2 & \ldots \\ m_{\mu_1} & m_{\mu_2} & \ldots \end{pmatrix}.$$

**Translators' note.* The original has μ beneath Π on the right-hand side of (16).

***Translators' note.* The original has $\{d_{\lambda}^{i}\}$.

Since the sets of lower elements do not intersect, each substitution is equivalent to a substitution of rank 1, from which it follows that all substitutions on $m_1, m_2, \ldots$ are equivalent. Substitutions on the elements $m_1, m_2, \ldots,$ were only an abridged designation for transformations on the set M' sending $M_1, M_2, \ldots,$ where $m_i \in M_i$, to $m_{i_1}, m_{i_2}, \ldots,$ respectively. In this way, every substitution on the elements of M' having this form is equivalent to the substitution of rank 1 which sends all sets M_i to the element m_1. Since among the indicated substitutions there are some of rank ξ, by Lemma 2 all substitutions on M' are equivalent to a substitution of rank 1. Substitutions on M' were, in turn, only an abridged designation for substitutions on M having a specific form. Therefore we have proved the validity of all equivalences of the form

$$\begin{pmatrix} M_1 & M_2 & \ldots & R_1 & R_2 & \ldots \\ m_1 & m_2 & \ldots & r_1 & r_2 & \ldots \end{pmatrix} \sim \begin{pmatrix} M' & R_1 & R_2 & \ldots \\ m_1 & r_1 & r_2 & \ldots \end{pmatrix} \quad (M' = M_1 + M_2 + \ldots), \quad (17)$$

where $M_1, M_2, \ldots$ is any partition of the set M' into not more than ξ parts, $m_i \in M_i$, and $R_1, R_2, \ldots$ are fixed subsets, the number of which equals ρ, where $r_i \in R_i$. Now let arbitrary substitutions A_1 and B_1 of the set M be given, having exactly the same rank σ ($\sigma \leqslant \rho$) and differing at not more that ξ places. These substitutions can be written in the form

$$A_1 = \begin{pmatrix} P_1 & P_2 & \ldots & S_1 & S_2 & \ldots \\ p_1 & p_2 & \ldots & s_1 & s_2 & \ldots \end{pmatrix}, \quad B_1 = \begin{pmatrix} Q_1 & Q_2 & \ldots & S_1 & S_2 & \ldots \\ q_1 & q_2 & \ldots & s_1 & s_2 & \ldots \end{pmatrix},$$

where the number of elements s_i equals σ and the number of elements p_i and q_i does not exceed ξ. Multiplying (17) from the right by a substitution sending a suitable number of elements r_i to r_1, and taking a number of sets M_i equal to the number of elements p_i, we obtain an equivalence which, if multiplied from the left by the substitution sending the sets $P_1, P_2, \ldots$ and $S_1, S_2, \ldots$ to the elements $m_1, m_2, \ldots$ and $r_1, r_2, \ldots$ respectively and simultaneously multiplied from the right by a substitution sending $m_1, m_2, \ldots$ and $r_1, r_2, \ldots$ to $p_1, p_2, \ldots$ and $s_1, s_2, \ldots$ respectively, will give the equivalence

$$\begin{pmatrix} P_1 & P_2 & \ldots & S_1 & S_2 & \ldots \\ p_1 & p_2 & \ldots & s_1 & s_2 & \ldots \end{pmatrix} \sim \begin{pmatrix} M' & S_1 & S_2 & \ldots \\ p_1 & s_1 & s_2 & \ldots \end{pmatrix}.$$

If we compare it with a similar equivalence for the substitution B_1, we will have $A_1 \sim B_1$, which proves Lemma 5.

LEMMA 6. *If the distinct substitutions A and B are equivalent and have infinite rank ρ and differ at a finite number of places, then all substitutions of rank no more than ρ and differing at a finite number of places are equivalent.*

By repeating the previous argument, we obtain the equivalence

$$\begin{pmatrix} M_1 & \ldots & M_\alpha & R_1 & R_2 & \ldots \\ m_1 & \ldots & m_\alpha & r_1 & r_2 & \ldots \end{pmatrix} \sim \begin{pmatrix} M_1 & \ldots & M_\alpha & R_1 & R_2 & \ldots \\ m_{i_1} & \ldots & m_{i_\alpha} & r_1 & r_2 & \ldots \end{pmatrix},$$

where $m_i \in M_i$, $r_i \in R_i$ and the number of elements r_i equals ρ. If we consider the substitution

$$\begin{pmatrix} m_1 & \dots & m_\alpha & r_1 & \dots & r_\beta \\ a_1 & \dots & a_\alpha & b_1 & \dots & b_\beta \end{pmatrix}$$

on the elements of the set $m_1, \dots, m_\alpha, r_1, \dots, r_\beta$ (β arbitrary, finite and fixed) as representing the substitution

$$\begin{pmatrix} M_1 & \dots & M_\alpha & R_1 & \dots & R_\beta & R_{\beta+1} & \dots \\ a_1 & \dots & a_\alpha & b_1 & \dots & b_\beta & r_{\beta+1} & \dots \end{pmatrix}$$

on the given set, then from the equivalence

$$\begin{pmatrix} m_1 & \dots & m_\alpha & r_1 & \dots & r_\beta \\ m_1 & \dots & m_\alpha & r_1 & \dots & r_\beta \end{pmatrix} \sim \begin{pmatrix} m_1 & \dots & m_\alpha & r_1 & \dots & r_\beta \\ m_{i_1} & \dots & m_{i_\alpha} & r_1 & \dots & r_\beta \end{pmatrix}$$

we conclude on the basis of Lemma 3 that

$$\begin{pmatrix} m_1 & \dots & m_\alpha & r_1 & r_2 & r_3 & \dots & r_\beta \\ m_1 & \dots & m_\alpha & r_1 & r_1 & r_3 & \dots & r_\beta \end{pmatrix} \sim \begin{pmatrix} m_1 & \dots & m_\alpha & r_1 & \dots & r_\beta \\ m_1 & \dots & m_1 & m_1 & \dots & m_1 \end{pmatrix},$$

i.e.,

$$\begin{pmatrix} M_1 & \dots & M_\alpha & R_1' & R_3 & \dots \\ m_1 & \dots & m_\alpha & r_1 & r_3 & \dots \end{pmatrix} \sim \begin{pmatrix} M_1 + \dots + M_\alpha & R_1' & R_3 & \dots \\ m_1 & r_1 & r_3 & \dots \end{pmatrix}.$$

Hence, as in the proof of the previous lemma, we immediately deduce the entire assertion of Lemma 6.

From Lemmas 2–6 the validity of the theorem can now be deduced in an obvious way. Indeed, let some equivalence be given on the infinite symmetric groupoid C_M. If all substitutions from C_M are equivalent, then the equivalence coincides with the congruence with respect to the ideal $I_\nu = C_M$. Suppose not all substitutions are equivalent. Denote by η_1 the minimal rank of a substitution not equivalent to a substitution of rank 1. If η_1 is finite, then, by virtue of Lemma 4, the equivalence under consideration coincides with a congruence with respect to one of the normal divisors $\mathfrak{N}_{\eta_1}$. Therefore suppose that η_1 is infinite. If η_α is some cardinal, denote by ξ_α the least cardinal greater than the number of places at which any two equivalent substitutions of rank η_α differ. Lemmas 5 and 6 show that if $\eta_\alpha < \eta_\beta$, then $\xi_\alpha \geqslant \xi_\beta$. Let η_2 be the least cardinal η_α for which $\xi_\alpha < \xi_1$, η_3 the least cardinal η_β for which $\xi_\beta < \xi_2$, and so on. Since the cardinal numbers are well ordered with respect to magnitude, the process must cut off and we shall obtain the finite sequence

$$\xi_k < \dots < \xi_2 < \xi_1 \leqslant \eta_1 < \eta_2 < \dots < \eta_k. \tag{18}$$

Let A and B be arbitrary substitutions of rank η satisfying the relation $\eta_i \leqslant \eta < \eta_{i+1}$. If the number of places at which these substitutions differ is less than ξ_i, then, by definition of the sequence (18), they will be equivalent. Conversely, if A and B are equivalent, then the number of places at which they differ is less than ξ_i. Therefore the given equivalence coincides with the equivalence $I_{\eta_1}\mathfrak{N}_{\xi_1} + I_{\eta_2}\mathfrak{N}_{\xi_2} + \dots + I_{\eta_k}\mathfrak{N}_{\xi_k}$. The theorem is proved.

Following E. S. Ljapin [1], a system of elements K of an associative groupoid C is called a *normal subsystem* if there exists a homomorphic mapping of C onto a groupoid with identity under which K is the complete inverse image of the identity. From our theorem it is immediately obvious that the only normal subsystems of a finite symmetric groupoid C_M are the normal divisors of the symmetric group S_M. This result was obtained by Vorob'ev [2]. In the case of infinite symmetric groupoids our theorem shows that the only normal subgroupoids will be the set of substitutions moving less than ξ elements, where ξ is any given infinite cardinal.

BIBLIOGRAPHY

1. E. S. Ljapin, *Normal complexes of associative systems,* Izv. Akad. Nauk SSSR Ser. Mat. **14** (1950), 179–192. (Russian) MR **11**, 575.

2. N. N. Vorob'ev, *Normal subsystems of finite symmetric associative systems,* Dokl. Akad. Nauk SSSR **58** (1947), 1877–1879. (Russian) MR **9**, 330.

3. J. Schreier and S. Ulam, *Über die Permutationsgruppe der natürlichen Zahlenfolge,* Studia Math. **4** (1933), 134–141.

4. J. Schreier, *Über Abbildungen einer abstrakten Menge auf ihre Teilmengen,* Fund. Math. **28** (1936), 261–264.

5.* A. H. Clifford and G. B. Preston, *The algebraic theory of semigroups.* Vol. II, Amer. Math. Soc., Providence, R. I., 1967.

Translated by HEIDI SEMLA and
BOB SULLIVAN

---

*Added by translators.